Saliva in Health and Disease

Asta Tvarijonaviciute
Silvia Martínez-Subiela • Pia López-Jornet
Elsa Lamy
Editors

Saliva in Health and Disease

The Present and Future of a Unique Sample for Diagnosis

Springer

Editors
Asta Tvarijonaviciute
Interdisciplinary Laboratory of Clinical Analysis (Interlab-UMU), Regional Campus of International Excellence Campus Mare Nostrum
University of Murcia
Murcia, Spain

Pia López-Jornet
Department of Oral Medicine, Faculty of Medicine and Dentistry, Regional Campus of International Excellence Campus Mare Nostrum
University of Murcia
Murcia, Spain

Silvia Martínez-Subiela
Interlab-UMU
University of Murcia
Murcia, Spain

Elsa Lamy
MED - Mediterranean Institute for Agriculture Environment and Development, IIFA - Instituto de Investigação e Formação Avançada
University of Évora
Évora, Portugal

ISBN 978-3-030-37680-2 ISBN 978-3-030-37681-9 (eBook)
https://doi.org/10.1007/978-3-030-37681-9

This Springer imprint is published by the registered company Springer Nature Switzerland AG.
The registered company address is: Gewerbestrasse 11, 6330 Cham, Switzerland

Preface

In the past, saliva was considered just as a component of the digestive process with a main role of initiating the breakdown of lipids and starches. However, during the last years, the view of saliva in biological sciences has changed dramatically; it was shown to contain a variety of molecular and bacterial compounds that can change in local and systemic pathologies. Furthermore, a number of proteins that are present in saliva are absent in plasma, representing an opportunity to identify new molecular biomarkers of different conditions such as disease diagnosis and treatment monitoring. Nevertheless, currently, blood is the sample most frequently used for health assessment in human and veterinary medicine. However, blood sampling is physically intrusive and technically demanding. In contrast, the use of saliva supposes a reduction of stress associated with the sampling due to its noninvasive nature and a decrease in costs because there is no need for a specialized personal for the sampling. In the particular case of elderly persons, children and animals, the samplings could be performed at home without the need to go to the clinic or hospital to avoid stressful situations. However, the use of saliva in routine practice is still limited. This is mainly due to *the lack of knowledge*. Therefore, in order to explore the diagnostic potential of this biofluid, existing literature related to salivary biomarkers in both humans and animals were gathered together, giving place to the current form of this book. The chapters of it could be grouped into three: (1) *generalities*, which describe anatomy and physiology of salivary glands (Chap. 1), saliva and its biomarkers role in ingestive behaviour (Chap. 2), pros and cons of saliva usage as a diagnostic biofluid (Chap. 3), the methodologies for salivary biomarker identification and validation (Chap. 4) and main challenges restricting the use of saliva in a clinical settings from the scientific point of view and the possible ways to solve them (Chap. 15), (2) *oral and systemic pathologies* (Chaps. 5–12) and (3) *physical and psychological stress and welfare* studies (Chaps. 13 and 14).

It is important to highlight that each of the chapters was integrated not only by the knowledge existing in human medicine but also for animals. This was performed in agreement with One Health approach, stating that integration of knowledge leads to higher quality or larger quantity of relevant information, leading to more economically efficient research that drives to more accurate conclusions.

Murcia, Spain Asta Tvarijonaviciute
Silvia Martínez-Subiela
Pia López-Jornet
Evora, Portugal Elsa Lamy

Acknowledgements

We, the editors, would like to thank all authors who agreed to participate in this project, making the production of this book possible. Also, we are grateful to Dr. Gonzalo Cordova for his help in every moment and for his valuable advices.

This book would not be possible without the financial support given by the Seneca Foundation of Murcia Region through projects 20649/JLI/18 and 19894/GERM/15, the Ministry of Economy and Competitiveness of Spain through the postdoctoral programme "Ramon y Cajal" and the project AGL2016-79096-R, and COST (European Cooperation in Science and Technology) Action 16215 the European network for the promotion of portable, affordable and simple analytical platforms (PortASAP). Thanks are also due to the FCT-Portuguese Science Foundation under the project UID/AGR/00115/2013 and the research contract CEEIND to Elsa Lamy.

Contents

Part I
Saliva Basics

Chapter 1
Salivary Glands' Anatomy and Physiology

María D. Contreras-Aguilar and Francisco Gómez-García

Objectives
This contribution aims to show the anatomical and physiological characteristics of the salivary glands as entity for the production of saliva, and to present the composition of the saliva fluid as a protective medium for the mouth, the start of digestion and as diagnostic medium. All this from a comparative point of view between humans and animals.

1.1 Salivary Glands' Anatomy and Histology

The salivary glands are part of the digestive tract (Nater and Rohleder 2009), and their major physiological function is to secrete saliva into the oral cavity, which is essential for the lubrication, digestion, immunity, and overall maintenance of homeostasis within the body.

M. D. Contreras-Aguilar (✉)
Interdisciplinary Laboratory of Clinical Analysis (Interlab-UMU), Veterinary School, Campus of Excellence Mare Nostrum, University of Murcia, Murcia, Spain
e-mail: mariadolores.contreras@um.es

F. Gómez-García
Department of Dermatology, Stomatology, Radiology and Physic Medicine. Faculty of Medicine. Campus of Excellence Mare Nostrum, University of Murcia, Hospital Universitario Morales Meseguer, Murcia, Spain
e-mail: fjgomez@um.es

A. Tvarijonaviciute et al. (eds.), *Saliva in Health and Disease*,
https://doi.org/10.1007/978-3-030-37681-9_1

1.1.1 Types of Salivary Glands

Salivary glands are classified as major and minor salivary glands. The terms *major* and *minor* are associated to the anatomic size of the glands and to the volume of saliva produced (Edgar 1990).

The major salivary glands are bilateral paired glands and include parotid (PG), which are located opposite the maxillary first molars, and submandibular (SMG) and sublingual glands (SLG), which are found in the floor of the mouth (Ferreira and Hoffman 2013; Humphrey and Williamson 2001). Although they are distributed in a horseshoe shape from the midline to the external auditory canal, bordering the jaw in human (Velayos and Santana 2007); in domestic animals, not all are equally developed; neither its topographic localization is the same nor the salivary ducts arranged identically in all of them. In horses and swine, as well as in human, PG is the most developed, with an extension from the base of the ear to zone of external jugular vein conformation. The SMG in cows is the most developed and it is the one that brings more saliva secretion. In horses and humans, the SMG is divided in its cervical extreme placed in the cervico-facial transit, and in its rostral extreme placed in the inter-mandibular space. In human, SMG is the second largest salivary gland, extended from mandibular angle to the hyoid bone. The SLG in domestic animals are divided in polystomatic and monostomatic SLG, although only the polystomatic presented in the horse. In dogs, a differentiated gland appears, named Zigomatic gland, which derives from the minor dorsal buccal glands in relation with the pterygopalatine fossa (Vázquez-Autón et al. 2002).

The minor salivary glands are distributed in groups of hundreds in the upper digestive tract mucosa, eg, in the lower lip, tongue, palate, cheeks, and pharynx (Roth and Calmes 1981). However, in domestic animal, there are other minor salivary glands topographed in relation to the buccinator muscle, the dorsal and ventral buccal glands, although there are also intermediate buccal glands in ruminants (Sandoval 1998; Vázquez-Autón et al. 2002).

1.1.2 Histology

Salivary glands are exocrine and are comprised of different types of cells: acinar cells, various duct system cells, and myoepithelial cells (Humphrey and Williamson 2001) (Fig. 1.1). The glandular histophysiological unit is called sialone and comprises the acinar cells as a secretory piece similar to a cluster of grapes (adenomere), and the ductal portions performed by the duct system cells inside the lobules (intralobulal ducts). However, in the minor salivary glands, subdivision into lobules is not always complete and they have short, individual excretory ducts (Chimenos Kustner et al. 2012). Acinar cells secret firstly the saliva into the ductal lumen, which could be classified as serous, mucous, or mixed (Fig. 1.1). In fact, the quality of saliva

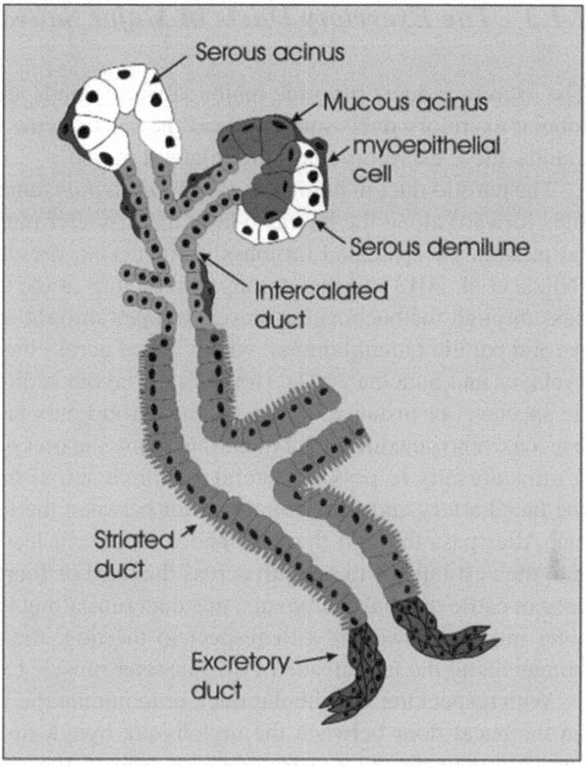

Fig. 1.1 Salivary glandular tissue ducts. (Figure taken from Edgar et al. 2012)

content varies according the histology of each salivary gland (Edgar 1990): the PG produces mainly serous secretions, minor glands mainly mucous secretions, and SLG and SMG produce mixed serous and mucous secretions. Duct system in the major salivary glands is complex and comprises intercalated, striated, and excretory ducts. The first two are shaped by a simple epithelium and are located inside the lobules, so they are called intralobular ducts. While intercalated cells are not involved in the modification of electrolytes, striated cells are responsible for electrolyte regulation. The final ducts called excretory or collector ducts, are shaped by a bilaminar epithelium with a large light surrounded by connective tissue and contribute in the modification of electrolytes as well. They are the last part of the duct network and lead saliva to the oral cavity (Fig. 1.1). With respect to the last cellular type, the myoepithelial cells, they are arranged intimately around acinar cells, being part of the basal layer. Its function is to contract to constrict the acini in order to make secreting or "squeezing out" accumulating fluid. This secreting process is purely neural (Chimenos Kustner et al. 2012; Edgar 1990; Roth and Calmes 1981). Finally, a large amount of adipocytes may be found in the stroma, mainly in PG from human, which with the age a large part of the functional parenchyma might be replaced by adipose tissue (Velayos and Santana 2007).

1.1.3 The Excretory Ducts in Major Salivary Glands

The excretory ducts from the major salivary glands are formed when several interlobular excretory ducts join and lead the saliva to the oral cavity. Like the salivary glands, these excretory ducts are bilateral paired.

The parotid duct in human, called *Stensen duct*, emerges rostral to the gland, and runs forward along the lateral side of the masseter muscle until crossing the buccal fat pad (an encapsulated fat mass in the cheek, presented in humans and porcine) (Niada et al. 2013), to then taking a steep turn at the border of the masseter and to pass through the buccinador muscle to open into the vestibule of the mouth on the parotid papilla (small bumps), which is lied across the second superior molar tooth (Velayos and Santana 2007). However, its layout is different with respect to domestic species. For instance, in horses, the parotid duct begins also rostral to the gland but runs ventromedial to the mandible along the pterygoid medial muscle, and then it turns dorsally to pass the lateral mandible across the mandibular notch together the facial artery and vein, running after between the facial vein and masseter muscle. After pass through the buccinador muscle, it leads on the parotid papilla lied into the vestibule of the mouth across the third or fourth maxillary premolar. In the case of cattle or small ruminants, this duct runs along the ventral surface of the masseter muscle; however, with respect to the dog, the parotid duct runs similar to human along the lateral side of the masseter muscle (Sandoval 1998).

With respect the mandibular duct, or in human the *Wharton duct*, its trajectory is on the bucal floor between the mylohyoid, hyoglosus and genioglosus muscles to run rostral to the lingual frenulum to lead in the sublingual caruncle (Sandoval 1998; Vázquez-Autón et al. 2002; Velayos and Santana 2007).

The SLG in human has several excretory ducts called *Walther's ducts* that lead in the sublingual caruncles. In the monostomatic SLG from domestic animals such as cattle or dogs, there is a major sublingual duct, which is similar to the largest *Walther's ducts* in human, called *Bartholino duct*; that runs alongside the mandibular duct discharging their excretion into the sublingual caruncles. However, the polystomatic SLG discharge salivary secretions through severals minor sublingual ducts located in the sublingual recess of the mucosal fold (Sandoval 1998; Vázquez-Autón et al. 2002; Velayos and Santana 2007).

Finally, and particularly in the dog, the zygomatic duct from the zygomatic gland leads into the vestibule of the mouth on the zygomatic papilla, caudal to the parotid papilla (Sandoval 1998).

1.2 Saliva Secretion

1.2.1 Regulation

The functional innervation of the human salivary glands is known from studies performed with animal models (Ferreira and Hoffman 2013). Acinar cells and their associated myoepithelial cells are both innervated by both the sympathetic and the parasympathetic branches of the autonomic nervous system (ANS), and there is no antagonism between the two branches (Emmelin 1987). Parasympathetic nerve impulses produce high-flow, low-protein saliva, whereas sympathetic impulses produce low-flow, high-protein saliva (Fig. 1.2). However, these are not absolutes. Parasympathetic stimuli can increase the exocytosis from salivary cells for releasing proteins in saliva, but seem particularly important for the secretion of mucins on mucous gland secretion; and adrenergic stimuli can invoke some salivary flow but not as part of the salivary reflex (Carpenter 2013; Proctor and Carpenter 2007). As result, the resting flow rate observed in human sampled by passive drooling ranges from 0.25 to 0.90 mL/min with a mean of approximately 0.4 mL/min, increasing when a sensory stimuli (visual, olfactory or taste) or parasympathetic activity happens (Beltzer et al. 2010; Chimenos Kustner et al. 2012; Thie et al. 2002). Although other factors may vary the resting salivary secretions, such as the age, number of teeth, sex, body weight the circadian rhythmic, or some medications; as well as some diseases such as hepatic disease, malnutrition, depression, neurological diseases, diabetes, chronic painful disorders, among other (Thie et al. 2002).

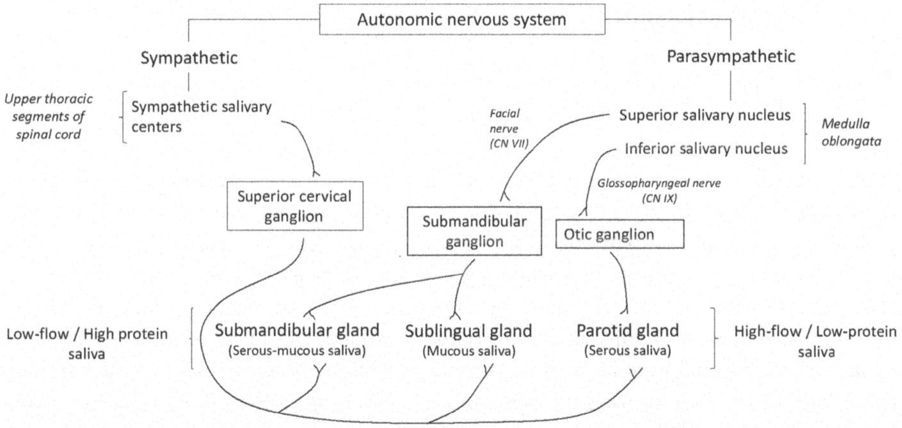

Fig. 1.2 Scheme on the regulation of salivary flow and protein production

1.2.1.1 Salivary Gland Innervation Routes

Preganglionic parasympathetic innervation of the PG comes from fibers of the IX cranial nerve (glossopharyngeal), which are conducted from the medullary region of the brainstem in the inferior salivatory nucleus (ISN) to its synapse in the otic ganglion (Fig. 1.2). Subsequently, the postganglionic fibers of the otic ganglion reach the PG, providing parasympathetic innervation for the secretion of serous-watery saliva. On the other hand, the innervation of both the SMG and the SLG is due to parasympathetic preganglionic fibers belonging to the facial nerve (or cranial nerve VII). These come from the superior salivary nucleus (SSN) to reach the submandibular ganglion, where short postsynaptic fibers leave the ganglion to innervate the SMG and SLG, which secrete serous-mucous and mucous saliva, respectively (Davis et al. 1956).

On the other hand, the sympathetic branches come from the sympathetic salivary centers located in the upper thoracic segments of the spinal cord. The paravertebral sympathetic trunk carries the ascending preganglionic fibers from the thoracic ganglion to synapse at the superior cervical ganglion (SCG). Postganglionic sympathetic fibers exit the SCG and the sympathetic fibers give off branches to reach all three pairs of major salivary glands through the external carotid artery plexus and its branches, as the facial artery (Kahle et al. 2010). These branches take charge of regulating peripheral blood flow, salivary secretion, and local inflammatory and immune mediators (Emilio Savastano et al. 2010; Mathison et al. 1994).

1.2.1.2 Neurotransmitters Released by Autonomic Nerves Innervating Salivary Glands

The major types of neurotransmitters released by the autonomic nervous fibers that innervate the salivary glands are the acetylcholine (ACh) and noradrenaline (NA), which provides cholinergic and adrenergic signaling responses, respectively (Fig. 1.2). ACh is mainly the neurotransmitter between preganglionic and postganglionic neurons in both the sympathetic and parasympathetic fibers, and between postganglionic parasympathetic neurons and salivary glands. Meanwhile, between postganglionic sympathetic neurons and effector salivary glands, NA is the major neurotransmitter. Calcitonin gene-related peptide (CGRP), Neuropeptide Y (NPY), neurokinin A (NKA), neuronal nitric oxide synthase (nNOS), pituitary adenylate cyclase activating peptide (PACAP), substance P (SP) and vasoactive intestinal peptide (VIP) are other non-adrenergic, non-cholinergic transmitters released from either parasympathetic, sympathetic or both autonomic nerves in salivary glands. All of them are neuropeptides with effects on the blood vessels and on the salivary cells which can modify the protein and/or fluid secretion (Ekström 1999; Ekström et al. 1989). For example, together with Ach, VIP is capable to cause vasodilation in the SMG to increase blood flow and salivary secretion (Lundberg et al. 1981). Or SP and/or CGRP can be found in sensory nerve fibers from the sympathetic and parasympathetic nerve bundles which target ducts and blood vessels in rat SMGs (Kobashi et al. 2005).

1.2.2 Pathways of Saliva Secretion

It must be emphasized that the effects of parasympathetic and sympathetic nerve impulses on protein and fluid secretion from salivary glands can differ between glands in the same species and between the same gland in different species (Proctor and Carpenter 2007). Acinar cells are responsible for the secretion of the primary fluid and of most of the proteins found in saliva (>85%), although duct cells secrete numerous proteins with important biological activities, e.g., nerve growth factor, epidermal growth factor, immunoglobulin (Ig) A, and kallikrein (Melvin et al. 2005). The different elements that compound the gland-derived salivary constituents are transported from the salivary glands' cells by:

- *Process of exocytosis* ➔ Most of the proteins secreted by salivary glands are derived from protein storage granules in acinar cells, which are then released by a process of exocytosis (Castle et al. 1975; Segawa and Yamashina 1998). Those proteins have to be firstly synthetized and packaging (Edgar et al. 2012). The exocytosis starts when neurotransmitters, mainly by a sympathetic stimulation, but also by a parasympathetic one; exert their activity at the cell membrane (Asking and Gjorstrup 1987; Nater and Rohleder 2009). They bind to specific receptor proteins on the basolateral membrane, causing a degranulation of the storage granules. Na^+ from sympathetic neurons binds to both α - and β-adrenergic receptors on the acinar cell. $\alpha 1$-receptor activation is linked to elevation of intracellular Ca^{2+}, which results in large-scale fluid and electrolyte transport, and modest exocytosis of stored protein; while $\beta 1$-receptor activation causes elevation of intracellular cyclic adenosine monophosphate (cAMP) followed by activation of protein kinase A and phosphorylation of endogenous proteins leading the fusion of secretory granules with the apical membrane of cells (Baum 1993; Castle and Castle 1998). In addition, neuropeptide VIP via parasympathetic activation can also acts on the exocytosis process through cAMP; and cholinergic stimuli or substance P can give rise to the release of protein by a coupling mechanism involving elevated intracellular Ca^{2+} and activation of protein kinase C (Ekström et al. 1989; Moller et al. 1996). For example, this is the way of excretion of the salivary alpha-amylase (sAA).
- *Vesicular protein transport (endocytosis)* ➔ Other way to protein secretion is the vesicular protein transport, as occurs with the secretory IgA. This is actively carried across acinar and ductal cells via a transporter protein, that in the case of the IgA in called the polymeric immunoglobulin receptor (pIgR). Although this process can be up regulated by neural activity due to a parasympathetic and sympathetic activation (Carpenter et al. 1998; Proctor and Carpenter 2002), it does not involve storage of protein within cells. In the case of IgA, which is made by plasma cells located in the connective tissues' gland, binds to the pIgR on acinar and ductal cells on the basolateral surface and it is endocytosed into the cell. Then, the vesicle is transported across the cell to the apical membrane, where the membrane receptor is cleaved to release secretory IgA (the secretory component is the cleaved part of the pIgR). The pIgR is specific to IgA, and so even though

there are equal numbers of IgG-, IgA-, and IgM-producing plasma cells within the gland (Mega et al. 1992), IgA becomes the single-most- abundant antibody in saliva because it is preferentially bound by the pIgR (Brandtzaeg 1998). IgG and IgM probably entry into the oral cavity via crevicular fluid (Eliaz Kaufman and Lamster 2000).

- Intracellular diffusion ➔ This is the case of cortisol, estriol and testosterone in saliva, which are transported passively (transcellular) from plasma as nonprotein-bound (unbound) fraction to saliva by diffusing through the cells of the salivary glands due to its solubility in the lipid-rich cell membranes (Vining et al. 1983). As result, its concentration in saliva does not depend on the rate of saliva production (Büttler et al. 2018; Vining et al. 1983). In the case of cortisol, it is only presented in saliva as free cortisol (unbound fraction) (Perogamvros et al. 2011).
- *Ultrafiltration* ➔ Ultrafiltration in salivary glands is the passive transport of compounds from plasma to saliva via the tight junctions between the acinar cells (paracellular) due to a positive hydrostatic pressure. It is only allowed for compounds with a relative molecular mass cutoff of approximately 100–300. This way of transport occurs in lipid-insoluble conjugated steroids, such as thyroxin or choriogonadotropin (Vining et al. 1983). This rout is by which water also crosses the cells of the salivary glands when a transepithelial osmotic gradient or higher hydrostatic pressure happen (Young et al. 1987). On this last case, the parasympathetic and sympathetic exercise on parenchymal cells short-term regulates the salivary gland blood flow increasing the hydrostatic pressure after an activation. E.g, the parasympathetic stimulus produce vasodilation within an integral salivary reflex (Anderson et al. 2006; Mizuta et al. 2000); while the sympathetic-mediated vasoconstriction, depends on a separate vasomotor control independent of the salivary reflex (Emmelin and Engstrom 1960). In addition, the myoepithelial cells, which surround glandular acini, are contracted by the stimulus of parasympathetic and sympathetic fibers (Tamarin 1966).
- *Water channels or aquaporins transport* ➔ Aquaporins are water channels that allow the transepithelial water movement. There are five aquaporins expressed in salivary glands (Aqp1, Aqp3, Aqp4, Aqp5, and Aqp8). However, Aqp5, which is highly expressed in the apical membranes of salivary acinar cells (Funaki et al. 1998; Matsuzaki et al. 1999), appears to be the only aquaporin to play a major role in salivation (Gresz et al. 2001; Melvin et al. 2005).
- *Active transport of ions by channels* ➔ Fluid secretion depends critically on the movement of Cl^-, K^+ and HCO^{-3} ions through Ca^{2+}-activated channels or cotransporters in the luminal membrane (transcellular) (Melvin et al. 2005). The release of ACh from parasympathetic nerves and its interaction with muscarinic cholinergic receptors (mAChRs) principally regulates fluid secretion on the salivary glands. Stimulation via mAChR receptors is coupled to secretion of inositol triphosphate (IP3) and diacylglycerol. The interaction of IP3 with its specific receptor on the endoplasmic reticulum causes release of stored Ca^{2+} (Baum and Wellner 1999). Rises in intracellular Ca^{2+} open, for example, apical membrane Cl^- channels and basolateral membrane K^+ channels in acinar cells. When Cl^- goes out through the acinar apical membrane and K^+ enters the interstitial fluid is

created a trans-epithelial potential difference. This electrical potential difference allows the passive exchange of cations across the tight intercellular junctions and the luminal accumulation of ions. This generates a transepithelial osmotic gradient that enables the movement of water via paracellular (ultrafiltration) or transcellular (water channels or aquaporins transport) to create the primary fluid secretion, which ionic composition in the interstitial fluid bathing the basolateral aspect of the acinar cells. In addition, the Na^+/HCO^{-3} cotransporter activity plays a potential role in salivary secretion due to the electrogenic HCO^{-3} conductance generated. However, it appears to be both gland- and species-specific (Kim et al. 2003; Luo et al. 2001; Melvin et al. 2005).

1.3 Saliva

1.3.1 Composition

Whole saliva consists of a mixture of oral fluids, including secretions of the major and minor salivary glands, in addition to constituents of non-salivary origin derived from blood and blood derivatives, as intraoral bleeding and gingival crevicular fluid (serum exudate and inflammatory cells as leukocytes); desquamated epithelial cells as keratinocytes; other fluids as expectorated bronchial and nasal secretions; extrinsic substances as food debris; as well as microbiota as bacteria and bacterial products, viruses and fungi (Kaufman and Lamster 2000).

In human, percentage contributions of the different salivary glands during unstimulated flow are as follows: 20% from PG, 65% from SMG, 7% to 8% from SLG, and less than 10% from numerous minor glands. Stimulated high flow rates drastically change percentage contributions from each gland, with the parotid contributing more than 50% of total salivary secretions (Edgar 1990; Matsuo 2000). In addition, although most proteins in saliva are secreted by the salivary glands, there are large differences between the glands as to which proteins they synthesize (Carpenter 2013). For example, PG secretes a serous secretion that contains no mucins but is rich in amylase and proline-rich proteins (PRPs, basic and acid). Mucins and cystatins are common to the SMG and SLG as well as most minor glands. Basic PRPs appear to be exclusive to the parotid glands, whereas acidic PRPs appear in submandibular and parotid glands. However, some proteins are universal to all glands, such as IgA (the main antibody in saliva). Moreover, the composition of saliva can be also affected by physiological situations (age, sex, body weight, circadian rhythmic, etc), systemic disorders or oral diseases, stress, exercise, etc. (Edgar et al. 2012; Greabu et al. 2009). Additionally, it is important to stress that the few studies that have targeted the saliva of other mammals suggest the existence of a different proteome composition among species (de Sousa-Pereira et al. 2015; Lamy et al. 2009).

The main gland-derived salivary constituents (Kaufman and Lamster 2000) are following:

- *Water and electrolytes* ➔ Salivary fluid is composed of 99% of water (Greabu et al. 2009). However, it is composed additionally of electrolytes derived from plasma (Humphrey and Williamson 2001; Kaufman and Lamster 2000), as chloride (Cl^-), bicarbonate (HCO_3^-), sodium (Na^+), potassium (K^+), calcium (Ca^{2+}), fluorid, thiocynate, iodine, magnesium (Mg^{+2}), phosphates, ammonia, and sulphates. Salivary fluid involves two stages. The primary fluid secreted by salivary acinar cells into the ductal lumen is a plasma-like, isotonic fluid (stage 1). Then, the striated duct cells contribute in resorbing Na^+ and Cl^- while the excretory duct cells continue the Na^+ resorption and secret K^+ and HCO_3^- (stage 2), subsequently modifying the isotonic fluid by a hypotonic one upon passage through the ducts. This is an energy rich process so that the striated duct cells have large numbers of mitochondria located in their basolateral part, leading to their description (Edgar 1990; Melvin et al. 2005). As the secretion of Na^+ and Cl^- by the acinar cells is upregulated by neural signal from the brain, but not its reabsorption by the duct cells, the stimulated saliva has a higher Na^+ and Cl^- concentration than resting saliva (Matsuo 2000).
- *Proteins*
  - *Mucin* ➔ There are two types, MG1 and MG2. MG1 is a high-molecular-weight (a molecular weight greater than 1000 kDa), highly glycosylated mucin, and forms heterotypic complexes with other salivary proteins such as amylase, PRPs, statherin, and histatins. However, the MG2 is a low-molecular-weight (with a molecular weight of 200–300 kDa), single-glycosylated peptide chain mucin (Iontcheva et al. 1997; Slomiany et al. 1996). They may be present in distinct amounts and types, and it might be the reason for the chief differences in the viscoelastic and antibacterial properties in saliva among species (de Sousa-Pereira et al. 2015).
  - *Enzymes* ➔ Saliva contains a variety of enzymes, such as lactoferrins, lysozymes, peroxidases (also known as sialoperoxidase or lactoperoxidase), glutathione-S-transferase P, sAA, esterases, among others (de Sousa-Pereira et al. 2015; Humphrey and Williamson 2001; Kaufman and Lamster 2000). All of them may be or not present in saliva from human and domestic animals, and at different concentrations. For example, the anhydrase carbonic (CA), an esterase, includes the 42-kDa CA secreted (CA VI, also named gustin), reported to be the main CA identified in human saliva, but also identified in dog, horse, cattle, sheep and rat (Asari et al. 2000; de Sousa-Pereira et al. 2015; Feldstein and Silverman 1984; Fernley et al. 1988). sAA (a-1,4-a-D-glucan 4-glucanohydrolase; EC 3.2.1.1) is one of the most important enzymes in human saliva (Nater and Rohleder 2009). Despite its concentration and activity being lower in other domestic animals, and even previously thought to be absent, it is also present in saliva from dog, horse, pig and sheep (Contreras-Aguilar et al. 2017; 2018a, b, c).

- *Immunoglobulins* ➔ Immunologic contents of saliva include secretory IgA, IgG, and IgM (Humphrey and Williamson 2001). IgA is the main antibody in saliva. It is actively carried from the connective tissues' gland across acinar and ductal cells that express the specific receptors labeled pIgR for polymeric IgA molecules, which are the prevalent form in mucosal secretions, while in serum is the monomeric form. Little or no diffusion of Igs into saliva occurs, except under conditions of inflammation or disease (Carpenter 2013).
- *Proline-rich proteins* ➔ These PRPs contribute 70% of all amino acids, with a strongly basic or acidic isoelectric point. They are highly polymorphic, with a huge variety of proteins not only between individuals but also within the same individual at different times of the day (Carpenter 2013). However, there is no evidence of them in some ruminants as sheep and goat (Amado et al. 2013; Lamy et al. 2009).
- *Cystatins, histatins and statherins* ➔ Cystatins and histatins are cysteine- rich proteins and histidine-rich proteins, respectively. Statherins are amphipathic molecules that contain both hydrophobic and hydrophilic domains, highly surface active (Carpenter 2013). Hitatins and statherins have not been found in saliva from dog, cattle, sheep, horse, rat and rabbit. However, it could be postulated that casein, observed in all the animal described above, except in human, could replace statherin and histatins as an adaptative mechanism (de Sousa-Pereira et al. 2015), since the genes encoding for this protein family belong to the same secretory gene cluster; and because its function in oral cavity is the same that cystatins, statherins, and histatins, by preventing adherence of salivary components and bacteria to enamel (Huq et al. 2005; Kawasaki and Weiss 2003).
- *Cathelicidins* ➔ Cathelicidins are antimicrobial peptides that contain a highly conserved signal sequence and cathelin domain but show substantial heterogeneity in the C-terminal dominion coding the mature active peptide (Murakami et al. 2002). They have been identified in saliva from cattle, dog, and rat (de Sousa-Pereira et al. 2015). Some studies highlight the presence of cathelicidins in human's saliva (Murakami et al. 2002), while in other they were not detected (de Sousa-Pereira et al. 2015).
- *Other proteins* ➔ In addition the main gland-derived salivary proteins, other proteins may be also found (Barranco et al. 2018; Contreras-Aguilar et al. 2019). One of these is albumin, which major route of entry into the oral cavity is via crevicular fluid (Henskens et al. 1993). Other protein that may be found in saliva is the Chromogranin A (CgA). CgA is an acidic, soluble protein, which is stored and co-released with catecholamines from granules of the adrenal medullary and sympathetic nerve chromaffin cells (Blaschko et al. 1967; Oconnor 1983); but is also produced and released from the serous acinar and ductal cells of the human SMG (Saruta et al. 2005), and also has been detected in salivary glands of rat and horse (Sato et al. 2002). In addition, there are species-specific proteins like the case of Latherin, a non-glycosylated, surface-active, detergent-like protein (Lindner et al. 2000; McDonald et al.

2009). Latherin is secreted by horses at unusually high concentrations in their sweat, to wet their hairs and facilitate the water flow for evaporative cooling, because their pelts are thick, hairy and waterproofed. Latherin has been also detected in saliva, probably as a result of an adaptative strategy to their needing to masticate and process large quantities of dry food material, and their ability to sustain high levels of exercise for long periods of time.

- *Small organic molecules* ➔ In saliva small organic molecules such as amino acids, creatinine, glucose, lipids, nitrogen, sialic acid, urea and uric acid can also be found (Eliaz Kaufman and Lamster 2000).
- *Hormones* ➔ Hormones, such as cortisol, estriol, estradiol, thyroxin and testosterone pass into saliva from plasma (Choe et al. 1983; Vining et al. 1983). All of them may reflect the levels in serum with different degree of correlation between its levels in saliva and serum. However, there are other hormones that could be produced directly by the salivary glands. For instance, there are evidences that salivary melatonin may have a local production in the salivary glands in addition to the free melatonin coming from plasma (Van Faassen et al. 2017). In addition, salivary glands during saliva production create an inactive hormone, cortisone, due to the inter-conversion of hormonally active cortisol (free serum cortisol) by the 11β-hydroxysteroid dehydrogenase type 2 (Shimojo et al. 1997). Cortisone is present at higher concentration than cortisol in saliva, and has previously been shown to have a linear relationship with serum cortisol (Debono et al. 2016). Although it has been suggested that a linear relationship exists between serum and saliva levels of insulin in humans (Marchetti et al. 1986), controversy remains as to the origin of insulin in saliva (Messenger et al. 2003) (See Chap. 8). Furthermore, other hormones related to the regulation of metabolism such as leptin, adiponectin and ghrelin, are also secreted by the salivary glands (Groschl et al. 2001, 2005; Cappai et al. 2016).

1.3.2 Function

The major functions of whole saliva are:

- *Lubrication and protection* ➔ The best lubricating components of saliva are mucins (Humphrey and Williamson 2001). Together PRPs, statherin, and histatins (Iontcheva et al. 1997), they make the saliva viscous since being of low solubility, high elasticity and strong adhesiveness. So mastication and swallowing are aided by the lubricating effects of mucins, increasing the resistance to shear at those moments (Humphrey and Williamson 2001).
- *Buffering action and clearance* ➔ In saliva, HCO_3^- is the most important buffering system. HCO_3^- is generated by the intracellular CA catalyzing the reversible reaction of water and CO_2 to form HCO_3^- and H^+ (Melvin et al. 2005). When

HCO_3^- diffuses into plaque, it acts as a buffer by neutralizing acids regulating the salivary pH: an higher flow rate will increase HCO_3^-concentrations in saliva, and salivary pH will thus be lowest (Edgar et al. 2012). In addition, HCO_3^- generates ammonia to form amines, which also serve as a buffer by neutralizing acids. After HCO_3^- activity, the major buffering ability of saliva is attributed to histatins (Mandel 1989). Urea, another buffer present in saliva, releases ammonia after being metabolized by urease activity from plaque bacteria and thus increases plaque pH (Edgar et al. 2012).

- Maintenance of tooth integrity ➔ MG1 adsorbs tightly to the tooth and thereby contributes to the enamel pellicle, which protects the tooth from acid challenges (Iontcheva et al. 1997). In addition, statherins stick tightly to the tooth surface, and together the acidic PRPs, bind to the high salivary concentrations of calcium and phosphate compared to hydroxyapatite (the main mineral component of teeth), to maintain the maturation and remineralization of enamel (Edgar et al. 2012). Therefore, they remain on the surface, bound to hydroxyapatite, to aid in controlling crystalline growth of the enamel by allowing the penetration of minerals into the enamel and by limiting mineral egress (Humphrey and Williamson 2001).
- Antibacterial activity ➔ The antibacterial activity of saliva may be specific (e.g. immunoglobulins) or non-specific (proteins, mucins, peptides, and enzymes), and it helps to control the oral microbiota (Edgar et al. 2012; Humphrey and Williamson 2001). IgA acts actively on mucosal surfaces, to neutralize viruses, serves as an antibody to bacterial antigens, and works to aggregate or clump bacteria, thus inhibiting bacterial attachment to host tissues (Humphrey and Williamson 2001; Kaufman and Lamster 2002). However, MG2 and IgA complex bind mucosal pathogens with greater affinity than either MG2 or IgA alone (Biesbrock et al. 1991). In the case of the heterotypic complexes of MG1 with sAA, PRPs, statherins, and histatins, attract the attachment of certain bacteria and providing a short-term nutrient source for bacteria (Iontcheva et al. 1997). In addition, cathelicidins act as antimicrobial peptides of the early host defenses of mammals against infection (Zanetti 2005). On the other hand, enzymes that act on the non-specific antibacterial activity are the lactoferrin, lysozymes or peroxidades (Edgar 1990; Humphrey and Williamson 2001).
- Taste and digestion ➔ The salivary hypotonicity enhances the tasting capacity of salty foods and nutrient sources (Humphrey and Williamson 2001) (see Chap. 2). Saliva has also the early, although limited, role in total digestion by beginning the breakdown of starch with sAA, the major component of parotid saliva in human that initially dissolves starch (Humphrey and Williamson 2001; Nater and Rohleder 2009). The major starch digestion results from pancreatic amylase (Butterworth et al. 2011). Other enzymes in saliva, such as lingual lipase also initiate fat digestion (Carpenter 2013; DeNigris et al. 1988). Finally, the bolus formation is another important function of saliva due to its wetting properties and its capacity to go into food to allow that the food particles to stick together once the food have been chewed and breakdown physically (Carpenter 2013).

1.4 Conclusions

The salivary gland system is a complex organ that produces saliva highly variable depending the specie and other factors related with its secretion. The secretory unit, called sialone, comprises an adenomere (secretory part) and the ductal portions that modify the product secreted by it. The regulation of salivary secretion (fluid and protein secretion) is carried out by the autonomic nervous system (sympathetic and parasympathetic) and depends on sensory, electrical and mechanical stimuli. However, nervous impulses on protein and fluid secretion from salivary glands can differ between glands in the same species and between the same gland in different species.

Whole saliva includes the secretions of the salivary glands, besides of non-salivary origin derived from blood and blood derivatives, desquamated epithelial cells, other organic fluids, extrinsic substances, as well as microbiota as bacteria and bacterial products, viruses and fungi. The main gland-derived salivary constituents are water and electrolytes, proteins (mucin, enzymes, immunoglobulins, among others), small organic molecules and hormones. The primary functions of saliva include comprise lubrication, protection, bufferin action, clearance, maintenance of tooth integrity, antimicrobial activity, taste and digestion.

References

Amado FML, Ferreira RP, Vitorino R (2013) One decade of salivary proteomics: current approaches and outstanding challenges. Clin Biochem 46(6, SI):506–517. https://doi.org/10.1016/j.clinbiochem.2012.10.024

Anderson LC, Martin DJ, Phillips DL, Killpack KJ, Bone SE, Rahimian R (2006) The influence of gender on parasympathetic vasodilatation in the submandibular gland of the rat. Exp Physiol 91(2):435–444. https://doi.org/10.1113/expphysiol.2005.032730

Asari M, Miura K, Ichihara N, Nishita T, Amasaki H (2000) Distribution of carbonic anhydrase isozyme VI in the developing bovine parotid gland. Cells Tissues Organs 167(1):18–24. https://doi.org/10.1159/000016762

Asking B, Gjorstrup P (1987) Synthesis and secretion of amylase in the rat parotid gland following autonomic nerve stimulation in vivo. Acta Physiol Scand 130(3):439–445. https://doi.org/10.1111/j.1748-1716.1987.tb08160.x

Barranco T, Tvarijonaviciute A, Tecles F, Carrillo JM, Sanchez-Resalt C, Jimenez-Reyes P et al (2018) Changes in creatine kinase, lactate dehydrogenase and aspartate aminotransferase in saliva samples after an intense exercise: a pilot study. J Sports Med Phys Fitness 58(6):910–916. https://doi.org/10.23736/S0022-4707.17.07214-0

Baum B (1993) Principles of saliva secretion. Ann N Y Acad Sci 20(694):17–23

Baum BJ, Wellner RB (1999) Receptors in salivary glands. In: Garrett JR, Ekstrom J, Anderson LC (eds) Neural mechanisms of salivary secretion, Frontiers in oral biology. Karger, Basel, pp 44–58

Beltzer EK, Fortunato CK, Guaderrama MM, Peckins MK, Garramone BM, Granger DA (2010) Salivary flow and alpha-amylase: collection technique, duration, and oral fluid type. Physiol Behav 101(2):289–296. https://doi.org/10.1016/j.physbeh.2010.05.016

Biesbrock A, Reddy M, Levine M (1991) Interaction of salivary mucin-secre- tory immunoglobulin A complex with mucosal pathogens. Infect Immun 59(10):3492–3497

Blaschko H, Comline RS, Schneider FH, Silver M, Smith AD (1967) Secretion of a chromaffin granule protein, chromogranin, from the adrenal gland after splanchnic stimulation. Nature 215:58–59

Brandtzaeg P (1998) Synthesis and secretion of human salivary immunoglobulins. In: Garrett J, Ekstrom J, Anderson LC (eds) Glandular mechanisms of salivary secretion. Karger, Basel, pp 167–199

Butterworth PJ, Warren FJ, Ellis PR (2011) Human α-amylase and starch digestion: an interesting marriage. Starch/Staerke 63(7):395–405. https://doi.org/10.1002/star.201000150

Büttler RM, Bagci E, Brand HS, den Heijer M, Blankenstein MA, Heijboer AC (2018) Testosterone, androstenedione, cortisol and cortisone levels in human unstimulated, stimulated and parotid saliva. Steroids 138:26–34. https://doi.org/10.1016/j.steroids.2018.05.013

Cappai MG, Dall'Aglio C, Sander SJ, Ratert C, Dimauro C, Pinna W, Kamphues J (2016) Different physical forms of one diet fed to growing pigs induce morphological changes in mandubular glands and local leptin (Ob) production and receptor (ObR) expression. J Anim Physiol Anim Nutr 100(6):1067–1072. https://doi.org/10.1111/jpn.12478

Carpenter GH (2013) The secretion, components, and properties of saliva. Annu Rev Food Sci Technol 4(February 2013):267–276. https://doi.org/10.1146/annurev-food-030212-182700

Carpenter GH, Garrett JR, Hartley RH, Proctor GB (1998) The influence of nerves on the secretion of immunoglobulin A into submandibular saliva in rats. J Physiol 512(2):567–573. https://doi.org/10.1111/j.1469-7793.1998.567be.x

Castle D, Castle A (1998) Intracellular transport and secretion of salivary proteins. Crit Rev Oral Biol Med 9(1):4–22. https://doi.org/10.1177/10454411980090010301

Castle J, Jamieson J, Palade G (1975) Secretion granules of the rabbit parotid gland. Isolation, subfractionation, and characterization of the membrane and content subfractions. J Cell Biol 64(1):182–210. https://doi.org/10.1083/jcb.64.1.182

Chimenos Kustner E, Ribera Uribe M, López López J (2012) Gerodontología (1ª; S. E. de Gerontología, ed.). Santiago de Compostela, España: La Ibérica

Choe JK, Khan-Dawood FS, Yusoff-Dawood M (1983) Progesterone and estradiol in the saliva and plasma during the menstrual cycle. Am J Obstet Gynecol 147(5):557–562. https://doi.org/10.1016/0002-9378(83)90016-9

Contreras-Aguilar MD, Tecles F, Martínez-Subiela S, Escribano D, Bernal L, Cerón JJ (2017) Detection and measurement of alpha- amylase in canine saliva and changes after an experimentally induced sympathetic activation. BMC Vet Res 13(266):1–6. https://doi.org/10.1186/s12917-017-1191-4

Contreras-Aguilar MD, Escribano D, Martín-cuervo M, Tecles F, Cerón JJ (2018a) Salivary alpha-amylase activity and cortisol in horses with acute abdominal disease: a pilot study. BMC Vet Res 14(156):1–7

Contreras-Aguilar MD, Escribano D, Martínez-subiela S, Martínez-Miró S, Cerón JJ, Tecles F (2018b) Changes in alpha-amylase activity , concentration and isoforms in pigs after an experimental acute stress model: an exploratory study. BMC Vet Res 14(256):1–8

Contreras-Aguilar M, Escribano D, Quiles A, López-Arjona M, Cerón J, Martínez-Subiela S et al (2018c) Evaluation of new biomarkers of stress in saliva of sheep. Animal 13(6):1–9. https://doi.org/10.1017/S1751731118002707

Contreras-Aguilar M, Escribano D, Martínez-Subiela S, Martín-Cuervo M, Lamy E, Tecles F, Cerón J (2019) Changes in saliva analytes in equine acute abdominal disease: a sialochemistry approach. BMC Vet Res 6:1–9

Davis R, Anson B, Budinger J, Kurth L (1956) Surgical anatomy of the facial nerve and parotid gland based upon a study of 350 cervicofacial halves. Surg Gynecol Obstet 102(4):384–412

de Sousa-Pereira P, Cova M, Abrantes J, Ferreira R, Trindade F, Barros A et al (2015) Cross-species comparison of mammalian saliva using an LC-MALDI based proteomic approach. Proteomics 15(9):1598–1607. https://doi.org/10.1002/pmic.201400083

Debono M, Harrison RF, Whitaker MJ, Eckland D, Arlt W, Keevil BG, Ross RJ (2016) Salivary cortisone reflects cortisol exposure under physiological conditions and after hydrocortisone. J Clin Endocrinol Metab 101(4):1469–1477. https://doi.org/10.1210/jc.2015-3694

DeNigris SJ, Hamosh M, Kasbekar DK, Lee TC, Hamosh P (1988) Lingual and gastric lipases: species differences in the origin of prepancreatic digestive lipases and in the localization of gastric lipase. Biochim Biophys Acta Lipids Lipid Metab 959(1):38–45. https://doi.org/10.1016/0005-2760(88)90147-6

Edgar WM (1990) Saliva and dental health. Clinical implications of saliva: report of a consensus meeting. Br Dent J 169(3–4):96–98. https://doi.org/10.1038/sj.bdj.4807284

Edgar M, Dawes C, O'Mullane D (2012) In: Edgar M, Dawes C, O'Mullane D (eds) Saliva and oral health, vol 56, 4th edn. https://doi.org/10.1590/S0104-42302010000100001

Ekström J (1999) Role of nonadrenergic, noncholinergic autonomic transmitters in salivary glandular activities in vivo. In: Garrett J, Ekström J, Anderson L (eds) Neural mechanisms of salivary gland secretion, Frontiers in oral biology, 11th edn. Karger, Basel, pp 94–130

Ekström J, Garrett J, Månsson B, Rowley P, G T (1989) Depletion of large dense-cored vesicles from parasympathetic nerve-terminals in rat parotid-glands after prolonged stimulation of the auriculotemporal nerve. Regul Pept 25(1):61–67. https://doi.org/10.1016/0167-0115(89)90248-6

Emilio Savastano L, Elizabeth Castro A, Rene Fitt M, Fredensborg Rath M, Eduardo Romeo H, Maris Munoz E (2010) A standardized surgical technique for rat superior cervical ganglionectomy. J Neurosci Methods 192(1):22–33. https://doi.org/10.1016/j.jneumeth.2010.07.007

Emmelin N (1987) Nerve interactions in salivary glands. J Dent Res 66(2):509–517. https://doi.org/10.1177/00220345870660022101

Emmelin N, Engstrom J (1960) Effect of sympathetic denervation on the sensitivity of the submaxillary gland to stimulating agents. J Physiol Lond 153(1):9–16. https://doi.org/10.1113/jphysiol.1960.sp006515

Feldstein JB, Silverman DN (1984) Purification and characterization of carbonic-anhidrase from the saliva of the rat. J Biol Chem 259(9):5447–5453

Fernley RT, Coghlan JP, Wright RD (1988) Purification and characterization of a high-Mr carbonic-anhydrase from sheep parotid-gland. Biochem J 249(1):201–207. https://doi.org/10.1042/bj2490201

Ferreira JN, Hoffman MP (2013) Interactions between developing nerves and salivary glands. Organogenesis 9(3):199–205

Funaki H, Yamamoto T, Koyama Y, Kondo D, Yaoita E, Kawasaki K et al (1998) Localization and expression of AQP5 in cornea, serous salivary glands, and pulmonary epithelial cells. Am J Phys Cell Phys 275(4):C1151–C1157

Greabu M, Battino M, Mohora M, Totan A, Didilescu A, Spinu T (2009) Saliva – a diagnostic window to the body, both in health and in disease. J Med Life 2(2):124–132

Gresz V, Kwon TH, Hurley PT, Varga G, Zelles T, Nielsen S et al (2001) Identification and localization of aquaporin water channels in human salivary glands. Am J Physiol Gastrointest Liver Physiol 281(1):G247–G254

Groschl M, Rauh M, Wagner R, Neuhuber W, Metzler M, Tamguney G et al (2001) Identification of leptin in human saliva. J Clin Endocrinol Metab 86(11):5234–5239. https://doi.org/10.1210/jc.86.11.5234

Groschl M, Topf HG, Bohlender J, Zenk J, Klussman S, Dotsch J et al (2005) Identification of ghrelin in human saliva: production by the salivary glands and potential role in proliferation of oral keratinocytes. Clin Chem 51(6):997–1006. https://doi.org/10.1373/clinchem.2004.040667

Henskens YMC, Vandervelden U, Veerman ECI, Amerongen AVN (1993) Protein, albumin and cystatin concentrations in saliva of healthy-subjects and of patients with gingivitis or periodontitis. J Periodontal Res 28(1):43–48. https://doi.org/10.1111/j.1600-0765.1993.tb01049.x

Humphrey SP, Williamson RT (2001) A review of saliva: normal composition, flow, and function. J Prosthet Dent 85(2):162–169

Huq NL, Cross KJ, Ung M, Reynolds EC (2005) A review of protein structure and gene organisation for proteins associated with mineralised tissue and calcium phosphate stabilisation encoded on human chromosome 4. Arch Oral Biol 50(7):599–609. https://doi.org/10.1016/j.archoralbio.2004.12.009

Iontcheva I, Oppenheim FG, Troxler RF (1997) Human salivary mucin MG1 selectively forms heterotypic complexes with amylase, proline-rich proteins, statherin, and histatins. J Dent Res 76(3):734–743. https://doi.org/10.1177/00220345970760030501

Kahle W, Frotscher M, Gerhard S (2010) Color atlas of human anatomy: nerve system and sensory organs, vol 3. Theime Medical, Stuttgart

Kaufman E, Lamster IB (2000) Analysis of saliva for periodontal sis of saliva may offer a cost-effective approach to assessment of periodontal disease in large populations. J Clin Periodontol C Munksgaard 27(27):453–465

Kaufman E, Lamster I (2002) The diagnosis application of saliva: a review. Crit Rev Oral Biol Med 13(2):197–212

Kawasaki K, Weiss KM (2003) Mineralized tissue and vertebrate evolution: the secretory calcium-binding phosphoprotein gene cluster. Proc Natl Acad Sci U S A 100(7):4060–4065. https://doi.org/10.1073/pnas.0638023100

Kim YB, Yang BH, Piao ZG, Oh SB, Kim JS, Park K (2003) Expression of Na+/HCO3- cotransporter and its role in pH regulation in mouse parotid acinar cells. Biochem BIiophys Res Commun 304(4):593–598. https://doi.org/10.1016/S0006-291X(03)00632-6

Kobashi M, Ichikawa H, Kobashi M, Funahashi M, Mitoh Y, Matsuo R (2005) The origin of sensory nerve fibers that innervate the submandibular salivary gland in the rat. Brain Res 1060(1–2):184–187. https://doi.org/10.1016/j.brainres.2005.08.012

Lamy E, da Costa G, Santos R, e Silva F, Potes J, Pereira A et al (2009) Sheep and goat saliva proteome analysis: a useful tool for ingestive behavior research? Physiol Behav 98(4):393–401. https://doi.org/10.1016/j.physbeh.2009.07.002

Lindner A, Marx S, Kissenbeck S, Mosen H (2000) Saliva collection and relationship between lactate concentration in blood and saliva of exercising horses. J Equine Vet Sci 20(1):52–54. https://doi.org/10.1016/S0737-0806(00)80322-6

Lundberg J, Anggård A, Fahrenkrug J (1981) Complementary role of vasoactive intestinal polypeptide (VIP) and acetylcholine for cat submandibular gland blood flow and secretion. II effects of cholinergic antagonists and VIP antiserum. Acta Physiol Scand 113(3):329–336. https://doi.org/10.1111/j.1748-1716.1981.tb06903.x

Luo X, Choi JY, Ko SBH, Pushkin A, Kurtz I, Ahn W et al (2001) HCO3- salvage mechanisms in the submandibular gland acinar and duct cells. J Biol Chem 276(13):9808–9816. https://doi.org/10.1074/jbc.M008548200

Mandel ID (1989) The role of saliva in maintaining oral homeostasis. J Am Dent Assoc 119(2):298–304. https://doi.org/10.14219/jada.archive.1989.0211

Marchetti P, Benzi L, Masoni A, Cecchetti P, Giannarelli R, Di Cianni G et al (1986) Salivary insulin concentrations in type 2 (non-insulin-dependent) diabetic patients and obese non-diabetic subjects: relationship to changes in plasma insulin levels after an oral glucose load. Diabetologia 29(10):695–698. https://doi.org/10.1007/BF00870278

Mathison R, Davison J, Befus A (1994) Neuroendocrine regulation of inflammation and tissue repair by submandibular gland factors. Immunol Today 15(11):527–532. https://doi.org/10.1016/0167-5699(94)90209-7

Matsuo R (2000) Role of saliva in the maintenance of taste sensitivity. Crit Rev Oral Biol Med 11(2):216–229. https://doi.org/10.1177/10454411000110020501

Matsuzaki T, Suzuki T, Koyama H, Tanaka S, Takata K (1999) Aquaporin-5 (AQP5), a water channel protein, in the rat salivary and lacrimal glands: immunolocalization and effect of secretory stimulation. Cell Tissue Res 295(3):513–521. https://doi.org/10.1007/s004410051257

McDonald RE, Fleming RI, Beeley JG, Bovell DL, Lu JR, Zhao X et al (2009) Latherin: a surfactant protein of horse sweat and saliva. PLoS One 4(5):1–12. https://doi.org/10.1371/journal.pone.0005726

Mega J, McGhee J, Kiyono H (1992) Cytokine- and Ig-producing T cells in mucosal effector tissues: analysis of IL-5- and IFN-gamma-producing T cells, T cell receptor expression, and IgA plasma cells from mouse salivary gland-associated tissues. J Immunol 148(7):2030–2039
Melvin JE, Yule D, Shuttleworth T, Begenisich T (2005) Regulation of fluid and electrolyte secretion in salivary gland acinar cells. Annu Rev Physiol 67(1):445–469. https://doi.org/10.1146/annurev.physiol.67.041703.084745
Messenger B, Clifford MN, Morgan LM (2003) Glucose-dependent insulinotropic polypeptide and insulin-like immunoreactivity in saliva following sham-fed and swallowed meals. J Endocrinol 177(3):407–412. https://doi.org/10.1677/joe.0.1770407
Mizuta K, Karita K, Izumi H (2000) Parasympathetic reflex vasodilatation in rat submandibular gland. Am J Physiol Regul Integr Comp Physiol 279(2):R677–R683
Moller K, Benz D, Perrin D, Soling HD (1996) The role of protein kinase C in carbachol-induced and of cAMP-dependent protein kinase in isoproterenol-induced secretion in primary cultured guinea pig parotid acinar cells. Biochem J 314(1):181–187
Murakami M, Ohtake T, Dorschner RA, Gallo RL (2002) Cathelicidin antimicrobial peptides are expressed in salivary glands and saliva. J Dent Res 81(12):845–850. https://doi.org/10.1177/154405910208101210
Nater UM, Rohleder N (2009) Salivary alpha-amylase as a non-invasive biomarker for the sympathetic nervous system: current state of research. Psychoneuroendocrinology 34(4):486–496. https://doi.org/10.1016/j.psyneuen.2009.01.014
Niada S, Ferreira LM, Arrigoni E, Addis A, Campagnol M, Broccaioli E, Brini AT (2013) Porcine adipose-derived stem cells from buccal fat pad and subcutaneous adipose tissue for future preclinical studies in oral surgery. Stem Cell Res Ther 4:148. https://doi.org/10.1186/scrt359
Oconnor DT (1983) Chromogranin - widespread immunoreactivity in polypeptide hormone producing tissues and in serum. Regul Pept 6(3):263–280. https://doi.org/10.1016/0167-0115(83)90145-3
Perogamvros I, Aarons L, Miller AG, Trainer PJ, Ray DW (2011) Corticosteroid-binding globulin regulates cortisol pharmacokinetics. Clin Endocrinol 74(1):30–36. https://doi.org/10.1111/j.1365-2265.2010.03897.x
Proctor G, Carpenter G (2002) Neural control of salivary secretory IgA secretion. In: Clow A, Hucklebridge F (eds) Neurobiology of the immune system. Academic, San Diego, pp 187–212
Proctor GB, Carpenter GH (2007) Regulation of salivary gland function by autonomic nerves. Auton Neurosci Basic Clin 133(1):3–18. https://doi.org/10.1016/j.autneu.2006.10.006
Roth G, Calmes R (1981) Salivary glands and saliva. In: Mosby C (ed) Oral biology. CV Mosby, St Louis, pp 196–326
Sandoval J (1998) Tratado de Anatomía Veterinaria. Tomo III: Cabeza y Sistemas viscerales, 3rd edn. Imprenta Sorles, León
Saruta J, Tsukinoki K, Sasaguri K, Ishii H, Yasuda M, Osamura YR et al (2005) Expression and localization of chromogranin A gene and protein in human submandibular gland. Cells Tissues Organs 180(4):237–244. https://doi.org/10.1159/000088939
Sato F, Kanno T, Nagasawa S, Yanaihara N, Ishida N, Hasegawa T, Iwanaga T (2002) Immunohistochemical localization of chromogranin A in the acinar cells of equine salivary glands contrasts with rodent glands. Cells Tissues Organs 172(1):29–36. https://doi.org/10.1159/000064389
Segawa A, Yamashina S (1998) The dynamics of exocytosis of preformed secretory granules from acini in rat salivary glands. In: Garrett J, Ekström J, Anderson L (eds) Glandular mechanisms of salivary secretion, Frontiers in oral biology, 10th edn. Karger, Basel, pp 80–100
Shimojo M, Ricketts ML, Petrelli MD, Moradi P, Johnson GD, Bradwell AR et al (1997) Type 2 in human mineralocorticoid target tissues: evidence for nuclear localization. Society 138(3):1305–1311
Slomiany BL, Murty VLN, Piotrowski J, Slomiany A (1996) Salivary mucins in oral mucosal defense. Gen Pharmacol 27(5):761–771. https://doi.org/10.1016/0306-3623(95)02050-0

Tamarin A (1966) Myoepithelium of rat submaxillary gland. J Ultrastruct Res 16(3–4):320–338. https://doi.org/10.1016/S0022-5320(66)80066-7

Thie NMR, Kato T, Bader G, Montplaisir JY, Lavigne GJ (2002) The significance of saliva during sleep and the relevance of oromotor movements. Sleep Med Rev 6(3):213–227. https://doi.org/10.1053/smrv.2001.0183

Van Faassen M, Bischoff R, Kema IP (2017) Relationship between plasma and salivary melatonin and cortisol investigated by LC-MS/MS. Clin Chem Lab Med 55(9):1340–1348. https://doi.org/10.1515/cclm-2016-0817

Vázquez-Autón J, Gil-Cano F, Latorre-Reviriego R, Ramírez-Zarzosa G, López-Albors O, Ayala-Florenciano M et al (2002) In: Marín D (ed) Manual de Prácticas de Anatomía Veterinaria: Sistemas Viscerales, 1st edn. Diego Marín, Murcia

Velayos J, Santana H (2007) Anatomía de la Cabeza para odontólogos (4[a]). Médica panamericana, Madrid

Vining RF, Mcginley RA, Symons RG (1983) Hormones in saliva: mode if entry and consequent implication for clinical interpretation. Clin Chem 29(10):1752–1756

Young J, Cook D, Lennep E, Roberts M (1987) Secretion by the major salivary glands. In: Johnson L (ed) Physiology of the gastrointestinal tract. Raven Press, New York, pp 773–816

Zanetti M (2005) The role of cathelicidins in the innate host defenses of mammals. Curr Issues Mol Biol 7:179–196

Chapter 2
Saliva in Ingestive Behavior Research: Association with Oral Sensory Perception and Food Intake

Elsa Lamy, Ann-Marie Torregrossa, Paula Midori Castelo, and Fernando Capela e Silva

Objectives

The present chapter aims to present what is known about the involvement of saliva in the way humans and animals perceive the sensory characteristics of diet. The chapter will start by presenting the biology of flavour perception and how it influences food acceptance and choice. We will then review the newest information on the participation of saliva in flavour and taste perception and the effect of pathologies in oral sensory perception.

E. Lamy (✉)
MED - Mediterranean Institute for Agriculture, Environment and Development, IIFA - Instituto de Investigação e Formação Avançada, University of Évora, Évora, Portugal
e-mail: ecsl@uevora.pt

A.-M. Torregrossa
Department of Psychology, University at Buffalo and University at Buffalo Center for Ingestive Behavior Research, Buffalo, NY, USA
e-mail: amtorreg@buffalo.edu

P. M. Castelo
Department of Pharmaceutical Sciences, Universidade Federal de São Paulo, Diadema, Brazil
e-mail: paula.castelo@unifesp.br

F. Capela e Silva
MED - Mediterranean Institute for Agriculture, Environment and Development, IIFA - Instituto de Investigação e Formação Avançada, University of Évora, Évora, Portugal

Department of Biology, University of Évora, Évora, Portugal
e-mail: fcs@uevora.pt

A. Tvarijonaviciute et al. (eds.), *Saliva in Health and Disease*,
https://doi.org/10.1007/978-3-030-37681-9_2

2.1 Introduction

There is large individual variation in oral conditions. People may vary in how they manipulate food inside the mouth, in the biology of their sensory receptors and in their salivary composition. It is expected that the same foods may results in variable reports of perception and acceptance. The role of saliva in food acceptance and ingestion is unsurprising. The oral cavity, the structures responsible for the sensory perception of food, and the constituents of consumed foods come into direct contact with saliva. As soon as food enters the mouth, saliva will begin to dissolve tastants and change food texture (De Wijk et al. 2004; Ferry et al. 2006). Moreover, salivary constituents interact with food molecules, at times altering the way that the molecules access sensory structures.

In the recent years, increasing evidence has emerged suggesting that saliva is a major contributor to both flavour (taste and aroma) and texture perception (Mosca and Chen 2017). Specific salivary proteins have been identified, which are related to individual sensitivity for basic tastes in humans (Dsamou et al. 2012; Rodrigues et al. 2017c) and recent studies using animal models reported changes in taste thresholds as well as acceptance after salivary changes (Martin et al. 2018, 2019b). In addition, variations in the dynamics of salivary protein secretion have been identified as a factor contributing to inter-individual differences in astringency sensitivity and acceptance (Dinnella et al. 2011).

In the context of ingestive behaviour, saliva composition studies are not only relevant because of the potential saliva has in influencing food sensory perception, but also due to the influence this fluid can have on the bioavailability of dietary compounds and in the subsequent effects these compounds can have in individual health and performance. Moreover, saliva secretion responds to diet, with changes in composition that can be more or less pronounced, and more or less prolonged in time, according to the different dietary constituents. For example, rodents submitted to increased levels of tannins in their diets respond with pronounced changes in salivary glands (Lamy et al. 2010b) and with changes in salivary protein composition, after few days [e.g. (Da Costa et al. 2008; Lamy et al. 2010a)]. These adaptive changes appear to protect animals from potential negative effects of these compounds, reversing the weight loss observed during the first days of consumption (Da Costa et al. 2008; Lamy et al. 2010b). Recent work has suggested that these changes in protein expression are also active at the sensory level. The presence of particular salivary proteins increases acceptance for bitter foods that would otherwise be rejected (Martin et al. 2019a).

Salivary composition is also variable between species, which may highlight the interplay of saliva and food. Saliva composition is substantially different between carnivore and herbivore or omnivore, reflecting the different dietary composition of these trophic groups (Sales Baptista et al. 2010). There are even significant differences in salivary proteome among herbivores with different feeding strategies (Lamy et al. 2008, 2009). While we believe food choices can be modulated by saliva composition, we should not forget that saliva composition also depends on food choices.

2.2 Oral Food Perception

2.2.1 Mechanisms of Basic Taste, Aroma and Oral Tactile Perception

Flavour is defined by the American Society for Testing and Materials (ASTM) as a "perception resulting from stimulating a combination of the taste buds, the olfactory organs, and chemesthetic receptors within the oral cavity" (Relating 2007). In fact, it is due to this combination of senses that we recognize each food. Any factor that can affect taste, olfactory or tactile perception will affect food flavour. For that reason, the potential contribution of saliva for the perception and acceptance of each of these senses will contribute to flavour acceptance.

2.2.1.1 Taste

Taste results from the activation of epithelial-derived taste cells, which are responsible for taste reception and signal transduction. These taste cells are organized into taste buds, which are distributed through the surface of papillae located mostly on the superior surface of the tongue. Among the different types of taste papillae, fungiform papillae are mainly distributed by the anterior two-thirds of the tongue, foliate papillae locate on the lateral sides of the tongue, and circumvallate papillae on the posterior part (Gravina et al. 2013). Taste reception occurs at the level of taste receptor cells, which are located in taste buds, being these last located in the papillae. For a deeper understanding on the types of cells present in taste buds, see the review by Gravina and colleagues (Gravina et al. 2013).

Sweet, bitter and umami tastes are transduced in similar way, which is dissimilar from salty and sour tastes transduction. In the case of sweet, bitter and umami, G protein-coupled receptors (GPCRs) located in the plasma membranes of Type II cells bind the taste stimuli. This binding causes in the conformational change of the protein, resulting in the activation of an intracellular heterodimeric G protein, composed of alpha, beta and gamma subunits. This activation will, in turn, result in the stimulation of multiple downstream intracellular pathways, depolarization of the taste cell and neurotransmitter release. Sweet and umami transduction are mediated by a small family of heterodimer GPCRs: T1R2 + T1R3 for sweet-tasting compounds and T1R1 + T1R3 for umami. T2R GPCR family respond to a diversity of bitter compounds (Roper 2007).

Salty and sour tastes are transduced through different processes. In the case of sour taste, it is assumed to be mainly elicited by acids, although it is known that sourness is not correlated simply with the concentration of protons in solution. One of the mechanisms appears to be through passage of extracellular protons through ion channels and ion exchangers, such as the sodium-hydrogen exchanger, at cell membrane. Other routes of entry for extracellular protons might also exist, such as amiloride-sensitive epithelial sodium channels (ENaC) [reviewed in (Roper 2007)]. In the case of salty taste, amiloride-sensitive and amiloride insensitive pathways were referred: 1) the amiloride-sensitive is mediated by the epithelial channels referred and respond selectively to sodium (and lithium) salts, whereas the amiloride-insensitive responds to a broad range of sodium- and non-sodium salts (Lewandowski et al. 2016).

2.2.1.2 Food Aroma Perception

Ingestive behaviour, oral sensory perception and food intake are also modulated by food odour. Inhalation of positive hedonic (pleasant smelling) scents has appetizing effects in both hungry and satiated states (Warwick et al. 1993; Zoon et al. 2016). To understand the participation of saliva in aroma perception, it is important to mention that odour molecules can reach the nasal cavity not only through orthonasal olfaction (direct inhalation in front of the nose), but also through retronasal route (this occurs during chewing or drinking, when the tongue pushes air to the back of the nasal cavity and the volatile molecules released reach the olfactory epithelium) (Genva et al. 2019). The retronasal olfactory pathway, is commonly associated with the sense of taste contrary to what happens with orthonasal olfaction (Small et al. 2005; Kakutani et al. 2017). It has been observed that odour can modulate taste perception. Kakutani and colleagues (Kakutani et al. 2017) demonstrated that retronasal odour of vanilla after drinking enhanced sweet taste, whereas orthonasal odour does not.

The transduction of olfactory information occurs in the olfactory epithelium, that lines approximately half of the nasal cavity, the remaining surface is lined by respiratory epithelium, which lacks neurons. The olfactory mucosa in adults is characterized by irregular boundaries and interspersed patches of respiratory epithelium (Escada and Escada 2013). The human olfactory mucosa consists of four cell types: ciliated bipolar olfactory receptors (ORs), sustentacular cells, microvillar cells, and basal cells (Escada and Escada 2013). The ciliated bipolar olfactory receptor cell is a true bipolar neuron that projects a single dendrite to the surface of the olfactory neuroepithelium where the olfactory receptor is located, and a single axon to the olfactory bulb, the primary olfactory centre (Escada and Escada 2013; Genva et al. 2019).

The lamina propria contains the Bowman glands, which are the main source of mucus in the olfactory region. These, among other functions, provide a suitable molecular and ionic medium for the olfactory transduction and detection of odours (Hajjar et al. 2006). This mucus has olfactory binding proteins (OBP), which allow the transport of the volatile molecules through the aqueous mucus layer and participate in the removal of odorants at the receptor for the cessation of olfactory signals (Hajjar et al. 2006).

Odour information is transmitted to glomeruli in the olfactory bulb (OB), and further to higher brain areas where odour perception is constructed (Genva et al. 2019). The actual number of olfactory stimuli that humans can discriminate is still unknown but estimates ranging from 400,000 to 1 trillion odours (Grabe and Sachse 2018). The ORs are a member of the G protein-coupled receptor superfamily. The olfactory system uses a combinatorial receptor-coding scheme to encode odour identities, which means that the vast majority of ORs recognize multiple odorants, and each odorant is recognized by different combinations of ORs. Moreover, changes in the concentration of an odorant can also change its receptor code and at higher concentrations, additional ORs were invariably recruited into the odour response (Niimura et al. 2014; Silva Teixeira et al. 2016).

2.2.1.3 Food Texture (Tactile) Perception

In addition to taste and smell, food and beverage texture and mouthfeel can influence the way in which multisensory flavours are experienced and for these reasons are critical to consumer choice and acceptability (Szczesniak 2002). Whereas texture is used mostly in reference to solid and semi-solid foods, mouthfeel includes all of the tactile (feel) properties of food or drinks perceived during the time they are in the mouth (Guinard and Mazzucchelli 1996). The tactile sensations elicited by foods can modulate the other oral sensations. Some results suggested that increased viscosity can reduce taste perception in sweet and salty solutions (Christensen 1980), or the intensity of an creamy odour (Weel et al. 2002).

Texture and mouthfeel can be perceived through: (1) mechanoreceptors in the superficial structures of the mouth such as the hard and soft palate, tongue and gums; (2) mechanoreceptors in the periodontal membrane surrounding the roots of the teeth; (3) and mechanoreceptors in the muscles and tendons involved in mastication (Guinard and Mazzucchelli 1996). All these mechanoreceptors have specialized nerve endings, which can be ‘free nerve endings’ (fine fibrils extending from the sensory nerves), or ‘encapsulated nerve endings’ (such as the larger Pacinian and Meissner corpuscles, Krause end bulbs and Ruffini cylinders) (Guinard and Mazzucchelli 1996). Detailed information on the sensory perception of texture and mouthfeel has been reviewed elsewhere [e.g. (Guinard and Mazzucchelli 1996; Engelen and Van Der Bilt 2008)].

To understand the sensory perception of texture and mouthfeel, we must also take into account: (1) that the oral and product temperature can influence this perception (Engelen et al. 2003); (2) the succession of perceptual events that happen in mouth during mastication (Lenfant et al. 2009); (3) and the effects of saliva composition (Engelen et al. 2007; Kupirovič et al. 2017). More detail about how saliva composition can affect texture and mouthfeel perception will be reviewed later in this chapter.

2.2.2 How Oral Food Perception Influences Food Acceptance and Choices

Taste has been referred as one of the main drivers of food consumption (Glanz et al. 1998). Food sensory qualities act as anticipatory signals of food nutritional content through both innate and learned processes (Bradbury 2004). Innately, sweet taste is connected to energy, conveyed by sugars (carbohydrates), whereas bitter is associated with poisonous and/or toxic compounds; salty, in turn, signals sodium chloride, a mineral necessary for almost all aspects of metabolism and neurological functioning. Even so, there is no denying that differences exist at the individual level, and there is variation in the acceptability of the different tastes and respective intensities.

Genetics contributes to taste sensitivity variation in people. One of the most well studied taste traits is the variation in the sensitivity to the bitter taste of the compounds phenyltiocarbamide (PTC) and 6-n-propylthiouracil (PROP). Although other factors can interfere, differences in the Single Nucleotide Polymorphisms (SNPs) in the TAR2R38 bitter taste receptor gene are linked to the ability to taste the bitterness of these compounds. Three SNPs result in altered transduction of amino acids (Pro49Ala, Ala262Val and Val296Ile), which in turn result in alterations in taste receptor function (Bufe et al. 2005). According to recent studies, the frequency of individuals with low sensitivity to bitterness (non-tasters) can vary from as low to 7%–40% (Cavallo et al. 2019). Although PTC and PROP are synthetic molecules, not found in nature, these polymorphisms correlate with perception of the bitterness of several vegetables (Shen et al. 2016). Although there is not yet a consensus, studies have reported differences in acceptance and consumption of bitter foods, like vegetables, by individuals with different SNPs for TAS2R38. AVI/AVI homozygotes, who experience low bitterness, have been reported as ingesting higher amounts of vegetables, compared to heterozygotes or homozygotes for PAV (high sensitivity individuals) (Sacerdote et al. 2007; Duffy et al. 2010; Sandell et al. 2014). Some of the controversy about the relationship between bitter taste sensitivity and bitter foods/drinks acceptance may be increased by the reaction people have to compounds like coffee, chocolate or alcoholic beverages, like beer. Despite the characteristic bitterness of these products, they also have psychoactive effects, which are associated with pleasure and enjoyment. These psychoactive effects can mediate the acceptance and preference of the associated bitter taste (Li et al. 2014; Masi et al. 2016; Fu et al. 2019).

Preferences can also be modified through life due to experience (Cowart et al. 2003). Repeated exposure and subsequent learning about the oral and post oral properties of foods can also drive preferences. Taste acceptance and preferences are determined very early, with some studies suggesting that taste learning occurs as early as foetal development. This issue has been focus of a recent review, which concluded that foetal flavour exposure increases the acceptance of similarly flavoured foods during infancy and potentially childhood (Spahn et al. 2019). Also, during breastfeeding, flavours from mothers' diets pass to milk, again increasing familiarity with these flavours and increasing their acceptance, leading to the hypothesis of these changes being also influenced by salivary protein expression. Increased acceptance of diverse foods is cited as one of the reasons breastfeeding is thought to prevent obesity development (Ventura 2017).

2.3 Saliva Composition and Food Sensory Perception

In the previous section, the basics of oral sensations, contributing to food oral perception were presented. Although food constituents are largely responsible for these sensory characteristics, saliva composition can alter the perception of foods. Mosca and Chen (2017) have recently reviewed how saliva-food interactions can influence different aspects of food oral perception, including flavour and texture. Saliva participates in food oral sensory perception at different levels. Saliva can act not only by interacting with food, but it also constitutes the medium that bathes the oral structures.

The influence of saliva on taste sensitivity was first suggested for bitter taste and astringency and much of the literature still focuses there. There is evidence in several animal models that bitter diets can drive changes salivary protein expression, an idea that will be covered in greater detail later in the chapter. Mehansho et al. (1987), demonstrated experimentally that animals that could produce proline rich salivary proteins (PRPs) in response to a tannin rich diet could maintain body mass on the diet, while those who were unable to produce the proteins could not maintain weight (Mehansho et al. 1987). This finding contributed to the idea that PRPs were a defence mechanism against the anti-nutritional action of tannin compounds. Recently these findings have been expanded to show that these bitter-diet regulated salivary proteins (BRSPs) can include more than PRPs, be induced by bitters other than tannin, and have distinct patterns of expression based on the diet. Glendinning (1992) later demonstrated that mice that were able to induce PRP in the saliva showed an increased preference for a tannic acid solution than mice who did not have these proteins (Glendinning 1992). This was an important finding because it suggested that the mice were not only more capable of consuming the food, as Mehansho showed, but that the food was more acceptable. Recently it has been demonstrated that the changes in acceptability of the diet extend past tannic acid. For example, the expression of BRSPs results in increased acceptance of a quinine, a proto-typical bitter, upon first exposure (Martin et al. 2019a), changes in taste-guided licking behaviours (Martin et al. 2018), taste nerve signalling (i.e., chorda tympani nerve recordings) (Martin et al. 2018), taste detection thresholds (Martin et al. 2019b), and behaviours driven by post oral feedback (e.g., meal size) (Martin et al. 2018). Together these findings suggest that the presence of these salivary proteins can fundamentally reduce the bitter taste of the diet. While identifying which proteins are the drivers for these changes is a work in progress the literature in humans offers some specific examples. A summary of the main evidences for the participation of salivary proteins in oral food perception is presented in Table 2.1.

Table 2.1 Summary of the salivary proteins described in literature as associated with oral food preferences

Salivary protein	Oral sensation	Type of association	References
α-Amilase	Taste – sweet	Negatively associated with sweet taste sensitivity	Rodrigues et al. (2017a)
	Aroma	Reduce the release of esters and ketones	Pagès-Hélary et al. (2014)
	Tactile	Affinity to bind and complex tannins (astringent compounds), and, as such, potentially involved in astringency development.	Da Costa et al. (2008) and Lamy et al. (2010a)
		Reduce the viscosity of starch-based products	De Wijk et al. (2004) and Mandel et al. (2010)
		Reduce the thickness of starch-based products	de Wijk and Prinz (2005) and Engelen et al. (2007)
Carbonic anhydrase VI (CA VI)	Taste	Contributes to growth and development of taste buds	Shatzman and Henkin (1981)
		Positively associated with sensitivity to bitter taste	Calò et al. (2011), Cabras et al. (2012) and Rodrigues et al. (2017b)
Cystatins	Taste – bitter	Negatively associated with bitterness	Dsamou et al. (2012) and Morzel et al. (2014)
		Association with bitter taste is dependent on BMI	Rodrigues et al. (2017b, 2019)
Proline-rich proteins (PRPs)	Taste – bitter	Associated with PROP bitter taste sensitivity	Cabras et al. (2012) and Melis et al. (2013)
		Reduce bitter taste perception and/or increase bitterness acceptance	Martin et al. (2018, 2019a, b)
	Tactile	Participate in the development of astringency	Soares et al. (2011, 2012, 2018)
Mucins	Tactile	Participate in astringency development and the incapacity of to maintain the levels increases the perception of this sensation	Dinnella et al. (2010) and Ployon et al. (2018)
	Aroma	Interfere with the retention of volatile compounds	Van Ruth et al. (2001)
Prolactin-inducible protein	Taste	Negatively associated with bitterness	Dsamou et al. (2012) and Rodrigues et al. (2017b)
		Negatively associated with sweetness	Rodrigues et al. (2017a)
Leptin	Taste	Negatively associated with sweet taste sensitivity	Rodrigues et al. (2017a)

2.3.1 Saliva and Basic Tastes

Despite saliva being in close contact with food and the structures in oral cavity, responsible for the sensory characteristics of food, there has been little work historically examining the role of saliva in sensory perception and acceptance. Much of the work on saliva focused on its relationship with oral cavity health. Only more recently has it started to gain attention as potentially involved in sensory perception. Studies have highlighted that individual variation in the composition of saliva can contribute to individual differences we see in diet acceptance and preference and even how food is sensed in oral cavity.

Carbonic Anhydrase VI (CA VI) CA VI was first identified as gustin, a major zinc-containing protein, making up 3% of the total protein in saliva (Shatzman and Henkin 1981). This protein, was linked to bitter taste sensitivity for several reasons. First it is secreted not only by the parotid, but also by the von Ebner serous minor salivary glands. Von Ebner glands are embedded in the tongue and secrete near the circumvallate papillae, a site where bitter taste receptors are at high density. Second, some studies implicated this protein in the growth and development of taste buds (Shatzman and Henkin 1981). Lastly, the levels of this protein were associated with the levels of zinc in saliva, and decreases in salivary zinc were correlated with decreased taste acuity (Shatzman and Henkin 1981).

Later it was demonstrated, with advances in molecular techniques that genetic differences at the level of salivary CA VI contribute to different bitter taste sensitivities between individuals. Padiglia and colleagues reported a link between bitter taste sensitivity (assessed through PROP taste status) and polymorphisms in the CA6 gene, which contains the sequence for this salivary protein (Padiglia et al. 2010). In other studies differences in the expression levels of CA VI were described between individuals with high and low sensitivities to bitter taste, namely increased levels were reported in PROP supertasters compared to those who were less sensitive (Cabras et al. 2012; Rodrigues et al. 2017b). Even so, the relationship between CA VI and bitter taste sensitivity is debatable, and factors such are Body Mass Index (BMI), sex or age can influence the way CA VI and bitterness associate (Rodrigues et al. 2019). This is an issue that needs to be explored more deeply.

Cystatins Cystatins are a class of cysteine protease inhibitors. Salivary cystatins are found in higher expression levels in non-stimulated saliva from men with low sensitivity to bitterness (Cabras et al. 2012; Dsamou et al. 2012) and children with higher levels of acceptance for bitterness (i.e. low sensitivity, Morzel et al. 2014). These data support a negative association between salivary cystatin levels and bitter taste sensitivity. However, this relationship was also recently observed to be dependent on BMI suggesting the relationship may be more complicated (Rodrigues et al. 2017b, 2019).

In addition to its potential role in bitter taste cystatins, have been shown to be increased in individuals with low sensitivity to saltiness. It has been suggested that increased cystatins inhibit the protease action needed for the epithelial sodium channel mediated NaCl transport (Stolle et al. 2017).

Proline Rich Protein Family (PRPs) Different lines of evidence have suggested that PRPs may interact with bitter taste stimuli. First, PRPs are produced not only in the parotid gland, but they are also produced in the Von Ebner's glands that are located around the circumvalate and foliate papillae where the majority of bitter taste receptors are located (Azen et al. 1990). Second, genes for PRPs and T2Rs, the purported bitter taste receptors, are interspersed on the same chromosome suggesting a possible link between PRPs and bitter taste (Adler et al. 2000). Third, gene linkage studies have implied a role for PRPs in human quinine acceptance (Reed et al. 2010). In mice, similar gene-linkage studies have implied a role for PRPs in tasting of the bitter taste stimuli: quinine, sucrose octaacetate, and cycloheximide (Lush and Holland 1988; Azen et al. 1989; Capeless et al. 1992). However, the human experimental data is the most compelling argument to date. Cabras and colleagues demonstrated that the basic salivary PRPs Ps-1 and Il-2 peptides are higher in individuals who find PROP bitterness to be more intense (Cabras et al. 2012). Their causal role in PROP bitter taste sensitivity was demonstrated by supplementing Ps-1 protein in individuals lacking it in saliva. Individuals with experimentally increased Ps-1 showed enhanced PROP responsiveness (Melis et al. 2013).

Salivary Amylase Recently studies of the salivary proteome has been extended to examine sweetness (Rodrigues et al. 2017a). Salivary amylase is one of the proteins associated with sweet taste sensitivity (Rodrigues et al. 2017a). It catalyses the hydrolysis of starch into sugars. Recently, it was demonstrated that the levels of this salivary protein are positively correlated with the levels of maltose produced, and therefore positively correlated with the sweetness produced by starch mastication (Aji et al. 2019). Nevertheless, the way salivary amylase contributes to sweetness perception may vary across different starch-based foods. A recent study indicated that individuals with higher levels of salivary amylase rated a bread with lower sweetness (Lamy et al., *submitted*). Although the causes for this apparently contradictory result need to be elucidated, it is possible that the other sensations elicited by the oral digestion of starch, including textural ones, can "mask" sweetness perception (Lapis et al. 2016, 2017).

Lipase Activity in Saliva There is not a consensus on whether or not fat qualifies as a basic taste or should be considered a somatosensory signal. However, variations in its perception also appears to be influenced by saliva composition. Several mechanisms have been suggested to explain the way saliva alters fat perception. In the case of "fat taste" it is thought that the lypolitic activity of saliva can influence the level of free fatty acids available for binding (Neyraud et al. 2012, 2017). However, some studies failed to find relevant lingual lipase activity during mastication of fatty foods casting doubt on the role of salivary lipase in fat oral perception (Kulkarni and Mattes 2014).

2.3.2 *Saliva and Tactile Sensations*

2.3.2.1 The Role of Saliva in Astringency Development

Astringency, the drying and rough sensation associated with some compounds (e.g. polyphenols, such as tannins) is an oral sensation which results from the interaction between these compounds and salivary proteins. According to the definition of American Society for Testing and Materials (ASTM), astringency refers to "the complex of sensations due to shrinking, drawing or puckering of the epithelium as a result of exposure to substances such as alums or tannins". When intense astringency acts as food deterrent. Despite the considerable number of studies working toward understanding astringency development, the mechanisms responsible for this tactile sensation are not completely elucidated.

Astringency is considered to be a tactile sensation, rather than a basic taste (chemical sensation). This is supported, on one hand, by the increase in astringency intensity perceived after repeated contact with the astringent compound (Dinnella et al. 2010), and on the other hand because this sensation is sensed on nongustatory mucosal surface, requiring tissue movement to be perceived. Nevertheless, it has been shown that, besides mechanoreceptors, chemosensory detection of astringent phenols can occur and this simultaneous stimulation of chemoreceptors and mechanoreceptors will create the entire sensation of astringency (Schöbel et al. 2014).

Astringency, produced by food items, involves mainly the interaction between polyphenols (particularly tannins) and proteins. This interaction between tannins and salivary proteins was first suggested by Bate-Smith, in 1954 (Bate-Smith 1954). Most of the studies relating salivary proteins with astringency are focused in the proline rich proteins (PRPs) family. This family contains three classes: basic (bPRPs), glycosylated (gPRPs) and acidic (aPRPs) (Manconi et al. 2016). Among these, bPRPs and gPRPs have been reported as having higher affinity to complex and precipitate tannins (Lu and Bennick 1998). In fact the binding of tannins, and acting as a defence against the potentially harmful, anti-nutritional, or toxic effects, was the main biological function attributed to bPRPs in humans and animals (Shimada 2006). More recently, besides bPRPs, some aPRPs and P-B peptide have been shown able to bind polyphenols and to contribute to the feeling of astringency (Brandão et al. 2014; Quijada-Morín et al. 2016; Soares et al. 2018).

Initially, astringency development was suggested to result from the formation of complexes between tannins and PRPs, which would precipitate, and these precipitates being sensed in oral cavity. In addition, it was proposed that tannins bind the proteins covalently linked to oral mucosal cells, which consist the layer surrounding the oral tissues, and that this complex causes increased friction, which results in the loss of lubrication in the oral cavity (Baxter et al. 1997). Later on, a 3-stage model of the interaction tannin-proteins was proposed (Charlton et al. 2002). In an initial stage, it is proposed that proteins bind to the surfaces of tannin aromatic rings through hydrophobic interactions, stabilization of these complexes is due to hydrogen binding between the hydroxyl groups of tannins and H-acceptor sites of proteins.

In a second stage, these tannin-protein complexes self-associate, via hydrogen bonding, aggregating in larger soluble protein-tannin complexes. These larger soluble complexes continue to grow until they are so large that they became insoluble and precipitate.

Ployon et al. (2018) did go deeper in understanding the participation of oral mucosal pellicle in astringency sensation using a cell-based model. MUC5B-tannin aggregates were observed, which increased in size by increasing tannin amounts in the protein-tannin mixture. However, it was observed that, in presence of bPRPs these structural alterations were reversed, suggesting a protective role of these salivary proteins suggesting they work to reduce astringency.

The influence of inter-individual salivary proteome variations in astringency perception was also observed. Dinnella and colleagues reported that individuals perceiving astringency with higher intensity presented higher levels of salivary PRPs, but were incapable of maintaining them after contact with astringent solutions (Dinnella et al. 2010). On the other hand, individuals less sensitive to astringency were capable of maintaining constant saliva characteristics. This different response, in saliva secretion, between astringent sensitive and low sensitive individuals was observed, with proteins other than PRPs as well, including salivary amylase (Lamy et al. 2016).

Despite the recognized participation of PRPs in astringency development, other salivary proteins, such as amylase and cystatins are observed in the precipitates resultant from the interaction between salivary proteins – tannins (e.g. (Lamy et al. 2016), and the involvement of these proteins in this oral sensation and/or polyphenol-rich food acceptance deserves further study.

2.3.2.2 How Can Saliva Influence Food Texture Perception in Mouth?

Texture evaluation is one of the most important factors in food appreciation. This is particularly evident in children, where the texture of vegetables is one the main characteristics influencing their acceptance (Appleton et al. 2016). Creaminess is particularly important, with a widespread consumer appeal of smooth texture of dairy and non-dairy or even starch-based semi-solid foods. In the food industry the incorporation of commercial ingredients, like aggregated protein microparticles and hydrocolloids is common to enhance creaminess. The interaction of these agents with salivary proteins is of major importance to defining their final properties. Saliva is also linked to viscosity ratings. Saliva appears to increase the viscosity of liquids with low viscosity, but decrease it for liquids with high viscosity [reviewed in (Kupirovič et al. 2017)].

Glycoproteins have been shown to induce flocculation of emulsions, which could suggest a role for them in the perception of texture attributes of foods (Vingerhoeds et al. 2005). Glycoproteins present in saliva, are thought to provide the lubricating properties of saliva and the largest group are the mucins. Mucins are the main organic constituents of the slimy visco-elastic layer that coats mucosal surfaces (Efremova et al. 2002). Mucins have a highly glycosylated central part accounting for up 80% of its molecular mass (Shi et al. 2000). Mucins may interact with food components affecting the way individuals perceive sensory and textural attributes. In fact, bovine submaxillary gland mucin was observed to be able to act as an emulsifier, stabilizing oil-in-water emulsions. This may facilitate distribution of food lipids in the mouth, acting at their textural perception level. Mucin interactions with food proteins were recently reviewed (Çelebioğlu et al. 2019).

Although mucins are the most abundant, there are other proteins that have been suggested to alter perception of food texture. Salivary proteins with molecular masses of 27 and 55 kDa have also been suggested as involved in emulsification process (Glumac et al. 2019) and α-amylase has been linked to perception of food texture too. α-Amylase initiates starch digestion in the mouth, by cutting starch at the alpha (1–4) binding sites. This action of α-amylase results in a decrease in the ability of starch to bind water, with the resultant reduction of food viscosity (De Wijk et al. 2004). Different levels of amylase among different individuals appear to be responsible for the variation in the textural perception of starch containing foods. High enzymatic activities of this protein were reported to be associated with lower reported thickness of starch-based semi-solid custards (de Wijk and Prinz 2005; Engelen et al. 2007). Individuals with high amylase levels report faster and higher decreases in starch viscosity, comparatively to the ones having low amylase levels (Mandel et al. 2010).

2.3.3 Saliva and Aroma

During mastication, volatile compounds are released, which will access olfactory cells receptors through retronasal route. Through mastication, food compounds are mixed with saliva. At this point, aroma compounds can interact with salivary proteins.

It has been observed that mucins decrease the retention of highly volatile compounds but increased the retention of hydrophobic compounds (Van Ruth et al. 2001). In another study, mucins and α-amylase were observed to reduce the release of the aroma compounds, specifically esters and ketones (Pagès-Hélary et al. 2014). It is interesting to note that in this study the type of aromas affected by human saliva were different from those affected by the use of artificial saliva, suggesting that the effect is due to proteins not just moisture (Pagès-Hélary et al. 2014).

A study, performed with cheese products, showed that both saliva sodium content and lipolytic activity impacted aroma perception (Guichard et al. 2017). The authors observed that salt congruent aroma perception was lower in individuals with high salivary sodium, on one hand, and that fat related aromas perception was increased in individuals with higher salivary lipolytic activity. In the same sense, aroma perception of wines vary among individuals with different BMIs and these differences were attributed to alterations in the salivary microbiome and/or protein profile (Piombino et al. 2014). Recently, the role of these salivary constituents was highlighted, by testing different wine matrixes with artificial or human saliva

(Piombino et al. 2019). The effect of adding each type of saliva in wine was that volatile release was different. Mucins and α-amylase were reported to be involved in the retention of the most hydrophobic volatiles as well as in the metabolization of some aromas.

BMI has also been implicated in variation in olive oil perception. Olive oil mixed with human saliva has a higher release of C6 compounds and a lower release of C5 when saliva is from overweight individuals compared to normal weight individuals (Genovese et al. 2018).

The importance of the interaction of saliva and aroma perception is highlighted by a study suggesting that individual characteristics of saliva composition affect the in-mouth aroma release of vegetables like raw cabbage. In this study, authors monitored the production of sulphur volatile resultant from the interaction between raw cabbage and human saliva *in vitro*. They found a tenfold difference in the sulphur volatile production across individuals (Frank et al. 2018). These data suggest that the differences in release will result in variable aroma perception among these individuals. The constituents of saliva responsible for these differences were not identified.

2.4 Changes in Saliva Composition in Response to Ingestion: Can This Affect the Sensory Perception of Subsequently Ingested Food?

Not only can saliva influence oral food perception and dietary choices, but food intake can also influence saliva composition. One of the best examples for understanding how sensory stimuli influence saliva secretion is the use of acid stimulation to induce saliva production.

The association between saliva composition and the type of food ingested is well understood when looking at the different animal species, thriving in different food environments. Carnivores, herbivores and omnivores present different salivary compositions due to the particular composition of foods usually ingested (Sales Baptista et al. 2010). Even for species belonging to the same trophic group, differences are apparent across different feeding strategies. For example, small ruminant species, which share pasture areas, but choose different plants and/or plant constituents have different salivary protein profiles (Lamy et al. 2009). Two decades ago, herbivore species started to be compared for the presence of salivary PRPs. PRPs, as mentioned previously, bind tannins, forming complexes that are stable under different pH conditions. This capacity of keeping tannins complexed through the entire gastrointestinal system, impeding their interaction with proteins from food or endogenous proteins from animals, was considered a defence against potential harmful effects of these plant secondary metabolites. Shimada reviewed the presence of tannin binding salivary proteins in the different animal species, the proteins were abundant in animals with diets rich in tannins, inducible in animals with mixed diets (periods of low vs. periods of high levels of tannins in diets) and absent in animals with diets low in these plant secondary metabolites (Shimada 2006). In rodents (rats and mice), the presence of salivary PRPs is low or even absent, when in standard diets, but induced after exposure to a tannin-rich diet (Shimada 2006; Lamy et al. 2010a). In these animals, tannin-enriched diets resulted in considerable increase in the size of acinar structures of major salivary glands (Da Costa et al. 2008; Lamy et al. 2010b). These changes in salivary glands and saliva composition were associated with recovery of animals from the weight loss animals had in the first 1–2 days of consumption (da Costa et al. 2008), reinforcing the role of saliva as defence against negative effects of dietary tannins. Although not at the level of PRPs, increases in dietary tannin levels also changed the saliva composition of species of ruminants e.g. (Lamy et al. 2011, 2019)].

The effect of diet composition in saliva secretion has been observed for other dietary constituents besides tannins. For example, the AMY1 gene, which is the gene that codes for salivary amylase, presents a higher copy number in populations that evolved under high-starch diets, comparatively to the ones from low-starch diets (Perry et al. 2007; Santos et al. 2012). This has been thought of as a selective advantage allowing individuals to thrive on starch-based diets. In line with this, it has been hypothesized that the amylase gene (AMY) underwent several gene copy number gains in humans, dogs, and mice, possibly along with increased starch consumption during the evolution of these species (Pajic et al. 2019).

The levels of salivary amylase are observed to increase when the levels of available starch decrease: in rats receiving a chow diet enriched with fat, the salivary amylase levels increased, even when the animals ingested similar amounts of energy (Rodrigues et al. 2015). Despite this needs further clarification, the lower percentage of starch ingested in the same amount of caloric content may be the reason for increased amylolytic activity.

The association of saliva with the type of diet consumed leads to the possibility that saliva acts as a non-invasive biomarker of food intake (Proctor et al. 2017). Although this possibility needs further studies to be elucidated, some researchers recently suggested that the qualitative aspects of diet, assessed by food frequency questionnaires, are linked to saliva global composition (Morzel et al. 2017).

Concerning the changes in saliva composition that diet induces in the short time, less is known. We recently showed that the immediate changes ingestion produces in saliva are dependent on the type of food ingested. Some salivary proteins change in response to multiple foods while other responses are food specific (*not published*). If saliva composition is linked to oral food perception, as is being discussed in the present chapter the ability to alter the salivary proteome with diet has huge implications. What you eat can change your experience with subsequent foods.

2.5 How Saliva Changes Under Pathological States Can Affect Oral Food Perception

As saliva has many functions of great importance, and some of them were discussed above like digestion, lubrication, gustatory function and maintenance of mucosal and teeth integrity (Marquezin et al. 2016; Freitas et al. 2017), it is expected that in cases of quantitative and qualitative changes in saliva secretion, chewing and taste perception will be impaired. Taste alterations related to diseases or specific health treatments have been recognized and, as stated by Zabernigg et al. (2010) in cases of cancer patients receiving chemotherapy, there is an urgent need for increased attention to the side effects of those therapies, both in research and clinical practice. Changes in salivary secretion related to cancer, dialysis and other chronic diseases (e.g. diabetes and obesity), and also to medication use, may impair chewing and gustatory functions (Buckner and Dwyer 2003; Marquezin et al. 2016; Walliczek-Dworschak et al. 2017; Wang et al. 2018) and, ultimately, food preferences and food intake.

Chewing is the first stage of the digestive process and results from a rhythmic pattern of mandibular movements in which food is crushed between the occlusal faces of the teeth under the influence of saliva: the fragmentation and humidification of food is the main function of mastication, but it also involves sensations related to the taste and pleasure of eating (Pereira et al. 2006). In addition, three functions were assigned to the sense of taste, namely a warning of danger, interpersonal communication and importance for eating and drinking (Stevenson 2009). For this reason, food, eating, and quality of life are intimately related and, when impaired, they will dramatically affect the individual's health-related quality of life (HRQOL) even before they affect nutritional status (Buckner and Dwyer 2003; Walliczek-Dworschak et al. 2017).

Saliva flow is important in dissolving non-volatile flavors in the oral cavity as well as it takes part in the maintenance of taste papillae and taste-sensing cells (Foster et al. 2011; Fábián et al. 2015). In the study of Walliczek-Dworschak et al. (2017), a correlation was observed between taste perception (evaluated by taste strips) and the salivary flow rate, probably because of variations in tastants substances as a result of a dilution effect. An increase in salivary flow may also change the concentration of electrolytes such as $Na+$, and these changes have been shown to influence taste perception (O'Mahony et al. 1982; Delwiche and O'Mahony 1996; Lugaz et al. 2005). In healthy children, a correlation between bitter taste sensitivity and unstimulated salivary flow rate was also observed (Marquezin et al. 2016). Thus, healthy conditions or the use of some medications that predispose to changes in salivary flow may compromise chewing function and taste perception.

Changes in salivary secretion may have many causes, including autoimmune exocrinopathy (e.g. primary Sjögren's syndrome), prescribed medication/polypharmacy, radiotherapy, poorly-controlled diabetes mellitus, among others (Mortazavi et al. 2014).

Of the most commonly reported complaints in the elderly is dry mouth, which has two aspects: xerostomia (symptom, subjective) and hyposalivation (sign, objective) (Thomson 2015; Turner 2016). This population is particularly vulnerable to gustatory alterations due to the use of many medications (polypharmacy), which may predispose to changes in the reception, transduction, propagation or perception of flavours and, also, in salivary secretion in terms of its physicochemical composition and volume. The global population is ageing and a reduced intake of food as a result of impaired physical function or difficulties in chewing, swallowing and poor appetite may lead to malnutrition. Ageing may also affect the ability to taste and smell due to diminished cognition and reduced chewing ability due to tooth loss (Xu et al. 2019). Poor appetite may be ascribed to impaired taste sensitivity and a past study showed an association between underweight and low taste sensitivity to bitter taste in middle- to old-aged nursing home residents (Fuchida et al. 2013).

In patients suffering from burning mouth syndrome, which is characterized by chronic pain and burning sensation of the tongue and oral mucosa, a reduced salivary flow rate and higher threshold for sourness were observed, which may justify the frequent simultaneous complaints of xerostomia and dysgeusia (Imura et al. 2016).

Many cancer patients receive chemotherapy as treatment during the course of their disease. A frequent side effect of this therapy is taste and smell impairments, experienced by a large proportion of this population, with variable duration that may persist even after the end of treatment (Zabernigg et al. 2010; IJpma et al. 2015, 2017). The most common symptom is the perception of a persistent metallic flavor and/or aftertaste with or without food intake (IJpma et al. 2015). As a consequence, those patients may suffer from loss of appetite, food avoidance/aversion, loss of weight, malnutrition, and depression, which may impact the prognosis of the treatment. According to a systematic review, caffeinated foods and drinks, red meat and citrus fruits or juices are frequently reported as aversive during chemotherapy (Boltong and Keast 2012), symptoms varying according to the chemotherapy regimen, tumour type and stage of treatment.

In patients receiving chemotherapy, the release of unbound and reactive forms of iron (Fe2+) in saliva secretion may result in the above mentioned persistence of metallic taste in the oral cavity (Toyokuni 2009); high Fe concentration may also be associated with neurodegenerative alterations, which was pointed out by Kamalinia et al. (Kamalinia et al. 2013). Another possible mechanism of taste/smell abnormality is the tissue damage caused by chemotherapy. By using a proteomic analysis, Wang et al. (2018) observed lower intensity of salivary α-amylase and low-abundance proteins in cancer patients comparing to healthy subjects which, according to the authors, might result in a difficulty to fully digest carbohydrates and a relative increase in the intensity of metallic taste. As sweet taste sensitivity and salivary amylase concentration seem to be associated (Marquezin et al. 2016; Rodrigues et al. 2017c), differences in amylase abundance may impair sweet taste perception.

A past study observed that patients suffering from taste disorders showed higher concentrations of total protein in saliva than healthy controls (Walliczek-Dworschak et al. 2017), although, according to Fabian et al. (2015), salivary proteins are likely to improve the availability of lipophilic polyphenol tastants for taste cells, which will ultimately increase taste perception. On the other hand, lower carbonic anhydrase VI concentrations were found in subjects with reduced taste and smell function (Shatzman and Henkin 1981; Henkin et al. 1999).

Infectious diseases, such as influenza-like infection (Henkin et al. 1975) or human immunodeficiency virus (HIV) (Graham et al. 1995), have been associated with taste abnormalities characterized by increased detection and recognition thresholds for various taste stimuli. At the oral cavity level, caries and periodontitis are two of the main frequent infectious diseases. It is known that these diseases are related with changes in the amount and types of microorganisms in mouth (Tezal et al. 2006), and this may ultimately change oral food perception.

Diabetes also seems to influence the composition of saliva (Abd-Elraheem et al. 2017) and saliva flow rate may be altered as a consequence of diabetic neuropathy. The subjects with higher mean blood glucose reported more taste disorders, probably as a result of the poor diabetic control (Dodds and Dodds 1997). The study of Yazla et al. (2018) demonstrated that diabetes type 2 was associated with olfactory and gustatory dysfunctions and as there was no difference between the diabetic patients with and without peripheral neuropathy, the authors hypothesized a central neuropathy. Hypogeusia may be an early sign of diabetic neuropathy which may be detrimental to the compliance of these patients to dietary recommendations (De Carli et al. 2018).

The relationship between overweight, salivary characteristics and taste/texture perception and preferences was little explored. Mosca et al. (2019) did not find differences in salivary parameters according to body mass index, while some changes in salivary pH, flow, electrolytic and protein composition of children and women with overweight were reported earlier (Pannunzio et al. 2010; Tremblay et al. 2012; Lamy et al. 2015; Rodrigues et al. 2019); these differences may have an influence on taste sensitivity (Overberg et al. 2012; De Carli et al. 2018) and preferences, especially for sweet taste, which needs further clarification. Mennella et al. (2014) observed higher salivary lipase and amylase activities in overweight subjects, in addition to a positive correlation between lipolysis activity and individual preference for high-fat foods and fat content consumption, suggesting an adaptive response to the low fat-taste perception related to a reduced salivary zinc concentration. Recently, increased levels of some isoforms of salivary cystatins was observed in children with obesity; in addition, cystatins were also related to bitter taste sensitivity: cystatin SN was found in higher levels in overweight children with low sensitivity to bitterness, and the opposite relationship was found in normal-weight ones (Rodrigues et al. 2019). Future studies are needed to clarify if those qualitative differences in salivary secretion are a cause and/or a consequence of overweight, which means that salivary secretion may influence taste perception and food intake and/or may serve as a source of biomarkers of fat accumulation, respectively.

2.6 Conclusion

More research is needed to elucidate the relationship between salivary proteins and taste, namely to establish if there is a causal relationship between proteins and taste perception. Moreover, understanding how variations in saliva can affect whole food perception can be of major interest for modulating food palatability and increase the acceptance of healthy, often bitter, food items. By reviewing the literature, it is clear that taste disorders have a substantial effect on various aspects of quality of life, and the loss of eating pleasure as a result of illness or therapies may be associated with weight loss and negative impact on social life. Therefore, the issue is worth of further investigation by nutritionists and/or health practitioners together with the food industry to develop interventions and food strategies involving taste enhancement and food flavouring considering the saliva variability of the different consumer groups.

References

Abd-Elraheem SE, saeed EL, mohammed A, Mansour HH (2017) Salivary changes in type 2 diabetic patients. Diabetes Metab Syndr Clin Res Rev 11:S637–S641. https://doi.org/10.1016/j.dsx.2017.04.018

Adler E et al (2000) A novel family of mammalian taste receptors. Cell 100(6):693–702. Available at: http://www.ncbi.nlm.nih.gov/pubmed/10761934. Accessed 16 Sept 2015

Aji GK, Warren FJ, Roura E (2019) Salivary α-amylase activity and starch-related sweet taste perception in humans. Chem Senses., Narnia 44(4):249–256. https://doi.org/10.1093/chemse/bjz010

Appleton KM et al (2016) Increasing vegetable intakes: rationale and systematic review of published interventions. Eur J Nutr 55(3):869–896. https://doi.org/10.1007/s00394-015-1130-8

Azen EA et al (1989) Prp (proline-rich protein) genes linked to markers Es-12 (esterase-12), Ea-10 (erythrocyte alloantigen), and loci on distal mouse chromosome 6. Genomics 5(3):415–422. https://doi.org/10.1016/0888-7543(89)90004-9

Azen EA et al (1990) mRNAs for PRPs, statherin, and histatins in von Ebner's gland tissues. J Dent Res 69(11):1724–1730. https://doi.org/10.1177/00220345900690110401

Bate-Smith EC (1954) Astringency in foods. Food 23:124–135

Baxter NJ et al (1997) Multiple interactions between polyphenols and a salivary proline-rich protein repeat result in complexation and precipitation. Biochemistry 36:5566–5577. https://doi.org/10.1021/bi9700328

Boltong A, Keast R (2012) The influence of chemotherapy on taste perception and food hedonics: a systematic review. Cancer Treat Rev 38:152–163. https://doi.org/10.1016/j.ctrv.2011.04.008

Bradbury J (2004) Taste perception: cracking the code. PLoS Biol 3. https://doi.org/10.1371/journal.pbio.0020064

Brandão E et al (2014) In vivo interactions between procyanidins and human saliva proteins: effect of repeated exposures to procyanidins solution. J Agric Food Chem 62(39):9562–9568. https://doi.org/10.1021/jf502721c

Buckner S, Dwyer J (2003) Do we need a nutrition-specific quality of life questionnaire for dialysis patients? J Ren Nutr 13:295–302. https://doi.org/10.1053/S1051-2276(03)00119-5

Bufe B et al (2005) The molecular basis of individual differences in phenylthiocarbamide and propylthiouracil bitterness perception. Curr Biol 15:322–327. https://doi.org/10.1016/j.cub.2005.01.047

Cabras T et al (2012) Responsiveness to 6-n-propylthiouracil (PROP) is associated with salivary levels of two specific basic proline-rich proteins in humans. PLoS One 7(2):e30962. https://doi.org/10.1371/journal.pone.0030962

Calò C et al (2011) Polymorphisms in TAS2R38 and the taste bud trophic factor, gustin gene co-operate in modulating PROP taste phenotype. Physiol Behav 104(5):1065–1071. https://doi.org/10.1016/j.physbeh.2011.06.013

Capeless CG, Whitney G, Azen EA (1992) Chromosome mapping of Soa, a gene influencing gustatory sensitivity to sucrose octaacetate in mice. Behav Genet 22:655–663. https://doi.org/10.1007/BF01066636

Cavallo C et al (2019) Consumers' perceptions and preferences for bitterness in vegetable foods: the case of extra-virgin olive oil and brassicaceae—a narrative review. Nutrients 11. https://doi.org/10.3390/nu11051164

Çelebioğlu HY, Lee S, Chronakis IS (2019) Interactions of salivary mucins and saliva with food proteins: a review. Crit Rev Food Sci Nutr 60:64–83. https://doi.org/10.1080/10408398.2018.1512950

Charlton AJ et al (2002) Polyphenol/peptide binding and precipitation. J Agric Food Chem 50:1593–1601

Christensen CM (1980) Effects of solution viscosity on perceived saltiness and sweetness. Percept Psychophys 28:347–353. https://doi.org/10.3758/BF03204394

Cowart BJ, Beauchamp GK, Mennella JA (2003) Development of taste and smell in the neonate. In: Fetal and neonatal physiology: third edition. https://doi.org/10.1016/B978-0-7216-9654-6.50180-6

Da Costa G et al (2008) Salivary amylase induction by tannin-enriched diets as a possible countermeasure against tannins. J Chem Ecol 34(3):376–387. https://doi.org/10.1007/s10886-007-9413-z

De Carli L et al (2018) Impaired taste sensation in type 2 diabetic patients without chronic complications: a case–control study. J Endocrinol Investig 41:765–772. https://doi.org/10.1007/s40618-017-0798-4

de Wijk RA, Prinz JF (2005) The role of friction in perceived oral texture. Food Qual Prefer 16:121–129. https://doi.org/10.1016/j.foodqual.2004.03.002

De Wijk RA et al (2004) The role of α-amylase in the perception of oral texture and flavour in custards. Physiol Behav 83:81–91. https://doi.org/10.1016/j.physbeh.2004.07.014

Delwiche J, O'Mahony M (1996) Changes in secreted salivary sodium are sufficient to alter salt taste sensitivity: use of signal detection measures with continuous monitoring of the oral environment. Physiol Behav 59:605–611. https://doi.org/10.1016/0031-9384(95)02120-5

Dinnella C et al (2010) Temporary modification of salivary protein profile and individual responses to repeated phenolic astringent stimuli. Chem Senses 35(1):75–85. https://doi.org/10.1093/chemse/bjp084

Dinnella C et al (2011) Individual astringency responsiveness affects the acceptance of phenol-rich foods. Appetite 56(3):633–642. https://doi.org/10.1016/j.appet.2011.02.017

Dodds MW, Dodds AP (1997) Effects of glycemic control on saliva flow rates and protein composition in non-insulin-dependent diabetes mellitus. Oral Surg Oral Med Oral Pathol Oral Radiol Endod 83(4):465–470

Dsamou M et al (2012) Salivary protein profiles and sensitivity to the bitter taste of caffeine. Chem Senses 37(1):87–95. https://doi.org/10.1093/chemse/bjr070

Duffy VB et al (2010) Vegetable intake in college-aged adults is explained by oral sensory phenotypes and TAS2R38 genotype. Chemosens Percept 3(3–4):137–148. https://doi.org/10.1007/s12078-010-9079-8.

Efremova NV et al (2002) Direct measurement of interactions between tethered poly(ethylene glycol) chains and adsorbed mucin layers. Langmuir 18:836–845. https://doi.org/10.1021/la011303p
Engelen L, Van Der Bilt A (2008) Oral physiology and texture perception of semisolids. J Texture Stud 39:83–113. https://doi.org/10.1111/j.1745-4603.2007.00132.x
Engelen L et al (2003) The effect of oral and product temperature on the perception of flavor and texture attributes of semi-solids. Appetite 41:273–281. https://doi.org/10.1016/S0195-6663(03)00105-3
Engelen L et al (2007) The effect of saliva composition on texture perception of semi-solids. Arch Oral Biol 52(6):518–525. https://doi.org/10.1016/j.archoralbio.2006.11.007
Escada P, Escada P (2013) Localization and distribution of human olfactory mucosa in the nasal cavities. Acta Medica Port 26:200–207. https://doi.org/10.20344/amp.4238
Fábián TK et al (2015) Molecular mechanisms of taste recognition: considerations about the role of saliva. Int J Mol Sci 16:5945–5974. https://doi.org/10.3390/ijms16035945
Ferry ALS et al (2006) In-mouth amylase activity can reduce perception of saltiness in starch-thickened foods. J Agric Food Chem 54:8869–8873. https://doi.org/10.1021/jf061607n
Foster KD et al (2011) The role of oral processing in dynamic sensory perception. J Food Sci 76:R49–R61. https://doi.org/10.1111/j.1750-3841.2010.02029.x
Frank D, Piyasiri U, Archer N, Jenifer J, Appelqvist I (2018) Influence of saliva on individual in-mouth aroma release from raw cabbage (Brassica oleracea var. capitata f. rubra L.) and links to perception. Heliyon 4: e01045
Freitas CN et al (2017) Educational strategies and atraumatic restorative treatment effect on salivary characteristics: a controlled clinical trial. Oral Dis 23:1116–1126. https://doi.org/10.1111/odi.12706
Fu D et al (2019) Complex relationship between TAS2 receptor variations, bitterness perception, and alcohol consumption observed in a population of wine consumers. Food Funct 10:1643–1652. https://doi.org/10.1039/c8fo01578c
Fuchida S et al (2013) Association between underweight and taste sensitivity in middle- to old-aged nursing home residents in Sri Lanka: a cross-sectional study. J Oral Rehabil 40:854–863. https://doi.org/10.1111/joor.12099
Genovese A, Rispoli T, Sacchi R (2018) Extra virgin olive oil aroma release after interaction with human saliva from individuals with different body mass index. J Sci Food Agric 98:3376–3383. https://doi.org/10.1002/jsfa.8848
Genva M et al (2019) Is it possible to predict the odor of a molecule on the basis of its structure? Int J Mol Sci 20(12):3018. https://doi.org/10.3390/ijms20123018
Glanz K et al (1998) Why Americans eat what they do: taste, nutrition, cost, convenience, and weight control concerns as influences on food consumption. J Am Diet Assoc 98:1118–1126. https://doi.org/10.1016/S0002-8223(98)00260-0
Glendinning JI (1992) Effect of salivary proline-rich proteins on ingestive responses to tannic acid in mice. Chem Senses 17:1–12. https://doi.org/10.1093/chemse/17.1.1
Glumac M et al (2019) Saliva could act as an emulsifier during oral processing of oil/fat. J Texture Stud 50:83–89. https://doi.org/10.1111/jtxs.12375
Grabe V, Sachse S (2018) Fundamental principles of the olfactory code. Biosystems 164:94–101. https://doi.org/10.1016/j.biosystems.2017.10.010
Graham CS et al (1995) Taste and smell losses in HIV infected patients. Physiol Behav 58:287–293. https://doi.org/10.1016/0031-9384(95)00049-O
Gravina SA, Yep GL, Khan M (2013) Human biology of taste. Ann Saudi Med 33:217–222. https://doi.org/10.5144/0256-4947.2013.217
Guichard E et al (2017) Model cheese aroma perception is explained not only by in vivo aroma release but also by salivary composition and oral processing parameters. Food Funct 8:615–628. https://doi.org/10.1039/c6fo01472k
Guinard JX, Mazzucchelli R (1996) The sensory perception of texture and mouthfeel. Trends Food Sci Technol 7:213–219. https://doi.org/10.1016/0924-2244(96)10025-X

Hajjar E et al (2006) Odorant binding and conformational dynamics in the odorant-binding protein. J Biol Chem 281:29929–29937. https://doi.org/10.1074/jbc.M604869200
Henkin RI, Larson AL, Powell RD (1975) Hypogeusia, dysgeusia, hyposmia, and dysosmia following influenza-like infection. Ann Otol Rhinol Laryngol 84:672–682. https://doi.org/10.1177/000348947508400519
Henkin RI, Martin BM, Agarwal RP (1999) Decreased parotid saliva gustin/carbonic anhydrase VI secretion: an enzyme disorder manifested by gustatory and olfactory dysfunction. Am J Med Sci 318:380–391. https://doi.org/10.1016/S0002-9629(15)40663-9
IJpma I et al (2015) Metallic taste in cancer patients treated with chemotherapy. Cancer Treat Rev 41:179–186. https://doi.org/10.1016/j.ctrv.2014.11.006
IJpma I et al (2017) Changes in taste and smell function, dietary intake, food preference, and body composition in testicular cancer patients treated with cisplatin-based chemotherapy. Clin Nutr 36:1642–1648. https://doi.org/10.1016/j.clnu.2016.10.013
Imura H et al (2016) Characteristic changes of saliva and taste in burning mouth syndrome patients. J Oral Pathol Med 45:231–236. https://doi.org/10.1111/jop.12350
Kakutani Y et al (2017) Taste of breath: the temporal order of taste and smell synchronized with breathing as a determinant for taste and olfactory integration. Sci Rep 7:8922. https://doi.org/10.1038/s41598-017-07285-7
Kamalinia G et al (2013) Enhanced brain delivery of deferasirox-lactoferrin conjugates for iron chelation therapy in neurodegenerative disorders: in vitro and in vivo studies. Mol Pharm 10:4418–4431. https://doi.org/10.1021/mp4002014
Kulkarni BV, Mattes RD (2014) Lingual lipase activity in the orosensory detection of fat by humans. Am J Phys Regul Integr Comp Phys 306:R879–R885. https://doi.org/10.1152/ajpregu.00352.2013
Kupirovič UP et al (2017) Effect of saliva on physical food properties in fat texture perception. Crit Rev Food Sci Nutr 57:1061–1077. https://doi.org/10.1080/10408398.2013.766787
Lamy E et al (2008) Comparison of electrophoretic protein profiles from sheep and goat parotid saliva. J Chem Ecol 34(3):388–397. https://doi.org/10.1007/s10886-008-9442-2
Lamy E et al (2009) Sheep and goat saliva proteome analysis: a useful tool for ingestive behavior research? Physiol Behav 98(4):393–401. https://doi.org/10.1016/j.physbeh.2009.07.002
Lamy E, Graça G et al (2010a) Changes in mouse whole saliva soluble proteome induced by tannin-enriched diet. Proteome Sci 8:65. https://doi.org/10.1186/1477-5956-8-65
Lamy E, Baptista ES et al (2010b) Morphological alterations in salivary glands of mice (Mus musculus) submitted to tannin enriched diets: comparison with sialotrophic effects of sympathetic agonists stimulation. Arquivo Brasileiro de Medicina Veterinaria e Zootecnia 62(4):837. https://doi.org/10.1590/S0102-09352010000400012
Lamy E et al (2011) Effect of condensed tannin ingestion in sheep and goat parotid saliva proteome. J Anim Physiol Anim Nutr 95(3):304. https://doi.org/10.1111/j.1439-0396.2010.01055.x
Lamy E et al (2015) Changes in the salivary protein profile of morbidly obese women either previously subjected to bariatric surgery or not. J Physiol Biochem 71(4):691. https://doi.org/10.1007/s13105-015-0434-8
Lamy E, Mowe M, Pinheiro C, Rodrigues L, Lopes O, Capela e Silva F (2016) Oral food perception and polyphenol-rich foods acceptance – the importance of knowing individuals saliva characteristics for promoting consumption. J Int Soc Antioxid 3(4). https://doi.org/10.18143/JISANH_v3i4_1338
Lamy E et al (2019) Changes in salivary protein composition of lambs supplemented with aerial parts and condensed tannins: extract from Cistus ladanifer L.—a preliminary study. Agrofor Syst. https://doi.org/10.1007/s10457-019-00386-4
Lapis TJ, Penner MH, Lim J (2016) Humans can taste glucose oligomers independent of the hT1R2/hT1R3 sweet taste receptor. Chem Senses 41:755–762. https://doi.org/10.1093/chemse/bjw088
Lapis TJ et al (2017) Oral digestion and perception of starch: effects of cooking, tasting time, and salivary α-amylase activity. Chem Senses 42:635–645. https://doi.org/10.1093/chemse/bjx042

Lenfant F et al (2009) Perception of oral food breakdown. The concept of sensory trajectory. Appetite 52:659–667. https://doi.org/10.1016/j.appet.2009.03.003

Lewandowski BC et al (2016) Amiloride-insensitive salt taste is mediated by two populations of type III taste cells with distinct transduction mechanisms. J Neurosci 36:1942–1953. https://doi.org/10.1523/jneurosci.2947-15.2016

Li B, Hayes JE, Ziegler GR (2014) Interpreting consumer preferences: physicohedonic and psychohedonic models yield different information in a coffee-flavored dairy beverage. Food Qual Prefer 36:27–32. https://doi.org/10.1016/j.foodqual.2014.03.001

Lu Y, Bennick A (1998) Interaction of tannin with human salivary proline-rich proteins. Arch Oral Biol. Pergamon 43(9):717–728. https://doi.org/10.1016/S0003-9969(98)00040-5

Lugaz O et al (2005) Time-intensity evaluation of acid taste in subjects with saliva high flow and low flow rates for acids of various chemical properties. Chem Senses 30:89–103. https://doi.org/10.1093/chemse/bji004

Lush IE, Holland G (1988) The genetics of tasting in mice: V. Glycine and cycloheximide. Genet Res 52:207–212. https://doi.org/10.1017/S0016672300027671

Manconi B et al (2016) The intriguing heterogeneity of human salivary proline-rich proteins: short title: salivary proline-rich protein species. J Proteome. Elsevier 134:47–56. https://doi.org/10.1016/J.JPROT.2015.09.009

Mandel AL et al (2010) Individual differences in AMY1 gene copy number, salivary α-amylase levels, and the perception of oral starch. PLoS One 5(10):e13352. https://doi.org/10.1371/journal.pone.0013352

Marquezin MCS et al (2016) Descriptive analysis of the masticatory and salivary functions and gustatory sensitivity in healthy children. Acta Odontol Scand 74:443–448. https://doi.org/10.1080/00016357.2016.1191085

Martin LE et al (2018) Salivary proteins alter taste-guided behaviors and taste nerve signaling in rat. Physiol Behav 184:150–161. https://doi.org/10.1016/j.physbeh.2017.11.021

Martin LE et al (2019a) Altering salivary protein profile can increase acceptance of a novel bitter diet. Appetite 136:8–17. https://doi.org/10.1016/j.appet.2019.01.011

Martin LE, Kay KE, Torregrossa A-M (2019b) Bitter-induced salivary proteins increase detection threshold of quinine, but not sucrose. Chem Senses. Narnia 44(6):379–388. https://doi.org/10.1093/chemse/bjz021

Masi C et al (2016) Caffeine metabolism rate influences coffee perception, preferences and intake. Food Qual Prefer 53:97–104. https://doi.org/10.1016/j.foodqual.2016.06.002

Mehansho H et al (1987) Induction of proline-rich proteins in hamster salivary glands by isoproterenol treatment and an unusual growth inhibition by tannins. J Biol Chem 262:12344–12350

Melis M et al (2013) Marked increase in PROP taste responsiveness following oral supplementation with selected salivary proteins or their related free amino acids. PLoS One 8(3):e59810. https://doi.org/10.1371/journal.pone.0059810

Mennella I, Fogliano V, Vitaglione P (2014) Salivary lipase and α-amylase activities are higher in overweight than in normal weight subjects: influences on dietary behavior. Food Res Int 66:463–468. https://doi.org/10.1016/j.foodres.2014.10.008

Mortazavi H et al (2014) Xerostomia due to systemic disease: a review of 20 conditions and mechanisms. Ann Med Health Sci Res 4:503. https://doi.org/10.4103/2141-9248.139284

Morzel M et al (2014) Salivary protein profiles are linked to bitter taste acceptance in infants. Eur J Pediatr 173(5):575–582. https://doi.org/10.1007/s00431-013-2216-z

Morzel M et al (2017) Associations between food consumption patterns and saliva composition: specificities of eating difficulties children. Physiol Behav 173:116–123. https://doi.org/10.1016/j.physbeh.2017.02.005

Mosca AC, Chen J (2017) Food-saliva interactions: mechanisms and implications. Trends Food Sci Technol 66:125–134. https://doi.org/10.1016/j.tifs.2017.06.005

Mosca AC et al (2019) How are macronutrient intake, BMI, ethnicity, age, and gender related to the composition of unstimulated saliva? A case study. J Texture Stud 50:53–61. https://doi.org/10.1111/jtxs.12362

Neyraud E et al (2012) Variability of human saliva composition: possible relationships with fat perception and liking. Arch Oral Biol 57(5):556–566. https://doi.org/10.1016/j.archoralbio.2011.09.016

Neyraud E et al (2017) The basal free fatty acid concentration in human saliva is related to salivary lipolytic activity. Sci Rep 7. https://doi.org/10.1038/s41598-017-06418-2

Niimura Y, Matsui A, Touhara K (2014) Extreme expansion of the olfactory receptor gene repertoire in African elephants and evolutionary dynamics of orthologous gene groups in 13 placental mammals. Genome Res 24:1485–1496. https://doi.org/10.1101/gr.169532.113

O'Mahony M et al (1982) Salt taste sensitivity and stimulus volume: effect of stimulus residuals. Perception 11:347–357. https://doi.org/10.1068/p110347

Overberg J et al (2012) Differences in taste sensitivity between obese and non-obese children and adolescents. Arch Dis Child. BMJ Publishing Group Ltd 97(12):1048–1052. https://doi.org/10.1136/archdischild-2011-301189

Padiglia A et al (2010) Sensitivity to 6-n-propylthiouracil is associated with gustin (carbonic anhydrase VI) gene polymorphism, salivary zinc, and body mass index in humans. Am J Clin Nutr 92(3):539–545. https://doi.org/10.3945/ajcn.2010.29418

Pagès-Hélary S et al (2014) Retention effect of human saliva on aroma release and respective contribution of salivary mucin and α-amylase. Food Res Int 64:424–431. https://doi.org/10.1016/j.foodres.2014.07.013

Pajic P et al (2019) Independent amylase gene copy number bursts correlate with dietary preferences in mammals. elife 8. https://doi.org/10.7554/eLife.44628

Pannunzio E et al (2010) Analysis of the stimulated whole saliva in overweight and obese school children. Rev Assoc Méd Bras 56:32–36. https://doi.org/10.1590/s0104-42302010000100012

Pereira LJ, Duarte Gaviao MB, Van Der Bilt A (2006) Influence of oral characteristics and food products on masticatory function. Acta Odontol Scand 64:193–201. https://doi.org/10.1080/00016350600703459

Perry GH et al (2007) Diet and the evolution of human amylase gene copy number variation. Nat Genet 39:1256–1260. https://doi.org/10.1038/ng2123

Piombino P et al (2014) Saliva from obese individuals suppresses the release of aroma compounds from wine. PLoS One 9(1):e85611. https://doi.org/10.1371/journal.pone.0085611

Piombino P, Moio L, Genovese A (2019) Orthonasal vs. retronasal: studying how volatiles' hydrophobicity and matrix composition modulate the release of wine odorants in simulated conditions. Food Res Int 116:548–558. https://doi.org/10.1016/j.foodres.2018.08.072

Ployon S et al (2018) Mechanisms of astringency: structural alteration of the oral mucosal pellicle by dietary tannins and protective effect of bPRPs. Food Chem 253:79–87. https://doi.org/10.1016/j.foodchem.2018.01.141

Proctor GB et al (2017) The SALAMANDER project: SALivAry bioMarkers of mediterraneAN diet associated with long-tERm protection against type 2 diabetes. Nutr Bull 42:369–374. https://doi.org/10.1111/nbu.12298

Quijada-Morín N et al (2016) Effect of the addition of flavan-3-ols on the HPLC-DAD salivary-protein profile. Food Chem 207:272–278. https://doi.org/10.1016/j.foodchem.2016.03.118

Reed DR et al (2010) The perception of quinine taste intensity is associated with common genetic variants in a bitter receptor cluster on chromosome 12. Hum Mol Genet 19:4278–4285. https://doi.org/10.1093/hmg/ddq324

Relating ST (2007) Standard terminology relating to sensory evaluation of materials and products 1. Annual Book of ASTM Standards

Rodrigues L et al (2015) Effects of high fat diet on salivary α-amylase, serum parameters and food consumption in rats. Arch Oral Biol 60(6):854. https://doi.org/10.1016/j.archoralbio.2015.02.015

Rodrigues L et al (2017a) Association between salivary leptin levels and taste perception in children. J Nutr Metab 2017:1–7. https://doi.org/10.1155/2017/7260169

Rodrigues L, da Costa G, Cordeiro C, Pinheiro CC et al (2017b) Relationship between saliva protein composition and 6-n-Propylthiouracil bitter taste responsiveness in young adults. J Sens Stud 32(4):e12275. https://doi.org/10.1111/joss.12275

Rodrigues L, Costa G et al (2017c) Salivary proteome and glucose levels are related with sweet taste sensitivity in young adults. Food Nutr Res 61(1):1389208. https://doi.org/10.1080/16546628.2017.1389208

Rodrigues L et al (2019) Comparison of salivary proteome of children with different sensitivities for bitter and sweet tastes: association with body mass index. Int J Obes 43:701–712. https://doi.org/10.1038/s41366-018-0289-5

Roper SD (2007) Signal transduction and information processing in mammalian taste buds. Pflugers Arch - Eur J Physiol 454:759–776. https://doi.org/10.1007/s00424-007-0247-x

Sacerdote C et al (2007) Lactase persistence and bitter taste response: instrumental variables and mendelian randomization in epidemiologic studies of dietary factors and cancer risk. Am J Epidemiol 166(5):576–581. https://doi.org/10.1093/aje/kwm113

Sales Baptista E et al (2010) Variation in salivary protein composition related to feeding behavior and its ecological implications. Behav Chem Ecol

Sandell M et al (2014) Genetic variation in the hTAS2R38 taste receptor and food consumption among Finnish adults. Genes Nutr 9:433. https://doi.org/10.1007/s12263-014-0433-3

Santos JL et al (2012) Copy number polymorphism of the salivary amylase gene: implications in human nutrition research. J Nutrigenet Nutrigenomics 5:117–131. https://doi.org/10.1159/000339951

Schöbel N et al (2014) Astringency is a trigeminal sensation that involves the activation of G protein-coupled signaling by phenolic compounds. Chem Senses 39:471–487. https://doi.org/10.1093/chemse/bju014

Shatzman AR, Henkin RI (1981) Gustin concentration changes relative to salivary zinc and taste in humans. Proc Natl Acad Sci U S A 78(6):3867–3871. Available at: http://www.pubmedcentral.nih.gov/articlerender.fcgi?artid=319674&tool=pmcentrez&rendertype=abstract. Accessed 15 Sept 2014

Shen Y, Kennedy OB, Methven L (2016) Exploring the effects of genotypical and phenotypical variations in bitter taste sensitivity on perception, liking and intake of brassica vegetables in the UK. Food Qual Prefer 50:71–81. https://doi.org/10.1016/j.foodqual.2016.01.005

Shi L et al (2000) Mucin coating on polymeric material surfaces to suppress bacterial adhesion. Colloids Surf B: Biointerfaces 17:229–239. https://doi.org/10.1016/S0927-7765(99)00121-6

Shimada T (2006) Salivary proteins as a defense against dietary tannins. J Chem Ecol 32(6):1149–1163. https://doi.org/10.1007/s10886-006-9077-0

Silva Teixeira CS, Cerqueira NMFSA, Silva Ferreira AC (2016) Unravelling the olfactory sense: from the gene to odor perception. Chem Senses 41:105–121. https://doi.org/10.1093/chemse/bjv075

Small DM et al (2005) Differential neural responses evoked by orthonasal versus retronasal odorant perception in humans. Neuron 47:593–605. https://doi.org/10.1016/j.neuron.2005.07.022

Soares S et al (2011) Reactivity of human salivary proteins families toward food polyphenols. J Agric Food Chem 59(10):5535–5547. https://doi.org/10.1021/jf104975d

Soares S, Mateus N, de Freitas V (2012) Interaction of different classes of salivary proteins with food tannins. Food Res Int 49(2):807–813. https://doi.org/10.1016/j.foodres.2012.09.008

Soares S et al (2018) Study of human salivary proline-rich proteins interaction with food tannins. Food Chem. Elsevier 243:175–185. https://doi.org/10.1016/J.FOODCHEM.2017.09.063

Spahn JM et al (2019) Influence of maternal diet on flavor transfer to amniotic fluid and breast milk and children's responses: a systematic review. Am J Clin Nutr 109:1003S–1026S. https://doi.org/10.1093/ajcn/nqy240

Stevenson RJ (2009) An initial evaluation of the functions of human olfaction. Chem Senses 35:3–20. https://doi.org/10.1093/chemse/bjp083

Stolle T et al (2017) Salivary proteome patterns affecting human salt taste sensitivity. J Agric Food Chem 65:9275–9286. https://doi.org/10.1021/acs.jafc.7b03862

Szczesniak AS (2002) Texture is a sensory property. Food Qual Prefer 13:215–225. https://doi.org/10.1016/S0950-3293(01)00039-8
Tezal M et al (2006) Supragingival plaque may modify the effects of subgingival bacteria on attachment loss. J Periodontol 77:808–813. https://doi.org/10.1902/jop.2006.050332
Thomson WM (2015) Dry mouth and older people. Aust Dent J 60:54–63. https://doi.org/10.1111/adj.12284
Toyokuni S (2009) Role of iron in carcinogenesis: cancer as a ferrotoxic disease. Cancer Sci 100:9–16. https://doi.org/10.1111/j.1349-7006.2008.01001.x
Tremblay M et al (2012) Salivary pH as a marker of plasma adiponectin concentrations in women. Diabetol Metab Syndr 4:4. https://doi.org/10.1186/1758-5996-4-4
Turner MD (2016) Hyposalivation and xerostomia. Etiology, complications, and medical management. Dent Clin N Am 60:435–443. https://doi.org/10.1016/j.cden.2015.11.003
Van Ruth SM et al (2001) Interactions between artificial saliva and 20 aroma compounds in water and oil model systems. J Agric Food Chem 49:2409–2413. https://doi.org/10.1021/jf001510f
Ventura AK (2017) Does breastfeeding shape food preferences? Links to obesity. Ann Nutr Metab 70:8–15. https://doi.org/10.1159/000478757
Vingerhoeds MH et al (2005) Emulsion flocculation induced by saliva and mucin. Food Hydrocoll. Elsevier 19(5):915–922. https://doi.org/10.1016/j.foodhyd.2004.12.005
Walliczek-Dworschak U et al (2017) Differences in the density of fungiform papillae and composition of saliva in patients with taste disorders compared to healthy controls. Chem Senses 42:699–708. https://doi.org/10.1093/chemse/bjx054
Wang A et al (2018) Effect of lactoferrin on taste and smell abnormalities induced by chemotherapy: a proteome analysis. Food Funct 9:4948–4958. https://doi.org/10.1039/c8fo00813b
Warwick ZS et al (1993) Taste and smell sensations enhance the satiating effect of both a high-carbohydrate and a high-fat meal in humans. Physiol Behav 53:553–563. https://doi.org/10.1016/0031-9384(93)90153-7
Weel KGC et al (2002) Flavor release and perception of flavored whey protein gels: perception is determined by texture rather than by release. J Agric Food Chem 50:5149–5155. https://doi.org/10.1021/jf0202786
Xu F, Laguna L, Sarkar A (2019) Aging-related changes in quantity and quality of saliva: where do we stand in our understanding? J Texture Stud 50:27–35. https://doi.org/10.1111/jtxs.12356
Yazla S et al (2018) Evaluation of olfaction and taste function in type 2 diabetic patients with and without peripheral neuropathy. Diabetes Metab Res Rev 34:e2973. https://doi.org/10.1002/dmrr.2973
Zabernigg A et al (2010) Taste alterations in cancer patients receiving chemotherapy: a neglected side effect? Oncologist 15:913–920. https://doi.org/10.1634/theoncologist.2009-0333
Zoon H, de Graaf C, Boesveldt S (2016) Food odours direct specific appetite. Foods 5. https://doi.org/10.3390/foods5010012

Chapter 3
Saliva as a Non-invasive Sample: Pros and Cons

Lorena Franco-Martínez and Candela Castillo-Felipe

Objectives
In the last decades, the use of saliva for the evaluation of healthy and diseased individuals has been widely employed. This biofluid offers several advantages when compared to other invasive and non-invasive matrixes, although it also possesses some drawbacks and limitations. The present chapter briefly describes the most employed saliva collection methods, the principal uses of saliva, and the main benefits and inconveniences of the use of this biofluid in human and animal research and medicine.

3.1 Collection Methods

Two types of saliva are distinguished:

1. *Partial saliva*: composed by the secretion of only one gland.
2. *Total saliva*: composed of the mixture of saliva derived from all glands, microorganisms, crevicular fluids (derived from the gingival sulcus), remains of epithelial cells, food debris, sebaceous cells secretions and other particles. Total saliva is preferred over partial saliva for biomarkers determination due to its easy collection.

L. Franco-Martínez (✉)
Interdisciplinary Laboratory of Clinical Analysis (Interlab-UMU),
Campus of Excellence Mare Nostrum, University of Murcia, Espinardo, Spain
e-mail: lorena.franco2@um.es

C. Castillo-Felipe
Faculty of Medicine and Odontology, Hospital Morales Meseguer,
University of Murcia, Murcia, Spain

A. Tvarijonaviciute et al. (eds.), *Saliva in Health and Disease*,
https://doi.org/10.1007/978-3-030-37681-9_3

Thus, the collection method will depend on the type of saliva needed and the purpose of the sample (Humphrey and Williamson 2001). In all cases, the following recommendations should be fulfilled in order to not alter or mask concentrations of target salivary biomarker:

- Not to eat, brush teeth or chew gum 2 h before the test
- Not to smoke at least 10 min before the test
- Not use lipstick or other chemical compounds

3.1.1 Methods for Obtaining Total Unstimulated Saliva

Passive or unstimulated saliva has the advantage that analyte concentrations are not altered due to alterations in flow rate (Carpenter 2013). Drainage and spiting are the most commonly used methods for collecting saliva for analytical purposes although in cases of non-collaborating patients, the cotton technique is usually used.

3.1.1.1 Drainage

The saliva collection should be performed in a calm environment to avoid external stimulation. The patient's head should be placed slightly inclined forward, dropping the salivary flow into a funnel that will be connected to a graduated tube (Table 3.1). It is important to make clear to the patient that all saliva produced must be dropped without swallowing any of it. This method is used to calculate the amount of saliva produced in mL per minute to assess the flow and diagnose possible saliva flow alterations, hyper and hypo-salivation (López-Jornet 2000; Bagan and Jimenez 2010).

3.1.1.2 Spitting

Spitting is a variation of the drainage, the only difference is that the subject remains with the lips closed gathering saliva in the oral cavity and spitting it every so often in a graduated container (Table 3.1) (López-Jornet 2000). Position and situation of the subject should be the same as during drainage procedure. It is important to notice that using this technique there is a higher risk that the subject accidentally swallows the saliva. Furthermore, expectoration can occur resulting in stimulation of salivary flow. For these reasons, some authors recommend not to exceed 3–5 min (López-Jornet 2000). The amount of saliva obtained can be measured in mL per min.

Table 3.1 Most used collection methods (advantages and disadvantages)

Method		Collection tool	Advantages	Disadvantages
Drainage		Salivary flow into a funnel that will be connected to a graduated sterile tube	Lower risk of sample contamination, increased reliability of analytes by not stimulated saliva	Analytes alteration due to the presence of lipstick or other chemical substances on the lips
				Difficult in patients with dementia, severe disability, or any disease that prevents proper collection of the sample
Spitting	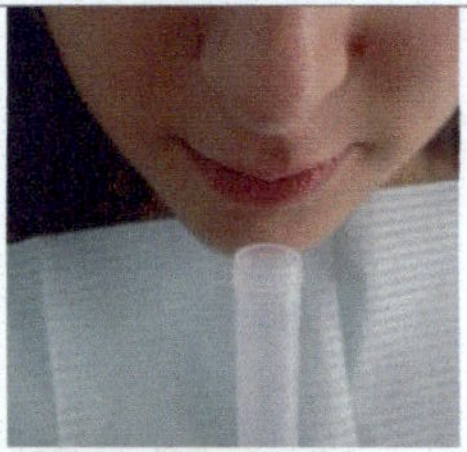	Saliva flow is accumulated for a certain time and then spat out in the collection vessel	Lower risk of sample contamination, increased reliability of analytes by not stimulated saliva	Analytes alteration due to the presence of lipstick or other chemical substances
				Difficult in patients with dementia, severe disability, or any disease that prevents proper collection of the sample
Ejector	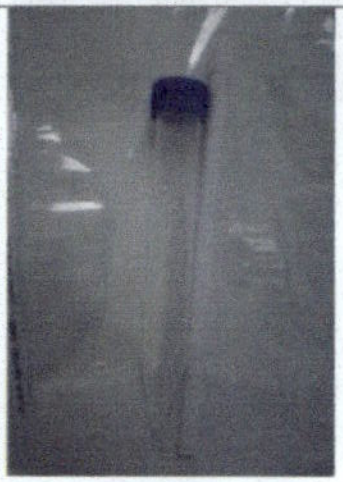	An ejector (plastic or glass) connected to a vacuum pump stores saliva into a sterile graduated tube	Greater patient comfort because does not need collaboration, it is fast and not painful	Greater risk of contamination and analyte alteration by interposing more materials until the collection container is reached and the degree of stimulation of glands
			It can be performed in patients with disabilities, dementia, etc	More expensive equipment

(continued)

Table 3.1 (continued)

Method		Collection tool	Advantages	Disadvantages
Syringe collection	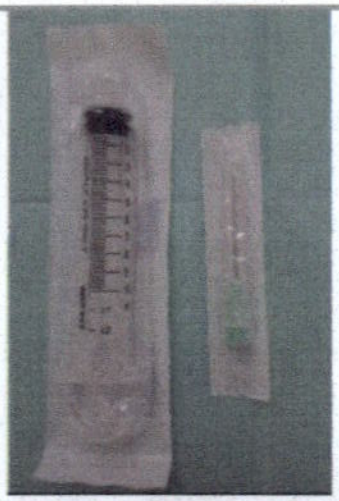	It is collected by sterile syringe and needle between the patient's lips	Little alteration of analytes	Difficult in patients with a low amount of saliva
Cotton	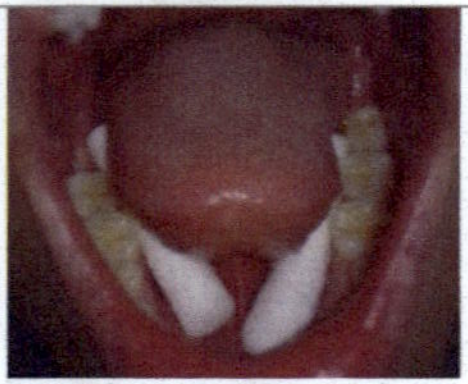	Cotton rolls are placed in the vestibular and sublingual area for 2 min	Easy. Useful in patients with difficulty to cooperate	Alteration of analytes absorbed by the cotton

3.1.1.3 Ejector (Saliva Aspiration)

An ejector (plastic or glass pipette) connected to a vacuum pump is used to collect the saliva. For this reason, this method can be considered as invasive. The ejector is placed under the tongue and then slides through the halls, jugal and lips mucosa to collect saliva that reaches a graduated tube. During this procedure, it is not necessary for the subject to remain with the head tilted forward. It should be highlighted that the use of ejector results in a certain degree of stimulation of the salivary glands producing higher volumes of saliva as compared to drainage or spitting (López-Jornet 2000) and, therefore, some concentrations of analytes can be altered (Humphrey and Williamson 2001). Salivary composition can be also altered if not sterile materials are used (Ngamchuea et al. 2017a). Saliva quantity is measured in mL per min (López-Jornet 2000).

3.1.1.4 Syringe Collection

The procedure consists on the extraction of global saliva by means of a sterile 5 mL glass or plastic syringe, in which a 2-inch needle with a rounded tip to avoid sharp areas is inserted between the lips and saliva is suctioned and measured in mL per min during a controlled period of time (López-Jornet 2000).

3.1.1.5 Cotton

Two or three rolls of dental cotton 1.5 centimetres long are used in this procedure. The cottons are weighed and placed in the sublingual and vestibular areas, or just in the sublingual zone (Table 3.1). The placed cottons are maintained for 2 min. Afterwards, soaked cottons are removed and weighed and the amount of absorbed saliva is calculated by subtracting the weight of the dry cotton. The amount of saliva is measured in g per min (López-Jornet 2000). It is important to highlight that cotton based sampling may affect some of the salivary composition levels because of the binding of analytes to cotton. In order to avoid the binding of analytes to the collection devices, non-cotton based materials such as polystyrene and polyester are preferable (Ngamchuea et al. 2017b).

3.1.2 Methods for Obtaining Total Stimulated Saliva

The methods for saliva collection involving stimulation of salivary glands are important since allow assessing the response of the glands to a stimulus. The stimulation can be carried out mechanically, chemically, pharmacologically, with electrostimulation or with a low power laser (Fedele et al. 2008a; Lončar et al. 2011a).

3.1.2.1 Mechanical Stimulation Techniques

- Chewing paraffin. A paraffin sheet of 0.5–1.5 g is introduced 30–60 s in the mouth (time that the paraffin takes to soften). Since the paraffin does not have any flavour and do not alter electrolytic composition of saliva, its use was recommended for studies aiming to measure ions in saliva. It is important to control two variables during the procedure: time and number of chews per minute (López-Jornet 2000).
- Saxon test. This test consists in measuring the difference in weight of a sponge when chewed vigorously for 2 min. The sponge should have 5 cm x 5 cm size (López-Jornet 2000; Li-Hui et al. 2016). As in case of cotton, use of sponge can result in altered salivary composition due to analyte binding to the sponge.

3.1.2.2 Chemical Stimulation Techniques

Chemical stimulation is mainly used to assess the response of salivary glands to a stimulant, and allows obtaining more saliva.

When chemical stimulation is used, four aspects should be considered:

1. Type of chemical substance used. The most commonly chemical substances used for salivary stimulation are acids, especially citric acid (2–10% concentration) and fresh lemon juice (López-Jornet 2000; Li-Hui et al. 2016). Weak acids (citric and acetic) stimulate secretion better than strong acids such as hydrochloric, yet, establishing a response pattern to the stimulus is complicated since it will always be conditioned to individual variability between subjects (gustatory perception, reaction capacity of the glands…) (Li-Hui et al. 2016). The use of citric acid lessens the pH of samples to below 3; so it is apparent that the use of citric acid stimulation is not suitable for obtaining samples for pH measurement in saliva and could interfere with methodologies that use determined pH for target biomarker determinations such as testosterone or electrolytes (Li-Hui et al. 2016). Furthermore, citric acid apart of increasing salivary flow it also increases salivary α-amylase activity, total proteins, and pH (Li-Hui et al. 2016). Besides the use of acids, there are other techniques such as the sugar cube test to stimulate saliva production. For this, a cube of sugar is placed on the lingual dorsum and the time elapsed until its dissolution by salivary secretion is counted. A cube of sugar technique is used for assessing xerostomia in patients undergoing radiotherapy (López-Jornet 2000; Bagan and Jimenez 2010) (Fig. 3.1a).
2. Stimulus application area. Chemical stimulus is usually applied on the lateral surface or on the anterior and dorsal section of the tongue (López-Jornet 2000; Li-Hui et al. 2016) (Fig. 3.1b).
3. Quantity. Usually a pipette is used in order to calculate the exact drops (2 to 3 drops) to apply to each patient.

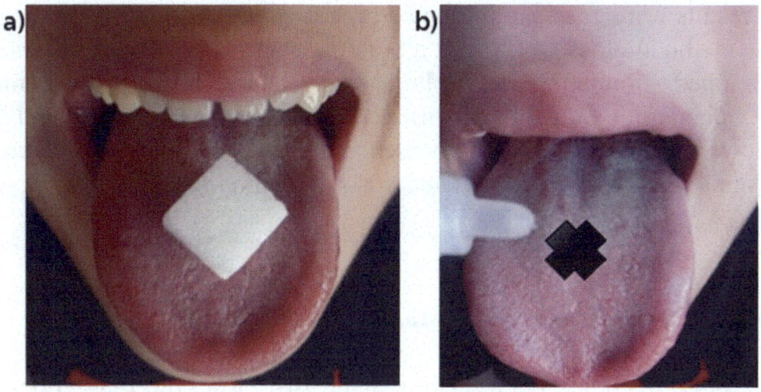

Fig. 3.1 (**a**) Sugar cube test to stimulate saliva production. (**b**) Pipette to stimulate saliva production. The "x" area correspond to the main zone of stimulation

4. Stimulus application frequency. It is suggested to renew the stimulus every 15–60 s since acids are rapidly buffered by saliva. The stimulant is applied before the sample collection, either for its biochemical study or to quantify the salivary flow at a clinical level.

The above variables must be taken especially into account if they are part of a clinical study to ensure that all study subjects are under the same conditions and in this way avoiding possible bias in the results.

3.1.2.3 Stimulation with Pharmacological Substance

Pilocarpine is the most commonly used systemic drug stimulator, it is a parasympathomimetic drug that stimulates exocrine gland secretions. Up to now it has been used in severe xerostomy due to radiotherapy or Sjogren's syndrome (López-Jornet 2000). Nevertheless, it is preferable for local action systems that achieve glandular stimulation since aprox. 30% of patients suffer side effects such as diarrhoea, dyspepsia, abdominal pain, nausea, vomiting, hypertension (López-Jornet 2000; Bagan and Jimenez 2010). For these reasons and because pilocarpine may interfere with other drugs commonly used in patients presenting xerostomy, a special care should be taken when using it (López-Jornet 2000; Bagan and Jimenez 2010).

3.1.2.4 Electrostimulation

In 1988 the use of electrostimulation (Biosonics®) (López-Jornet 2000) was approved for saliva secretion stimulation. The technique consisted in the stimulation of the lingual dorsum through a mouthpiece connected to a control centre. The obtained results were good but presented high cost and discomfort for the patients. Thus new methodologies working at a mucosa level of the lower third molar have been developed stimulating the sublingual and buccal nerve up to 5 mm away (Saliwel GenNarino; Saliwel Crown) (López-Jornet 2000; Bagan and Jimenez 2010). However, electrostimulation is used as a promising method for the treatment of xerostomia (Fedele et al. 2008b) and not as a method for obtaining bigger levels of saliva.

3.1.2.5 Low-Power Laser Stimulation

There are very few studies, but it seems that the use of low power laser does not only produce an increase in salivary flow but it also favours glandular regeneration. More studies are still necessary to clarify the dose and time of application (Lončar et al. 2011b).

3.1.2.6 Acupuncture

Some acupuncture techniques have been used obtaining better results than the use of pilocarpine and avoiding their side effects (Assy and Brand 2018).

No data about possible alterations in salivary biomarkers have been described while using the above described techniques (acupuncture, low-power laser stimulation, electrostimulation and pharmacological stimulation). Thus, further studies are needed to fill this gap of knowledge.

3.1.3 Methods for Collection of Partial Saliva

- Cannulas. Thin polyethylene tubes with diameters ranging from 0.5 to 1.5 mm are introduced directly into the excretory duct of the target gland. This technique must be carried out carefully since the walls of the ducts are very thin and can be easily damaged (López-Jornet 2000). Testing the sublingual and submandibular glands by this methodology is very complicated (López-Jornet 2000).
- Lashley capsule. This technique was popularized by Lashley in 1916 (López-Jornet 2000). The aim of this technique was to obtain saliva principally from the parotid gland in a safe and comfortable way for the patient. The original device consisted of a disk with two completely separate concentric chambers generating two compartments. Each of them was connected to a tube that comes out and goes into the mouth. The disc is placed in the exit of the Stenon canal, and through the suction of the external chamber it is kept in the oral mucosa at the same time as the saliva is collected in the inner chamber (López-Jornet 2000).
- Schneyer segregator device. This device was developed for specific collection of the submandibular and sublingual glands secret. Schneyer segregator is divided

in 3 sections: the central one is placed in the caruncles of the wharton duct, and the two laterals at the exit of the sublingual ducts. Saliva passes through polyethylene tubes into a graduated container out of the mouth. The reliability of the procedure depends on the obtaining of a good sealing of the different parts. However, it is time-consuming and efforts-demanding technique so it is not recommended for use in field work. Even though modifications have been carried out, it still presents certain instability that makes it difficult to seal the sections correctly. Instead, other methods such as direct cannulation or the use of micropipettes are recommended (López-Jornet 2000; Bagan and Jimenez 2010).

– Periotron absorbent strips. These strips are mainly used for saliva collection from minor salivary glands located above all on the lips. Periotron® absorbent strips are chromatography strips of 6 x 16 mm diameter. Before their use, it is recommended to isolate the areas of internal mucosa of the lips with cotton rolls and dry them, then place the strips for 30 s and measure the wet part with Periotron® calibrator (López-Jornet 2000).

From a practical point of view, these sampling procedures for partial saliva collection are slow and complex, and require a special knowledge and formation for collecting the specimens. For these reasons, currently, these collection methods are not widely used.

3.1.4 Saliva Collection Techniques in Animals

Not all the saliva collection methods used for human are available in animals due to several limitations such as animal size and/or behaviour. Although the collection method would differ between the different species and objectives, some methods such as the cannulation of salivary glands or mechanical stimulation by chewing are the most commonly used in veterinary medicine and research.

In dogs and cats, cotton swabs have been used by being rubbed on each side of the mouth and pressed against the upper dental arcade (Dodds 2017, 2019). Another way to obtain saliva from dogs and farm animals is pressing the corner of a sponge with a clamp while the animal chews it (Spiesberger et al. 2019) (Fig. 3.2). Meanwhile, the most common methods for salivary collection in laboratory animals such as mice, rats or rabbits may require anaesthesia or sedation and chemical stimulation, which may cause alteration in salivary flow and composition (Kang et al. 2008). In rodents, for whole saliva collection, a sponge can be place in the animal' mouth (Bagavant et al. 2018; Rasinhas et al. 2018) or it can be aspirated with a syringe (Blanchard et al. 2015). Alternatively, animals can be trained to tolerate the application of a pipette tip under the tongue and a gently drawing of small volumes of saliva, as described for mice (Torregrossa et al. 2014; Martin et al. 2018). Similarly than for humans for partial saliva collection, the cannulation of the desired salivary gland provides the most accurate and pure saliva, although it is a technically

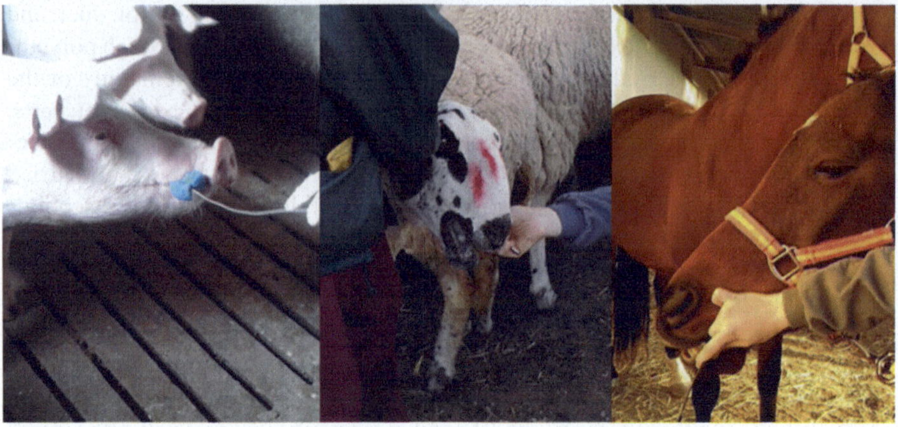

Fig. 3.2 Saliva sampling in different species using sponges or cotton swabs

Table 3.2 Advantages of saliva compared to other samples

<table>
<tr><th></th><th>Blood</th><th>Feces</th><th>Urine</th></tr>
<tr><td rowspan="8">SALIVA ADVANTAGES</td><td>Not painful</td><td colspan="2">Less uncomfortable</td></tr>
<tr><td>No need of specific training</td><td></td><td></td></tr>
<tr><td colspan="2">Ready for its immediately use</td><td></td></tr>
<tr><td colspan="3">Easier transport and storage</td></tr>
<tr><td>Not invasive</td><td></td><td></td></tr>
<tr><td>No stress</td><td></td><td></td></tr>
<tr><td>No fear</td><td></td><td></td></tr>
<tr><td>Less risk of infections</td><td></td><td></td></tr>
</table>

challenging procedure and may cause injury to the animal if performed incorrectly (Lamy et al. 2009, 2011; Bagavant et al. 2018).

Even arthropod saliva can be collected and analysed, being of high importance in the research of vector-borne diseases. In these cases, the mostly used method consist in promoting saliva expectoration during feeding, and recollecting it with filter paper cards (Hall-Mendelin et al. 2010).

3.2 Advantages and Disadvantages of Using Saliva

3.2.1 Advantages

Saliva has different advantages when compared to other analytical samples (Table 3.2) (Prasad et al. 2016). One of the most important advantages in comparison to blood and other specimens is that saliva is a minimally-invasive sample, meaning that the sampling does not requires breaching the skin or other biological barriers. Saliva is being widely studied due to the ease of obtaining the sample for both the medical staff and the patient. Thus, specific training is not necessary to obtain the sample, as it is the case with blood samples, which allows to almost everyone to adequately sampling saliva after a very short explanation of the protocol. This ease in sampling allows the patient to self-sampling, which are of high importance for consecutive samplings, in-home assays, or when no professional personnel are available. Furthermore, and in comparison to other non-invasive specimens such as urine and faeces, saliva is more comfortable to sample providing with better patient compliance. Lastly, the pain for blood sampling often causes stress, fear or nervousness in the patient which can produce the alteration of some parameters or may cause the patient to postpone the analysis or to reject to participate in studies. Since saliva sampling is completely painless the possibilities of these drawbacks would be reduced.

Another advantage of saliva is that it does not require the use of instruments such as needles, which reduces the risk of injuries or infections in this way (Yoshizawa et al. 2013).

In addition, saliva transport and storage is easier than in other samples such as blood or non-invasive samples such as faeces, which usually requires additional processing or complicated treatments before their analysis (Yoshizawa et al. 2013; McKenna 2014).

Finally, the protein content, complexity and varying composition of saliva is lower than serum (Alessandro et al. 2011; Liu and Duan 2012). Saliva has been also proposed as more cost-effective than blood, being of special interest for the monitoring of large populations (Liu and Duan 2012).

3.2.2 *Disadvantages*

Saliva represents many advantages, although it also possesses a series of limitations or disadvantages when compared with other biofluids, as summarized in Table 3.3.

Whereas salivary sampling is very easy and comfortable in humans and in some animals such as farm animals and dogs, in others cases sampling may be difficult or even invasive, depending on the species and animal behaviour. While in well behaved animals saliva can be placed by simply collecting the mouth flow into a large beaker or by allowing the animal to chew a sponge attached to a wire (Lamy and Mau 2012), this approach is rarely tolerated by conscious cats and wild animals, and cannot be used in others such as rodents due to their small mouth size. Additionally, and depending on the subject and species, saliva volume collected may be scarce, making difficult to perform a complete analysis. In these animals, saliva has to be sampled by other methods such as plastic cannulas in salivary glands, which are often invasive, or through stimulation by some of the methods described above.

Table 3.3 Main disadvantages of saliva as diagnostic fluid and possible solutions for each case

Category	Problem	Possible solutions
Sampling	Some methods may be inapplicable in some species	Alternative saliva sampling methods such as plastic canules
		Familiarise animals with sampling
		Use stimulating methods for higher sampling volume
Changes in composition	Inter- and intra-subject variability	Increment of sample number for each gender, age range, and physiological status
		Perform samplings at the same time of day/ year to minimize circadian/ circannual effect
		Protocolize sampling method and devices
		Analyse the potential influence of oral health
		Perform samplings at the same time of day/ year to minimize circadian/ circannual effect
		Avoid drink, eat, teeth brushing, smoke or drugs intake 2 h prior to sampling
		If possible, use the same individual as his/her own control
	Inter-laboratory variability	Use standardised protocols of saliva sampling processing
	Blood contamination	Gentle sampling
		Measure possible blood contamination
Processing	Low concentration of target analyte	Consider to use high sensible methods for analytes which are in low concentration in saliva
		Depletion of analytes of higher abundance which may interfere
		Discard analyte losses due to processing
	No assays or reference material available for saliva measurements	Analytical adaptation from assays in other specimens
		Development of novel assays for saliva
		Analytical and clinical validation of each novel analyte in saliva
Data analysis	Not consistent results	Consider to study the influence of flow rate, protein content, or osmolarity in the analyte levels

Saliva consistency and composition can vary greatly between and within subjects (Aps and Martens 2005), which poses a challenge for reproducible salivary measurements. Additionally, changes in saliva can be observed depending on the sampling method and processing (Khurshid et al. 2016). Different saliva composition has been observed when saliva is sampled by stimulation with citric acid or other compounds when compared with unstimulated saliva (Li-Hui et al. 2016; Gomar-Vercher et al. 2018; Lucena et al. 2018). Similarly, the use of passive drool for saliva sampling showed some differences in saliva composition when compared to active methods based on chewing or spitting (Michishige et al. 2006). If saliva is collected by chewing, different materials such as cotton or parafilm absorb different analytes, which can affect the results (Gröschl et al. 2008; Topkas et al. 2012). As stated above, saliva composition can differ from one salivary gland to others, which has to be taken into consideration if saliva is taken for a specific gland. Finally, although saliva processing in the laboratory is usually easy and does not requires specific equipment, differences in processing protocols such as freezing-thawing cycles, or the removal of contaminants, viscous components or certain components may influence saliva composition (Gröschl et al. 2008). For these reasons, the use of validated and reproducible sampling and processing protocols, and the knowledge of possible interactions with sampling devices is desirable in order to minimize these biases (Alessandro et al. 2011).

Although certain analytes such as α-amylase and phosphates are at higher concentration in saliva than in blood, most blood components are in much less concentration or not presented in saliva (Chiappin et al. 2007; Sanchez Martinez 2007). As examples, total protein level in plasma is up to 100 fold higher than in saliva, while steroid hormones such as cortisol present up to ten fold higher concentrations in serum (Turpeinen and Hämäläinen 2013). However, there are a number of different proteins in higher amounts in saliva than in blood, namely proline-rich-proteins, α-amylase, albumin, mucins and secretory immunoglobulin A. These proteins can pose up to 80% of total protein content (Sanchez Martinez 2007). Thus, the remaining components are frequently found in very small concentrations and require high sensible methods for their measurement. Additionally, special attention should be posed in possible blood contamination of saliva, since abnormally elevated levels of certain analytes can be caused due to relative small blood leakages (Behr et al. 2017).

Saliva composition is highly influenced by species, gender, age, physiological status, drugs, and time, among other factors (Sanchez Martinez 2007). For example, in animals, salivary composition is closely related to the species and food preferences: while in humans α-amylase is one of the predominant proteins (Noble 2000), in other species such as dogs and cats α-amylase concentration is much lower (Contreras-Aguilar et al. 2017). Gender and age-related differences in salivary composition have been observed in humans and animals (Franco-Martínez et al. 2019; Ruis et al. 1997; Sun et al. 2014), and several analytes in saliva have probed to show a circadian or circannual cycle (Alessandro et al. 2011; Stanton et al. 2011). Drugs and altered physiological states can affect saliva composition and secretion rate (Aps and Martens 2005; Nieuw Amerongen et al. 2007). In humans, sympathetic simulation promotes a more viscous saliva due to the decrease in secretion ratio and the increase on levels of α-amylase and mucins (Sanchez Martinez 2007). Similarly, drink, eat, teeth brushing or smoke – among other actions- are known to alter salivary composition (Maran et al. 2018; Engel et al. 2019). Thus, for more accurate results in research, it is recommended to prevent these activities for at least 1 or 2 h prior to saliva sampling.

There is still no consensus regarding the more appropriate way to report salivary levels of certain analytes, which could influence the interpretation. For example, the levels of a certain analyte can be expressed as absolute concentration or activity, or as a ratio to salivary total protein concentration, as well as corrected by saliva flow rate or osmolarity, among others parameters. This is due to the great variability that can be observed in these parameters between different subjects or time points, which can masks the real behaviour of the analyte in saliva (Lindsay and Costello 2016). Salivary protein content is an indicator of the degree of saliva concentration or dilution. Salivary flow rate can be defined as the millilitres of saliva produced per minute, which can be closely related to salivary protein content. Salivary flow may influence salivary concentration, especially in those molecules with high molecular weight, polar or ionic, or transported by active mechanisms. Some collection devices may induce incorrect volume measurement due recovery-loss of saliva, causing considerable errors into the analyses (Crouch 2005). Saliva osmolarity is lower than plasma's, and it is reduced at higher salivary secretion rates; thus, correcting the data by salivary osmolarity has been proposed for certain analytes (Sari-Sarraf et al. 2007). Therefore, ideally, the evaluation of the influence of the ways of reporting the results under different conditions should be made for each analyte.

As stated in Chap. 1, the whole saliva term includes saliva, gingival crevicular fluids, mucus from the nasal cavity and pharynx, mucosal transudate and other products such as bacteria, food debris, and desquamated epithelial cells, among other compounds (Farnaud et al. 2010). Although many components in saliva are an ultra-filtrate of blood and, others are locally produced and therefore may not be correlated with blood levels (Lindsay and Costello 2016).

Finally, when compared to blood, there is a scarcity of commercial analytical methods and reference material designed for it use in saliva. This makes saliva research more difficult and requires developing or adaptation of assays designed for serum or plasma. While some assays can be easily adapted from other biofluids to saliva measurements, others are difficult or inapplicable due to several factors such as low concentration of analyte, changes in analyte structure such as glycosylation, or differences in pH –among others- that may make the assay adaptation to saliva challenging.

Taking overall, when compared with other biofluids, saliva offers several advantages such as pain-free, easy and fast sampling, besides to be safer for the patient and the sanitary professionals. That encourages its use in research and in clinical practice in both humans and animals. However, several limitations have to been taken into account prior to the implementation of saliva-based method for diagnosis, prognosis or follow-up of several conditions. For this reason, the use of standardization methods and the evaluation of the best analyte for each condition and the potential biases that could affect our results should be performed before working with saliva.

References

Alessandro L et al (2011) Reference intervals for saliva analytes collected by a standardized method in a physically active population. Clin Biochem 44:1440–1444. https://doi.org/10.1016/j.clinbiochem.2011.09.012

Aps JKM, Martens LC (2005) Review: the physiology of saliva and transfer of drugs into saliva. Forensic Sci Int. https://doi.org/10.1016/j.forsciint.2004.10.026

Assy Z, Brand HS (2018) A systematic review of the effects of acupuncture on xerostomia and hyposalivation. BMC Complement Altern Med 18(1):57. https://doi.org/10.1186/s12906-018-2124-x

Bagan JV, Jimenez Y (2010) Técnicas de exploración salival: Sialometría. In JV Bagan, Y Jimenez (eds). Fisiopatología de las glándulas salivales

Bagavant H et al. (2018) A method for the measurement of salivary gland function in mice. J Vis Exp: JoVE. (131). https://doi.org/10.3791/57203. MyJoVE Corporation

Behr GA et al. (2017) A statistical method to calculate blood contamination in the measurement of salivary hormones in healthy women. Clin Biochem 50(7–8): 436–439. https://doi.org/10.1016/j.clinbiochem.2016.12.005. Elsevier

Blanchard AA et al. (2015) Towards further defining the proteome of mouse saliva. Proteome Sci 13(1): 10. https://doi.org/10.1186/s12953-015-0068-3. BioMed Central

Carpenter GH (2013) The secretion, components, and properties of saliva. Annu Rev Food Sci Technol 4:267–276. https://doi.org/10.1146/annurev-food-030212-182700

Chiappin S et al (2007) Saliva specimen: a new laboratory tool for diagnostic and basic investigation. Clin Chim Acta:30–40. https://doi.org/10.1016/j.cca.2007.04.011

Contreras-Aguilar MD et al. (2017) Detection and measurement of alpha-amylase in canine saliva and changes after an experimentally induced sympathetic activation. BMC Vet Res 13(1): 266. https://doi.org/10.1186/s12917-017-1191-4. BioMed Central

Crouch DJ (2005) Oral fluid collection: the neglected variable in oral fluid testing. Forensic Sci Int. https://doi.org/10.1016/j.forsciint.2005.02.028

Dodds JW (2017) Diagnosis of canine food sensitivity and intolerance using saliva: report of outcomes. J Am Holistic Vet Med Assoc (JAHVMA) 49. Available at: https://nutriscan.imatrixbase.com/documents/DODDS-_1.PDF. Accessed 10 Sept 2019

Dodds WJ (2019) Diagnosis of Feline Food Sensitivity and Intolerance Using Saliva: 1000 Cases. Animals 9(8): 534. https://doi.org/10.3390/ani9080534. Multidisciplinary Digital Publishing Institute

Engel S et al. (2019) Demographic, sampling- and assay-related confounders of endogenous oxytocin concentrations: a systematic review and meta-analysis. Front Neuroendocrinol: 100775. https://doi.org/10.1016/j.yfrne.2019.100775

Farnaud SJC et al. (2010) Saliva: physiology and diagnostic potential in health and disease. Sci World J. https://doi.org/10.1100/tsw.2010.38

Fedele S et al. (2008a) Neuroelectrostimulation in treatment of hyposalivation and xerostomia in Sjögren's syndrome: a salivary pacemaker. J Rheumatol 35(8): 1489–1494. Available at: http://www.ncbi.nlm.nih.gov/pubmed/18671323. Accessed 12 Sept 2019

Fedele S et al (2008b) Neuroelectrostimulation in treatment of hyposalivation and xerostomia in Sjögren's syndrome: a salivary pacemaker. J Rheumatol 35(8):1489–1494

Franco-Martínez L, Hernández JMG, Horvatić A, Guillemin N, Cerón J, Martínez-Subiela S, Sentandreu MÁ, Brkljačić M, Mrljak V, Tvarijonaviciute A, Reyes PJ (2019) Differences on salivary proteome at rest and in response to an acute exercise in men and women: A pilot study. J Proteome:103629

Gomar-Vercher S et al. (2018) Stimulated and unstimulated saliva samples have significantly different bacterial profiles. PLOS ONE 13(6): e0198021. https://doi.org/10.1371/journal.pone.0198021. Edited by R. K. Aziz

Gröschl M et al. (2008) Evaluation of saliva collection devices for the analysis of steroids, peptides and therapeutic drugs. J Pharm Biomed Anal. https://doi.org/10.1016/j.jpba.2008.01.033

Hall-Mendelin S et al. (2010) Exploiting mosquito sugar feeding to detect mosquito-borne pathogens. Proc Nat Acad Sci U S A 107(25): 11255–11259. https://doi.org/10.1073/pnas.1002040107. National Academy of Sciences

Humphrey SP, Williamson RT (2001) A review of saliva Normal composition, flow, and function. Humphrey, Williamson. 2001. Journal of prosthetic dentistry.Pdf. J Prosthet Dent 85(2):162–169. https://doi.org/10.1067/mpr.2001.113778.

Kang JG et al (2008) A prospective, randomized comparison of the effects of inhaled Sevoflurane Anesthesia and Propofol/Remifentanil intravenous Anesthesia on salivary excretion during laryngeal microsurgery. Anesth Analg 106(6):1723–1727. https://doi.org/10.1213/ane.0b013e3181730063

Khurshid Z et al. (2016) Human saliva collection devices for proteomics: an update. Int J Mol Sci 17(6). https://doi.org/10.3390/ijms17060846. Multidisciplinary Digital Publishing Institute (MDPI)

Lamy E et al (2009) Sheep and goat saliva proteome analysis: a useful tool for ingestive behavior research? Physiol Behav 98(4):393–401. https://doi.org/10.1016/j.physbeh.2009.07.002

Lamy E et al. (2011) Effect of condensed tannin ingestion in sheep and goat parotid saliva proteome. J Anim Physiol Anim Nutr 95(3): 304–312. https://doi.org/10.1111/j.1439-0396.2010.01055.x. Blackwell Publishing Ltd

Lamy E, Mau M (2012) Saliva proteomics as an emerging, non-invasive tool to study livestock physiology, nutrition and diseases. J Proteome:4251–4258. https://doi.org/10.1016/j.jprot.2012.05.007

Li-Hui W et al (2016) Gender differences in the saliva of young healthy subjects before and after citric acid stimulation. Clin Chim Acta 460:142–145. https://doi.org/10.1016/j.cca.2016.06.040

Lindsay A, Costello JT (2016) Realising the potential of urine and saliva as diagnostic tools in sport and exercise medicine. Sports Med 47(1): 1–21. https://doi.org/10.1007/s40279-016-0558-1. Springer International Publishing

Liu J, Duan Y (2012) Saliva: a potential media for disease diagnostics and monitoring. Oral Oncol 48(7): 569–577. https://doi.org/10.1016/J.ORALONCOLOGY.2012.01.021. Pergamon

Lončar B et al (2011a) The effect of low-level laser therapy on salivary glands in patients with Xerostomia. Photomed Laser Surg 29(3):171–175. https://doi.org/10.1089/pho.2010.2792

Lončar B et al (2011b) The effect of low-level laser therapy on salivary glands in patients with Xerostomia. Photomed Laser Surg 29(3):171–175. https://doi.org/10.1089/pho.2010.2792

López-Jornet P (2000) Recuerdo morfológico de las glándulas salivales. In A Bermejo Fenoll (ed.) Medicina Bucal Vol I Enfermedades mucocutáneas y de las glándulas salivales

Lucena S et al (2018) The effect of breed, gender, and acid stimulation in dog saliva proteome. Biomed Res Int 2018:7456894. https://doi.org/10.1155/2018/7456894

Maran S et al (2018) Evaluation of changes in salivary pH after intake of different eatables and beverages in children at different time intervals. Int J Clin Pediatr Dent 11(3):177–182. https://doi.org/10.5005/jp-journals-10005-1507

Martin LE et al (2018) Salivary proteins alter taste-guided behaviors and taste nerve signaling in rat. Physiol Behav 184:150–161. https://doi.org/10.1016/j.physbeh.2017.11.021

McKenna PB (2014) Further studies on the necessity or otherwise of multiple pre-treatment groups in faecal egg count reduction tests in sheep. Vet Parasitol 200(1–2):212–215. https://doi.org/10.1016/j.vetpar.2013.11.009

Michishige F et al (2006) Effect of saliva collection method on the concentration of protein components in saliva. J Med Investig. https://doi.org/10.2152/jmi.53.140

Ngamchuea K et al (2017a) Chemical analysis in saliva and the search for salivary biomarkers - a tutorial review. Analyst 143(1):81–99. https://doi.org/10.1039/c7an01571b.

Ngamchuea K et al (2017b) Chemical analysis in saliva and the search for salivary biomarkers - a tutorial review. Analyst 143(1):81–99. https://doi.org/10.1039/c7an01571b.

Nieuw Amerongen AV, Ligtenberg AJM, Veerman ECI (2007) Implications for diagnostics in the biochemistry and physiology of saliva. In Annals of the New York Academy of Sciences. https://doi.org/10.1196/annals.1384.033

Noble RE (2000) Salivary alpha-amylase and lysozyme levels: a non-invasive technique for measuring parotid vs submandibular/sublingual gland activity. J Oral Sci 42(2):83–86. Available at: http://www.ncbi.nlm.nih.gov/pubmed/10989590. Accessed: 30 July 2019

Prasad S, Tyagi AK, Aggarwal BB (2016) Detection of inflammatory biomarkers in saliva and urine: potential in diagnosis, prevention, and treatment for chronic diseases. Exp Biol Med 241(8):783–799. https://doi.org/10.1177/1535370216638770

Rasinhas ADC et al. (2018) First detection of dengue virus in the saliva of immunocompetent murine model. Mem Inst Oswaldo Cruz 113(4): e170208. https://doi.org/10.1590/0074-02760170208. Instituto Oswaldo Cruz

Ruis MAW et al. (1997) The circadian rhythm of salivary cortisol in growing pigs: effects of age, gender, and stress. Physiol Behav 62(3): 623–630. https://doi.org/10.1016/S0031-9384(97)00177-7. Elsevier

Sanchez Martinez P (2007) La saliva como fluido diagnóstico. Enfermedades Hepáticas Autoinmunes, 7(Unidad de Inmunología, Facultad de Medicina, Universidad de Granada, Granada), pp 44–52

Sari-Sarraf V et al. (2007) The effects of single and repeated bouts of soccer-specific exercise on salivary IgA. Arch Oral Biol. https://doi.org/10.1016/j.archoralbio.2006.11.016

Spiesberger, K. et al. (2019) The effects of play behavior, feeding, and time of day on salivary concentrations of sIgA in calves. Animals 9(9): 657. https://doi.org/10.3390/ani9090657. Multidisciplinary ssssDigital Publishing Institute

Stanton SJ, Mullette-Gillman OA, Huettel SA (2011) Seasonal variation of salivary testosterone in men, normally cycling women, and women using hormonal contraceptives. Physiol Behav 104(5):804–808. https://doi.org/10.1016/j.physbeh.2011.07.009

Sun S et al. (2014) Analysis of age and gender associated N-glycoproteome in human whole saliva. https://doi.org/10.1186/1559-0275-11-25

Topkas E et al. (2012) Evaluation of saliva collection devices for the analysis of proteins. Clin Chim Acta 413(13–14): 1066–1070. https://doi.org/10.1016/J.CCA.2012.02.020. Elsevier

Torregrossa A-M et al. (2014) Induction of salivary proteins modifies measures of both orosensory and postingestive feedback during exposure to a tannic acid diet. PloS one 9(8): e105232. https://doi.org/10.1371/journal.pone.0105232. Edited by J. I. Glendinning

Turpeinen U, Hämäläinen E (2013) Determination of cortisol in serum, saliva and urine. Best Pract Res Clin Endocrinol Metab 27:795–801. https://doi.org/10.1016/j.beem.2013.10.008

Yoshizawa JM et al. (2013) Salivary biomarkers: toward future clinical and diagnostic utilities. Clin Microbiol Rev 26(4): 781–791. doi: https://doi.org/10.1128/CMR.00021-13. American Society for Microbiology

Chapter 4
Methodology Assays for the Salivary Biomarkers' Identification and Measurement

Lorena Franco-Martínez, Camila Peres Rubio, and María D. Contreras-Aguilar

Objectives
In the last decades, the development of assays for the identification and quantification of biomarkers of health and disease has been immense. The increase of sensitivity of those assays allowed their use in saliva samples, although it may require several adaptations. This chapter aims to offer a schematic view of the most important methods to discover and measure salivary biomarkers in human and veterinary species. For academic purposes, the different methodologies have been classified into spectrophotometric assays, immunoassays, and "omics"; however, several techniques may belong to different categories. Finally, examples of the applications of these techniques in saliva are given.

4.1 The Concept of Biomarker and the Potential of Saliva as a Diagnostic Fluid

A biomarker is a measurable indicator of a specific biological state, which reflects the interaction between a biological system and a potential hazard (Rifai et al. 2006; Aitio et al. 1993), or a positive situation (Mitsui et al. 2011). Though historically, a biomarker has been often considered a physical trait or physiological metric (chemical or biological), the term is now typically shorthand for a molecular biomarker (Rifai et al. 2006).

Molecular biomarkers can take many forms and they may come from diverse nature, including antibodies (Ac), deoxyribonucleic acids (DNA), ribonucleic acids (RNA), lipids, metabolites, microorganisms or proteins (Yoshizawa et al. 2013).

L. Franco-Martínez · C. P. Rubio (✉) · M. D. Contreras-Aguilar
Interdisciplinary Laboratory of Clinical Analysis (Interlab-UMU), Veterinary School, Campus of Excellence Mare Nostrum, University of Murcia, Campus Espinardo, Murcia, Spain
e-mail: camila.peres@um.es

A. Tvarijonaviciute et al. (eds.), *Saliva in Health and Disease*,
https://doi.org/10.1007/978-3-030-37681-9_4

Presence or changes in their concentration, structure, function or action could be related to the apparition, progress, response to, or even regression of a certain process (Wagner et al. 2004). For instance, transcriptional profiling and DNA methylation studies have shown strong potential for biomarker discovery in cancer (Ramaswamy and Perou 2003), metabolomics approaches are beginning to show promise for metabolic disease or anxiety-related disorders (Fernie et al. 2004; Puurunen et al. 2016), and proteins are likely the most affected molecules in diseased conditions, response and recovery. The identification of early biomarkers of diagnosis is essential in order to improve survival rates. However, it is important to keep in mind that one single biomarker may not be able to point out a specific condition; thus, a battery of multiple sensible and specific biomarkers may be needed for screening purposes (Chattopadhyay and Panda 2019).

Saliva is one of the body fluids readily available for biomarker discovery and assessment (Drake et al. 2005; Yan et al. 2009). There is a growing interest in the use of salivary secretions in a clinical setting because it is easy to collect on a non-invasive and routine basis without any need for special training, and the procedure is economic (Lamy and Mau 2012; Yoshizawa et al. 2013). Therefore, the use of saliva as a sample allows repeated collections of a large number of samples even at short-time intervals (Lamy and Mau 2012; Schulz et al. 2013). In addition, scientific evidence indicate that saliva could be useful to reflect the body's health and well being (Schulz et al. 2013). Saliva contains biomolecules present in blood and also biomolecules produced directly by the salivary glands (Pedersen et al. 2002) that change after some diseases or external conditions. Nevertheless, the widespread implementation of saliva as a diagnostic fluid is scarce because there are some barriers (Schulz et al. 2013; Yoshizawa et al. 2013). From the methodological point of view, low concentration in saliva with respect to blood makes challenging the salivary biomarker discovery with clinical sensitivity and specificity; thus, the development of highly sensitive technologies capable of quantifying low concentration of biomolecules and their accurate validation is required (Chiappin et al. 2007; Schulz et al. 2013).

4.2 Methodology Assays

A wide variety of techniques has been satisfactorily adapted for the discovery and measurement of biomarkers in saliva. In many cases, the type and biochemical characteristics of each biomarker will determine the methods of discovery and measurement that can be employed (Cerón et al. 2005). In addition, although some assays for measurement can be performed manually, automatization is preferred since it allows obtaining better precision and higher processing speed leading to a minimum error of measurement due to manipulation, especially when a high number of samples are analyzed, and to reduced costs (Franco-Martínez et al. 2016).

In this section, the most important methodology techniques used to discover and quantify or semi-quantify biomolecules in saliva will be described, explaining the advantages or disadvantages of each one, with examples from both human and veterinary medicine.

4.2.1 Spectrophotometric Assays

Spectrophotometry is the quantitative measurement of the interaction of light with a material at a selected wavelength (Evenson 2001) and is one of the most used methods of analysis. The relationship between absorbance and concentrations is given by the Bouguer-Lambert-Beer law, which states that the concentration of a solute is proportional to the absorbance (Chance 1972; Swinehart 1962). Different spectrophotometric assays (as described below) and some examples of their use to detect salivary biomarkers can be found in Table 4.1.

4.2.1.1 Ultraviolet and Visible (UV-Vis) Absorption Spectroscopy

The UV-Vis portion of the electromagnetic spectrum normally yields little structural information but is very useful for quantitative measurements. The interactions between matter and electromagnetic radiation in the visible spectral region manifest themselves as colour (Perkampus 1992), therefore, usually called colorimetry. The UV-Vis spectrophotometry is a very fast and easy technique, moderately specific and applicable to compounds in small quantities and, therefore, it is widely used in all analytical chemistry issues.

Enzymes, metabolites, proteins, antioxidants and oxidants, are all examples of compounds that have been measured in saliva samples using UV-Vis spectrophotometry (Table 4.1). Just for instance, the determination of uric acid in saliva through UV-Vis spectrophotometry has been useful for monitoring chronic kidney disease progression and the efficacy of dialysis (Bibi et al. 2008).

4.2.1.2 Atomic Spectrophotometry

Atomic spectrophotometry is based on the measurement of the absorption, emission or fluorescence spectra of elementary atoms or ions. Free atoms, predominantly generated thermally, can absorb or emit radiation due to distinct transitions of the valence electrons of the outer shell of the atom. Hence element identification is feasible based on the atomic structure and ranges from wavelengths somewhat below 200 nm to around 850 nm (Stoeppler 1992). When sample interacts with light, absorption process occurs.

Atomic absorption is the absorption of light by atoms. In the atomic absorption spectrophotometry, the vapour phase atoms absorb energy radiations corresponding

Table 4.1 Spectrophotometric techniques and examples of biomarkers determined in saliva

Technique	Biomarker	References
Ultraviolet and visible absorption	Adenosine deaminase	Rai et al. (2011)
	Aspartate aminotransferase	Zappacosta et al. (2007)
	Arginase	Özmeriç et al. (2000)
	Total esterase	Tvarijonaviciute et al. (2017)
	Creatine kinase	Di Lenardo et al. (2019)
	Albumin	Metgud and Patel (2014)
	Uric acid	Diab-Ladki et al. (2003)
	Trolox equivalent antioxidant capacity	Tvarijonaviciute et al. (2016)
	Superoxide dismutase	Wei et al. (2010)
	Hydrogen peroxide	Barranco et al. (2019)
	Thiobarbituric acid reactive substances	Totan et al. (2015)
	Protein carbonyls	Nguyen et al. (2017)
Atomic absorption	Calcium	Shpitzer et al. (2007)
	Magnesium	
	Chromium	Olmedo et al. (2010)
	Cadmium	
	Manganese	
	Nickel	
	Lead	
Atomic emission	Sodium	Ben-Aryeh et al. (1996)
	Potassium	
Near infrared	Thiocyanate	Schultz et al. (1996)
	Immunoglobulin A	Khaustova et al. (2010, 2009)
	Cortisol	
	Salivary α-amylase	
	Urea	
	Phosphate	
Flow injection spectrophotometric analysis	Salivary α-amylase	Ohtomo et al. (2013)
	Lead	Luconi et al. (2001)

to their resonance lines (UV-Vis), in an amount proportional to their concentration. The technique is characterized by its simplicity, speed and selectivity. It is suitable for the determination of selected elements at the trace and partly at the ultra-trace level such as salivary Ca and Mg concentrations in patients with oral squamous cell carcinoma (Shpitzer et al. 2007).

Atomic emission spectrophotometry consists in the analysis of the radiation emitted after the excitation of atoms by sparks, arcs, flames or plasmas producing numerous transitions. Flame-emission spectrophotometry or flame photometry consists in the analysis of the radiation emitted after the atoms have been excited by the action of the flame (Berry et al. 1946; Holcombe and Hassell 1990; MacIntyre 1961). An advantage, especially in comparison to atomic absorption, is that the radiation intensity is directly proportional to the atom concentration (Stoeppler 1992). Instruments are very robust and relatively cheap offering high sample throughput. Moreover, the corresponding analytical methods are well established and validated (Fernández et al. 2018). Flame photometry is suitable for determining elementary compounds in saliva as only liquid samples may be used. This technique has been used to measure sodium and potassium concentrations in saliva of schizophrenic patients (Ben-Aryeh et al. 1996).

4.2.1.3 Near Infrared Spectrophotometry

The near infrared region of the electromagnetic spectrum extends from the red end of the visible spectrum, at a wavelength of 780 nm, to the onset of the microwave region at a wavelength approximately of 3000 nm (Shaw and Mantsch 2008). Near infrared spectroscopy is particularly useful in the characterization of rare earth and transition metal ions and quantitative determination of water, alcohols, phenols, amines, unsaturated hydrocarbons, and any molecule containing C-H, N- H, S-H or O-H bonds (Whetsel 1968). Near infrared spectrophotometry is very fast (1 min or less per sample), robust, non-destructive and non-invasive, suitable for in-line use, with minimum sample preparation demands. On the other hand, the equipment is expensive, a conventional and reference method must be available to provide the analytical results necessary for the modelling step of the spectral data, and the technique is not very sensitive, the limits of determination are in the range of tenths of a percentage (Pasquini 2003). Different authors used infrared spectrophotometry to quantify biochemical components in saliva such as α-amylase (sAA), cortisol, secretory IgA, urea, total protein and phosphates and provided real-time checking of response to stress without the use of reagents (Khaustova et al. 2009, 2010).

4.2.1.4 Flow Injection Spectrophotometric Analysis

The flow injection spectrophotometric technique involves the injection of the sample into a stream of a chromogenic reagent, which transports it to the detector. Virtually each colour reaction in a homogeneous medium can be adapted to flow injection processing creating a flow injection spectrophotometric method. This method is particularly interesting when only a small amount of sample is available (the analysis requires a small amount of sample, around 20 μl), when a rapid analysis is needed, and when the number of samples to be analysed is considerable (Łobiński and Marczenko 1992). Previous reports have shown the potential use of this technique with saliva samples in environmental, toxicological, medical and forensic studies (Luconi et al. 2001). In addition, although abundant reports have concerned the quantitative methods for sAA, the concentration of this analyte usually exceeds the determination ranges of these methods making the flow injection spectrophotometric a method of election for measuring this analyte in saliva samples (Ohtomo et al. 2013).

4.2.2 *Immunoassays (IAs)*

IAs are methods that measure the presence or concentration of biomolecules (from small molecules to macromolecules) in a solution, by the use of a specific antibody or an antigen as biorecognition agent which produce an analytical signal. The advantage of IAs methods is the high specificity due to the binding process between antibody-antigen (Ju et al. 2017a). A wide variety of IAs have been developed to provide the quantitative, semiquantitative, or qualitative detection of biomolecules (Vashist and Luong 2018a). The IAs techniques include labeling-free formats, in which observable detection signal is directly produced by the immunoreaction; and labeling formats, which need signal molecules to label antigens or antibodies that produces detectable analytical signals on the immunoreaction (Ju et al. 2017a). It is of high importance to notice that the correct interaction between antibody and target protein should always be evaluated, even between different batches of the same antibody (Muñoz-Prieto et al. 2017).

Currently, there is increasing interest in point-of-care technologies (POCT), defined as affordable, sensitive, specific, user-friendly, rapid and robust, equipment-free and deliverable to end-users (ASSURED) technologies (Govindaraju et al. 2019; Vashist and Luong 2018b). POCT often incorporate microfluidic IA protocols and biosensors resulting in fully automated IA formats that use minimal volume of reagents and samples, short turn-a-round time (<30 min) and permit wide range of determinations (Vashist and Luong 2018a, b, c).

In the following paragraphs, the most important IAs for salivary biomarkers and some examples (Table 4.2) will be briefly described.

4.2.2.1 Enzyme-Linked Immunoassays (EIAs)

EIAs are the techniques most widely used in clinical diagnostics. Over last decades, EIA techniques have been improved in terms of sensitivity, duration, procedure, cost-effectiveness, simplicity, and automation (Vashist and Luong 2018c), permitting their use with saliva samples. The EIAs techniques have two main formats:

Enzyme-linked immunosorbent assay (ELISA) is a heterogeneous immunoassay labeling format, where the enzyme-labeled immunocomplexes are bound to a solid substrate (e.g., microplate) to keep the molecules of interest while the unbound ones are removed by washing out of the system. Heterogenous assays require a longer run time, but are versatile, more sensitive, and more specific. They can be direct, indirect or sandwich ELISAs (Ju et al. 2017a; Vashist and Luong 2018c).

Table 4.2 Examples of immunoassays (IAs) used to quantificate salivary biomarkers

IAs	Biomarkers	Example	References
Enzyme-linked immunoassays	Adiponectin	Human Adiponectin ELISA, high sensitivity kit, bio-vendor–Laboratornimedicina	Tvarijonaviciute et al. (2014)
	Cortisone	Indirect ELISA	Al-Dujaili et al. (2012)
	C-reactive protein	Tridelta phase range canine CRP kit	Parra et al. (2005)
	Lactoferrin	Sandwich ELISA	Dipaola and Mandel (1980)
	Immunoglobulin A, M, G	ELISAQuantitation kit; Bethyl laboratories	Escribano et al. (2012b)
	Cortisol	Salimetrics, USA	Saiyudthong et al. (2010)
		Immunosensor is based on a direct competitive enzyme linked immunoassay	Kamarainen et al. (2018)
	C-reactive protein, Immunoglobulin E, myoglobin	AlphaLISA® kits (Perkin Elmer®, MA, USA)	Mohamed et al. (2012)
	D-dimer		Zhang et al. (2013)
Chemiluminescence immunoassays	Cortisol	Automated chemiluminescenceimmunoassay (Immulite 1000, Siemens)	Escribano et al. (2012a)
	Testosterone		Escribano and Cerón (2014)
	Lactate	Electrochemiluminescence-based disposable biosensor	Claver et al. (2009)
Fluoroimmunoassays	Chromogranin A	Time-resolved immunofluorometric assay (TR-IFMA)	Escribano et al. (2013)
	Salivary α-amylase		Contreras-Aguilar et al. (2017)
	C-reactive protein		Gutiérrez et al. (2009b)
	Haptoglobin		Gutiérrez et al. (2009a)
Radioimmunoassays	Cortisol	Spectria cortisol coated tube RIA using ^{125}I	Hansen et al. (2003)
	Estradiol	Diagnostic systems laboratories [^{125}I] double antibody test kit	Shirtcliff et al. (2000)
	Oxytocin	RIAgnosis, Regensburg, Germany	Martin et al. (2018)

(continued)

Table 4.2 (continued)

Non-labeled immunoassays	Immunoglobulin A	Particle-enhanced nephelo-metric immunoassay	Booth et al. (2009)
	Albumin	Nephelometry immunosassay	Booth et al. (2009)
	Immunoglobulin A, G	Turbidometric immunoassay	Patidar et al. (2011)
	Interleukin 8	BiacoreAB (Uppsala, Sweden) surface plasmon resonance (SPR) in a microfluidic channel	Yang et al. (2005)
	C-reactive protein	Plasmonic-photonic interaction in a gold-titanium dioxidegold metal-insulator-metal plasmonic nanocup array	Hu et al. (2019)
	Prostate specific antigen	Surface plasmon resonance (SPR) biosensors	Khan et al. (2018)
	Cortisol	Label-free immunosensor based on a single-walled, carbon nanotube-based chemiresistive transducer	Tlili et al. (2011)
Paper based-immunoassays	Cortisol	Label-free paper-based electrical biosensor chip	Pan et al. (2017)
		Lateral flow immunoassay (LFIA).	Miocevic et al. (2017)
Multiplex immunoassays	C-reactive protein, myoglobin, myeloperoxidase	Beadlyte technology (Millipore) in a Luminex IS-100	Floriano et al. (2009)
	Interleukin 8, 1β	Fluorokine MultiAnalyte profiling systems (Fluorokine MAP)	Arellano-Garcia et al. (2008)

Although the indirect ELISAs increase the specificity compared to the direct ones, the varying nonspecific binding in individual sera might lead to variability in IA results (Vashist and Luong 2018c).

Sandwich ELISAs are the most widely used EIA techniques because they are highly sensitive and permit automation. But the use of this format for very small biomarkers is limited due to the limited epitopes on their surface for the binding detection (Vashist and Luong 2018c).

Competitive EIAs are based on the ability of a substance to compete with the biomarker to be determined. Competitive EIAs are the most preferred formats to quantify small molecules such as steroids with low epitopes on their surface (Vashist and Luong 2018c). However, competitive immunoassays require more time and usually have lower sensitivity and specificity than the non-competitive ones (Pulli et al. 2005).

4.2.2.2 AlphaLisa

The current trend in IAs is in the development of wash-free IAs that consist in homogeneous immunoassay labeling formats, so it does not require the separation of the immunocomplexes from unbound immune reagents (Ju et al. 2017a). One wash-free ELISA-like is the AlphaLISA (amplified luminescent proximity homogeneous assay) technology. It is a non-radioactive technology consisting of two different types of neutral density beads, a donor and an acceptor, which in a bead-based proximity, a cascade of chemical reactions creates a highly amplified signal by the light emission proportional to the level of interaction. This technique reduces the time turn-around and provides higher sensitivity, efficiency and accuracy in comparison with ELISAs (Munoz-Prieto et al. 2019; PerkinElmer Inc. 2013; Vashist and Luong 2018c).

4.2.2.3 Chemiluminescence Immunoassays (CLIA)

CLIA are labeling IAs where a chemical reaction upon the immunoreaction happens producing luminescence (light) emission. The intensity of emission is directly proportional to the concentration of analytes in a sample. The mostly used chemiluminescent reagents are luminol and 4-methoxy-4-(3-phosphatephenyl)-spiro-(1,2-dioxetane-3,2-adamantane). CLIA methods are highly sensitive and with good specificity and with wide linear range (Ju et al. 2017b). They can be easily adapted to simple equipment and a wide range of applications. The most common format is the heterogeneous one, where a solid-phase material is required to immobilize capture antibody probes, since it presents much higher sensitivity in comparison with homogeneous format. Therefore, CLIA technology can be developed on conventional solid-phase formats, or on new solid-phase techniques such as the magnetics beads, or combined with flow injection analysis, capillary electrophoresis, electrochemiluminescence or immunosensors resulting in improved sensitivity (Hu and Xu 2010; Ju et al. 2017b, c; Wu and Liu 2009).

4.2.2.4 Fluoroimmunoassays (FIA)

FIA techniques are based on labeling immunoreactions with fluorescent probes. As in other IAs, FIA can be development in formats of heterogenous assays and homogeneous assays, depending if the separation step of the unbound immune reagents is or not needed. In addition, FIA methods can be based on labeled antigens or labeled antibodies, competition of antigens or sandwich FIA (Hemmila 1985; Parra et al. 2004). FIA methods offer high sensitivity, however, the high background of the fluorometric measurement by the non-specific bind of the fluorescent probes to other assay components limited the use of FIA for a long time (Hemmila 1985; Parra et al. 2004). But this residual fluorescence was solved with the development of the time-resolved immunofluorometric assay (TR-IFMA), where the signal is distinguished of the background because the fluorescence is not emitted immediately after the fluorescence compound's excitation, but a time later due to the use of long-lifetime probes, with a longer time in excited state than the average background duration. In TR-IFMA, the probes used are the lanthanide associated to chelates, such as the Eu-chelate label.

4.2.2.5 Radioimmunoassays (RIA)

RIA methods use radioactive isotopes, and traditionally, they are based on a competition between antigens for a limited amount of antibodies (Hemmila 1985; Liu et al. 2016; Yalow 1980). However, the use of radioisotopes offers problems derivative of radiation associated to the health hazards, waste disposal and the limited stability, restricting their use to specialized laboratories. Nevertheless, new metal stable isotope tags associated to mass spectrometry have been recently developed, which avoids radioactive problems and allows simultaneous detection of various biomarkers with wide dynamic ranges of measurement (Liu et al. 2016).

4.2.2.6 Non-labeled Immunoassays

Some examples of label-free immunoassays are nephelometry and turbidimetry methods, which determine the amount of analytes based in the effect of turbidity upon the light scattering or transmission, respectively, in a liquid with finely divided suspended particles previously formed. Label-free detection immunoassays have low sensitivity for detecting analytes at low concentrations compared to the labeled Lateral flow immunoassays (LFIAs), which use nanoparticles as signal labels (van Amerongen et al. 2018). The negative aspect of LFIAs is that they lack high analytical sensitivity comparable to other IAs formats (Vashist et al. 2015). However, new paper-based formats have been developed that incorporate electrochemical immunosensors resulting in improved limit of detection, linear range and limit of quantification (Mahato et al. 2018).

4.2.2.7 Multiplex Immunoassays

The multianalyte detection of biomarkers allows to improve assay efficiency, low sample and reagent consumption, and reduce overall cost per assay compared with parallel single-analyte IAs methods (Ju et al. 2017). Multiplexed detection of biomarkers are based in several formats, such as the multiplex bead-based Luminex IA (Arellano-Garcia et al. 2008; Siawaya et al. 2008) or multi-spot electrochemiluminescence-based IA (Chowdhury et al. 2009).

4.2.2.8 Liquid Biopsy

Liquid biopsy are tests that detect circulating tumor cells, exomes, tumor DNA, tumor RNA and proteins that are disseminated into bloodstream or saliva from the primary lesion, replacing the traditional tissue biopsy since the first is non-invasive, cheaper, non-time-consuming, and evaluate tumor heterogeneity in real time (Chen and Zhao 2019; Kaczor-Urbanowicz et al. 2017; Lousada-Fernandez et al. 2018). The development of real-time and high-sensitivity liquid biopsy POCT assays has allowed the identification of minimal residual disease. For all these characteristics, liquid biopsy can be considered a POC technology (Khan et al. 2017). Liquid biopsy using saliva can be used to diagnose and monitore local (such as oral cancer) and systemic (such as the Sjogren's syndrome) diseases (Khan et al. 2017).

4.2.3 "Omics" Techniques

The term '-omics' englobes different approaches for the study of biological molecules including DNA (genomic and epigenomics), RNA (transcriptomics), proteins (proteomics), metabolites (metabolomics), and others such as lipids and microbiome (lipidomics and microbiomics, respectively). The use of these approaches allows further increase of knowledge on entire pathways and to have a broad overview of the hierarchical linkage between genotype and phenotype (Guillemin et al. 2016), enabling a holistic view of many conditions while increasing exponentially the number of potential biomarkers (Horvatić et al. 2016).

One of the main advantages of omics is the possibility of analysing simultaneously hundreds of analytes, allowing detecting small but accurate differences in these molecules. Another important characteristic of omics is that they frequently offer high sensitivity and provides quantitative results. The huge amount of raw data generated by omics can be processed by bioinformatics and biostatistics programs in order to point out key markers or pathways of the condition of interest (Guillemin et al. 2016). This *in silico* modelling for omics has become an integral part of life sciences, pointing out the most relevant targets and their interactions in the context of any disease (Zeidán-Chuliá et al. 2012, 2013).

Since cell-free saliva contains mRNA, transcripts, metabolites, and thousands of proteins that play a wide variety of biological roles (Bonne and Wong 2012), saliva is a widely employed biofluid in omics technologies. Thus, the term 'salivaomics' was coined in 2008 to reflect the rapid development of knowledge regarding the information gathered about saliva by omics (Yan et al. 2008). These techniques have greatly contributed to the increasing value of saliva in medicine and research since one of its main limitations, the presence of many important biomolecules in very low concentration when compared to blood, is surmounting due to their high sensitivity. In saliva, the omics approach has led to complementary and powerful information related to early diagnosis, prognosis, treatment monitoring, recurrence prediction, and personalized therapeutic strategies, among other utilities (Kaczor-Urbanowicz et al. 2017; Wong 2007; Yang et al. 2016). For this, it is expected that salivaomics-derived biomarkers may overcome the use of other procedures (Schmidt et al. 2014).

It is desirable to evaluate saliva and other biofluids using different omics strategies because it is unusual to have a sole biomarker for the diagnosis of a certain condition. The use of different omic approaches in an adequate sample size may allow identifying a strong battery of biomarkers, because disease states may not be detectable by all techniques (Castagnola et al. 2017).

In the last years, the development of these novel techniques allowed the identification and validation of a wide range of salivary biomarkers that allow or will soon permit further use of saliva for clinical routines (Kaczor-Urbanowicz et al. 2017). Salivaomics has already been successfully applied in different fields including medicine, dentistry, epidemiology, pharmacotherapy and bioterrorism (Kaczor-Urbanowicz et al. 2017) (Fig. 4.1).

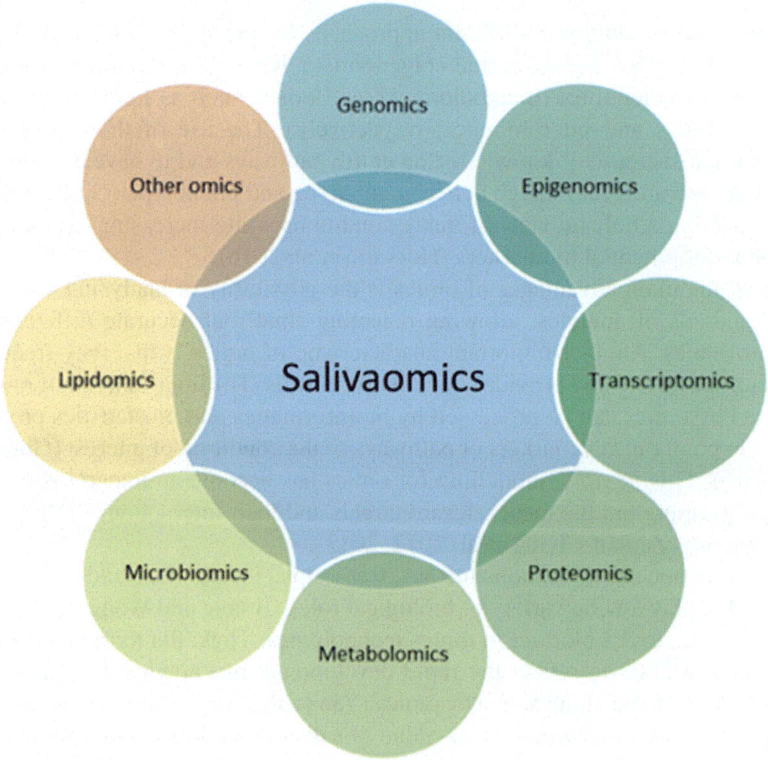

Fig. 4.1 The different and complementary components of salivaomics

4.2.3.1 Genomics

Genomics is focused on the study of the DNA structure and functions, and how these characteristics determine traits and phenotype expression. In saliva, the genome consists of both the individual (70%) and microbial DNAs (Rylander-Rudqvist et al. 2006). Of this, in humans, up to 96% of salivary DNA can be genotyped, 84% amplified and 67% sequenced and stored for long-term (Bonne and Wong 2012; Hansen et al. 2007; Looi et al. 2012).

The aim of genomics is to identify the association between a certain trait or phenotype and a specific genetic marker (a variation in DNA that can be easily identified). As an example, in cancer patients, for whom the early molecular characterization of the tumour may have an important impact on patient's prognosis, the study of circulating tumour DNA in body fluids can be used to monitor the tumour burden. Furthermore, circulating tumour DNA has been found in the saliva of the 100% oral cancer patients, being found in plasma only in 80% of cases (Chattopadhyay and Panda 2019). Similarly, GSTT1 gene is involved in the detoxification of active metabolites of tobacco, and the analysis of saliva showed that 24% of oral cancer patients showed GSTT1 null genotype versus 16% GSTT1 null genotype of healthy controls (Shah et al. 2011).

Aberrant DNA methylation is an epigenetic process that can change in response to several conditions such as environmental exposures (Bryan et al. 2013) or cancers (Carvalho et al. 2008), and thus it can be evaluated for the detection of these situations. In this sense, differences in salivary DNA-methylation patterns between pre- and postoperative patients with oral squamous cell carcinoma, and between those patients and healthy controls have been described (Viet and Schmidt 2008).

4.2.3.2 Transcriptomics

In contrast to genome, the transcriptome (RNA cell content) is very dynamic, and includes different types of RNA in order to transcript a gene into mRNA and translate it into a protein. Transcriptomics also provides with valuable information about the functional elements of the genome as well as the molecular components of cells and tissues (Wang et al. 2009). The salivary transcriptome was described in 2004, reporting a core of 180 RNAs (Li et al. 2004, 2006), and recent research has revealed more than 3000 species of mRNA and 300microRNA in saliva (Bentwich et al. 2005). Transcriptomes vary depending on the cell type and its status, and transcriptomics aims to quantitatively and qualitatively study these variations in RNA markers of the trait of interest. The molecular profiling of mRNA in saliva provides information regarding the gene transcription of the subject.

In saliva, mRNA and microRNA enters in the oral cavity from salivary glands, gingival crevicular fluid and the desquamation of oral epithelial cells, among other sources (Park et al. 2006). This RNA is mostly contained in exosomes in order to protect them from extracellular ribonucleases and, since it may be altered in diseases, it has been suggested that transcriptomic analysis of saliva could be of value for the study of health and diseases states (Bentwich et al. 2005).

Transcriptomics englobes different techniques, including the quantitative Real-Time Polymerase Chain Reaction (qPCR), in which a specific mRNA is quantified in a sample (Bustin 2000). This technique is widely used, and in saliva it has been employed for the detection of important diseases such as cancer (Lee et al. 2012; Zhang et al. 2010b, 2012), Zika virus RNA (Musso et al. 2015) or *Helicobacter pylori* infection, with several advantages compared to the traditional endoscopy or urea breath test (Ishihara et al. 2001; Young and Luo 2016). On the other side, microRNA (a group of small 19–25 nucleotides that are encoded by genes but not translated into proteins, known as non-coding RNAs) participates in important processes such as cell growth, differentiation, apoptosis, stress and immune response. Those microARN have been shown to be downregulated in cancer tissues when compared to healthy tissues. However, salivary microRNA are more stable (Gallo et al. 2012) and showed larger fold-differences between healthy controls and cancer patients (Jiang et al. 2005; Monticelli et al. 2005) than tissue microRNA, posing another advantage besides its non-invasive collection. Other technologies such as Affymetrix GeneChip Human Transcriptome Array 2.0 are commonly used, analysing all transcript isoforms of the human transcriptome, posing more than six million probes targeting coding and non-coding transcripts, and exon-exon splice junctions (Meghann et al. 2014). Similarly, RNA sequencing is been employed for the characterization of the full human transcriptome by deep-sequencing technologies (Kaczor-Urbanowicz et al. 2017). This approach is analytically more sensitive than microarrays, providing sequence information at each nucleotide position of specific genes (Spielmann et al. 2012; Wang et al. 2009).

Nowadays, transcriptomics in saliva has described hundreds of biomarkers of diagnosis for several systemic and oral diseases such as pancreatic (Zhang et al. 2010a), lung (Zhang et al. 2012), ovarian (Lee et al. 2012), and breast cancer (Zhang et al. 2010b), Sjögren's syndrome (Hu et al. 2010), or chronic periodontitis (Lakschevitz et al. 2013).

4.2.3.3 Proteomics

Proteomics studies are the large-scale screening of proteins, their expression, modifications, and their interactions through a high-throughput approach (Anderson and Anderson 1998). Currently, one of the most employed techniques are gel electrophoresis and mass spectrometry (Wong 2006a). Gel electrophoresis enables the separation of the different proteins according to their molecular weight, and additionally according to their isoelectric point in the case of two dimensional gel electrophoresis (Wong 2006a); while mass spectrometry analysis allows the qualitative and quantitative evaluation of proteins of a sample (Kossowska et al. 2009).

Although there are many different proteomics techniques that can be used, they can be firstly divided into 'bottom-up' and 'top-down' proteomics. While bottom-up proteomics analyses protein fragments obtained by sample pre-digestion, focused on one or more of their characteristics fragments for identification, top-down proteomics studies the intact naturally occurring structure of the proteins (Castagnola et al. 2011,

2017). Both approaches present several advantages and drawbacks, and the integration of the two in an adequate number of samples would result in the best way to carry out a robust protein biomarker identification, although this approach is not accessible to the majority of laboratories (Al-Tarawneh et al. 2011; Messana et al. 2013).

Only 27% of proteins presented in saliva are also found in plasma (Loo et al. 2010); thus, proteomic approaches in saliva have provided with a huge list of candidate biomarkers, being the actual tendency to focus on protein panels rather than unique proteins. However, the presence of several high-abundance proteins has been hampered the identification and quantification of low-sample proteins with valuable biomarker potential. For example, in human saliva, α-amylase, albumin and proline-rich proteins form >75% of the total salivary protein amount (Kaczor-Urbanowicz et al. 2017). For this, there are several methods for the removal of those high-abundance proteins, although they are not free of inconveniences such as the possible alteration of other salivary components.

Although the number of identified proteins would vary depending on the type of saliva, sample processing or species studied, normally hundreds to thousands of proteins are identified in the "salivaoma". The core salivary proteome was set in 1166 proteins (Denny et al. 2008), although more than 3000 proteins are commonly identified in saliva (Castagnola et al. 2017). For example, 487 proteins were identified in a study in the saliva of human patients with type 2 diabetes, being 65 of those differentially expressed when compared to healthy matched healthy controls (Rao et al. 2009). Using tandem mass tags technologies –a gel-free approach in which each sample is labelled individually- more than 500 proteins were identified and quantified in human saliva, being the combination of three of them namely cystatin B, triose-phosphate isomerase and deleted in malignant tumour 1 protein proposed as biomarkers of gastric cancer with 85% sensitivity, 80% specificity and 0.93 accuracy (Xiao et al. 2016). Additionally, the study of the changes in the salivary proteomic composition may point out the alteration in the regulation of some biological mechanisms relevant to a certain process prior to the development of any external clinical complication (Pappa et al. 2018). This approach has been recently applied in saliva of dogs for the discovery of potential biomarkers and to identify affected biological processes in parvoviral enteritis (Franco-Martínez et al. 2018) and leishmaniosis (Franco-Martínez et al. 2019). In humans, the salivary proteome has been characterised in several oral and systemic diseases such as Sjögren's syndrome, autoimmune disorders, schizophrenia, or Down's Syndrome (Bosello et al. 2016; Cabras et al. 2013; Iavarone et al. 2014; Katsiougiannis and Wong 2016).

4.2.3.4 Metabolomics

In metabolomics, the metabolites of living tissues –the end product of transcription and translation, including metabolic intermediates such as carbohydrates, lipids, nucleic acids, amino acids and hormones, among others- is studied (Wong 2007), allowing a better understanding of cellular functions (Chattopadhyay and Panda 2019). Some of the most common techniques for metabolomics are high-performance liquid chromatography-mass spectrometry (HPLC-MS) or two-dimensional gas chromatography MS and nuclear magnetic resonance spectroscopy (NMR) (Kaczor-Urbanowicz et al. 2017).

Although the metabolome can vary in seconds (Bonne and Wong 2012), metabolomics is essential for identifying metabolic alterations related to a specific condition, which can be used for diagnosis or monitoring (Ishikawa et al. 2017; Mikkonen et al. 2016; Ohshima et al. 2017; Sugimoto et al. 2010; Wei et al. 2011). For example, large amounts of lactic acid are often produced by tumor cells due to their excessive proliferation, which can be detected in saliva in the case of oral cancers (Wang et al. 2014). Additionally, salivary metabolomics has been successfully applied to hepatocellular, oral and colorectal cancers, periodontal diseases, or chronic renal diseases, among others conditions (Beger 2013; Mikkonen et al. 2016).

Similarly, several salivary glycoproteins levels were found to be higher in the saliva of oral cancer patients (Ohshima et al. 2017).

4.2.3.5 Microbiomics

The salivary microbiome is the significant microbial populations which colonize oral cavities (Chattopadhyay and Panda 2019), posing complex interactions between the microorganism and the host (Wong 2007). In humans, most salivary microbiomes are attached to exfoliated epithelial cells and their levels vary between saliva and subsites of the oral cavity such as supra- or sub-gingival plaque (Chattopadhyay and Panda 2019). Among the most common microbiome in humans, *Streptococci spp* is the dominant bacteria, which co-exist with *Veillonella, Gemella, Rothia, Fusobacterium*, and *Neisseria* species, among others.

Currently, newer technologies have been becoming available for the study of the salivary microbiomes, such as the study of microbial RNA or DNA sequencing (Kaczor-Urbanowicz et al. 2017). The oral microbiome has been related to diseases of the oral cavity and with the microbiomes of other microenvironments of the human body (Costalonga and Herzberg 2014). It is known that the characterization of the salivary microbiome would be of high utility in the diagnosis of several diseases (Nasidze et al. 2009), although it is very challenging due to its enormous diversity. To the date in humans, salivary microbiomics has been employed for early detection of influenza virus, herpes virus or Zika virus, among others (Wong 2006b). The characterization of specific oral microbiome of a certain condition might help in the development of strategies for its diagnosis, as described recently in children with autism spectrum disorder (Qiao et al. 2018; Rosenfeld 2015).

The detection of differences between the salivary microbiome in healthy and ill patients could provide useful information regarding the disease-associated changes at a certain time point (Schmidt et al. 2014). For example, the salivary microbiome has proven potentially useful for the detection of early resectable pancreatic cancer with 96.4% sensitivity and 82.1% specificity (AUC = 0.9) (Farrell et al. 2012). Smoking is known to affect the composition of the salivary microbiome, which may play important roles in the health status by inducing mutations in genes. Additionally, chronic inflammation caused by bacterial infection has been pointed out as responsible for tumorigenesis (Lim et al. 2017).

4.3 Validation

Of the hundreds of thousand discovered biomarkers with potential utility for clinical application, only a very low percentage have made it to clinical applications (Ransohoff 2005). This is partially due to the requirements that potential biomarkers must fulfil in order to be considered for clinical applications. Therefore, a full method validation should be performed, when introducing a new method (or even revised or based upon literature) in the laboratory. A complete validation should be achieved for each species and matrix concerned to demonstrate the reliability of this particular (new or not) method and to ensure high-quality laboratory services. That means a rigorously and impartially evaluation of its performance which includes consistently reliable results even when used by different operators and in different batches of reagents over a long period of time.

Thus, after the discovery of a biomarker, by using one of the techniques described above, the next step is to verify and validate it. For this, several different approaches may be used, depending on the characteristics of the biomarker, potential application, and techniques used. For verification, ideally, the discovered biomarker should be measured by a different technique with compatible results than the previously observed. After this, a preclinical validation followed by a definitive academic validation with the prospective-specimen collection – retrospective-blinded-evaluation design, and finally validation with multicentre studies (Wong 2007) should be ideally performed.

In this section, analytical validation of spectrophotometric and immunoassays and "Omics" validation will be differentiated:

4.3.1 Analytical Validation

The analytical performance of the method is related to several properties and should include, as a minimum, precision, accuracy, and sensitivity. To assess these characteristics, experimental studies are needed. It is important that the method validation be performed following principles of available Guidelines (European Medicine Agency 2012; Jensen and Kjelgaard-Hansen 2010; Karnes et al. 1991; Tiwari and Tiwari 2010). Thus, the findings should be evaluated and decision if the method performance is acceptable made.

Briefly, *precision* is expressed as the coefficient of variation (CV) and describes the closeness of agreement between a series of measurements (replicate) of the same sample. A*ccuracy* of an analytical method (also referred to as bias or systematic error), expressed in percentage, describes the closeness of the agreement between the measured value of an analyte and its true concentration and is frequently assessed using recovery and linearity tests, and comparison of analytical methods. *Sensitivity* is evaluated by the detection limit (the minimum concentration or quantity of an analyte that is detected with reasonable certainty for a given analytical method) and by the lower limit of quantification (the lowest amount that could be measured above the limit of detection with a CV <20%). In addition, the effects that a compound, or more than one, such as haemoglobin, lipids, bilirubin, may have on the accuracy of the analytical method are also included in a validation study. The stability of an analyte and other pre-analytical sources of variation such as anticoagulant, venous site of blood sampling, or medication may significantly affect test results so they should be included in the investigations of the performance study of the analytical method.

4.3.2 Clinical Validation

Finally, as part of this initial stage, a comparison of the test results in healthy populations to various groups of diseased subjects should be carried out.

One of the first desired requirements for a biomarker is to have a high sensitivity and specificity, meaning that most of the population with the condition of interest would have positive results, and most population without the condition would present negative test results. However, this often is difficult to achieve, and biomarkers with lower sensitivity or specificity may be also acceptable for some purposes. Biomarkers with high sensitivity and lower specificity (results that may be negative showed positive) may be employed for screening purposes, needing an additional confirmatory test in order to avoid unnecessary treatment to patients that are free of disease. Opposing this, biomarkers with low sensitivity and high specificity are used to confirm that the individual is disease-free, but could fail to detect the genuine cases of the condition of interest (Baum et al. 2011).

For clinical validation, it is important to keep in mind that is unlike that a single molecular biomarker will be able to discern disease from non-disease states due to the high heterogeneity presented in large populations (Baum et al. 2011).

4.3.3 Omics Validation

When omics approaches are employed, the clinically meaningful discoveries are normally hidden within millions of analyses due to the immense biological complexity, and hence, separating true signals from red herrings is challenging, and validation of proposed discoveries is essential (Ioannidis 2018; Ioannidis and Khoury 2011).

Potential biomarkers discovered by omics technologies have to be further validated before their implementation in routine analysis. Validation of candidate biomarkers identified by omics may require to check analytical validity (ideally, the accurately and repeatable measure the biomarker in different laboratories and techniques) (Ioannidis and Khoury 2011). Since omics are highly expensive and time consuming, it is normally required to validate those results in a larger cohort of sample size of hundreds or thousands using more economic and rapid techniques such as immunoassays or spectrophotometric assays (Mischak et al. 2015).

4.4 Conclusions

In the present chapter, a wide range of techniques that are used for the study of salivary composition have been described, from the most economical and easy to use (e.g. UV-Vis absorption spectroscopy) to the most sensible and advanced techniques (e.g. omics). In addition, it is essential to highlight the importance of validation of both, potential biomarkers identified by omics and analytical methods, before their usage in clinical settings.

We envision that, in parallel with the development of these methodologies, saliva would continue to gain increased attention for the diagnostic and monitoring of a wide range of conditions in both humans and animals.

References

Aitio A, Anderson D, Blain P, Bond J, Buratti M, Calder I, Chahoud I, Fowle J, Gerhardsson L, Henderson R, Koëter H, Nishikawa A, Thompson C, Zenick H (1993) Biomarkers and risk assessment: concepts and principles. In: World Health Organization (ed.), Environmental Health Criteria. United Nations Environment Programme, International Labour Organisation, World Health Organization, Génova, pp 1–82

Al-Dujaili EAS, Baghdadi HHS, Howie F, Mason JI (2012) Validation and application of a highly specific and sensitive ELISA for the estimation of cortisone in saliva, urine and in vitro cell-culture media by using a novel antibody. Steroids 77:703–709. https://doi.org/10.1016/j.steroids.2012.03.001

Al-Tarawneh SK, Border MB, Dibble CF, Bencharit S (2011) Defining salivary biomarkers using mass spectrometry-based proteomics: a systematic review. Omi A J Integr Biol. https://doi.org/10.1089/omi.2010.0134

Anderson NL, Anderson NG (1998) Proteome and proteomics: new technologies, new concepts, and new words. In: Electrophoresis. https://doi.org/10.1002/elps.1150191103

Arellano-Garcia MT, Hu S, Wang J, Henson B, Zhou H, Chia D, Wong DT (2008) Multiplexed immunobead-based assay for detection of oral cancer protein biomarkers in saliva. Oral Dis 14:705–712. https://doi.org/10.1111/j.1601-0825.2008.01488.x

Barranco T, Rubio CP, Tvarijonaviciute A, Rubio M, Damia E, Lamy E, Cugat R, Cerón JJ, Tecles F, Escribano D (2019) Changes of salivary biomarkers under different storage conditions: effects of temperature and length of storage. Biochem Med 29. https://doi.org/10.11613/BM.2019.010706

Baum BJ, Yates JR, Srivastava S, Wong DT, Melvin JE (2011) Scientific frontiers: emerging technologies for salivary diagnostics. Adv Dent Res 23:360–368. https://doi.org/10.1177/0022034511420433

Beger R (2013) A review of applications of metabolomics in Cancer. Meta. https://doi.org/10.3390/metabo3030552

Ben-Aryeh H, Jungerman T, Szargel R, Klein E, Laufer D (1996) Salivary flow-rate and composition in schizophrenic patients on clozapine: subjective reports and laboratory data. Biol Psychiatry 39:946–949. https://doi.org/10.1016/0006-3223(95)00296-0

Bentwich I, Avniel A, Karov Y, Aharonov R, Gilad S, Barad O, Barzilai A, Einat P, Einav U, Meiri E, Sharon E, Spector Y, Bentwich Z (2005) Identification of hundreds of conserved and nonconserved human microRNAs. Nat Genet 37(7):766–770. https://doi.org/10.1038/ng1590

Berry JW, Chappell DG, Barnes RB (1946) Improved method of flame photometry. Ind Eng Chem Anal Ed 18:19–24. https://doi.org/10.1021/i560149a005

Bibi G, Green Y, Nagler RM (2008) Compositional and oxidative analysis in the saliva and serum of predialysis chronic kidney disease patients and end-stage renal failure patients on peritoneal dialysis. Ther Apher Dial 12:164–170. https://doi.org/10.1111/j.1744-9987.2008.00564.x

Bonne NJ, Wong DTW (2012) Salivary biomarker development using genomic, proteomic and metabolomic approaches. Genome Med 4(10):82. https://doi.org/10.1186/gm383

Booth CK, Dwyer DB, Pacque PF, Ball MJ (2009) Measurement of immunoglobulin a in saliva by particle-enhanced nephelometric immunoassay: sample collection, limits of quantitation, precision, stability and reference range. Ann Clin Biochem 46:401–406. https://doi.org/10.1258/acb.2009.008248

Bosello S, Peluso G, Iavarone F, Tolusso B, Messana I, Faa G, Castagnola M, Ferraccioli G (2016) Thymosin β4 and β10 in Sjögren's syndrome: saliva proteomics and minor salivary glands expression. Arthritis Res Ther 18:229. https://doi.org/10.1186/s13075-016-1134-7

Bryan AD, Magnan RE, Hooper AEC, Harlaar N, Hutchison KE (2013) Physical activity and differential methylation of breast cancer genes assayed from saliva: a preliminary investigation. Ann Behav Med 45(1):89–98. https://doi.org/10.1007/s12160-012-9411-4

Bustin SA (2000) Absolute quantification of mRNA using real-time reverse transcription polymerase chain reaction assays. J Mol Endocrinol 25:169–193

Cabras T, Pisano E, Montaldo C, Giuca MR, Iavarone F, Zampino G, Castagnola M, Messana I (2013) Significant modifications of the salivary proteome potentially associated with complications of down syndrome revealed by top-down proteomics. Mol Cell Proteomics 12(7):1844–1852. https://doi.org/10.1074/mcp.M112.026708

Carvalho AL, Jeronimo C, Kim MM, Henrique R, Zhang Z, Hoque MO, Chang S, Brait M, Nayak CS, Jiang WW, Claybourne Q, Tokumaru Y, Lee J, Goldenberg D, Garrett-Mayer E, Goodman S, Moon CS, Koch W, Westra WH, Sidransky D, Califano JA (2008) Evaluation of promoter hypermethylation detection in body fluids as a screening/diagnosis tool for head and neck squamous cell carcinoma. Clin Cancer Res 14(1):97–107. https://doi.org/10.1158/1078-0432.CCR-07-0722

Castagnola M, Inzitari R, Fanali C, Iavarone F, Vitali A, Desiderio C, Vento G, Tirone C, Romagnoli C, Cabras T, Manconi B, Sanna MT, Boi R, Pisano E, Olianas A, Pellegrini M, Nemolato S, Heizmann CW, Faa G, Messana I (2011) The surprising composition of the salivary proteome of preterm human newborn. Mol Cell Proteomics 10(1). https://doi.org/10.1074/mcp.M110.003467

Castagnola M, Scarano E, Passali G, Messana I, Cabras T, Iavarone F, Di Cintio G, Fiorita A, De Corso E, Paludetti G (2017). Salivary biomarkers and proteomics. In: Acta Otorhinolaryngol Ital. pp 94–101. https://doi.org/10.14639/0392-100x-1598

Cerón JJ, Eckersall PD, Martínez-Subiela S (2005) Acute phase proteins in dogs and cats: current knowledge and future perspectives. Vet Clin Pathol 34:85–99. https://doi.org/10.1111/j.1939-165X.2005.tb00019.x

Chance B (1972) Principles of differential spectrophotometry with special reference to the dual wavelength method. Methods Enzymol 24:322–336. https://doi.org/10.1016/0076-6879(72)24080-0

Chattopadhyay I, Panda M (2019) Recent trends of saliva omics biomarkers for the diagnosis and treatment of oral cancer. J Oral Biosci 61:84–94. https://doi.org/10.1016/J.JOB.2019.03.002

Chen M, Zhao H (2019) Next-generation sequencing in liquid biopsy: cancer screening and early detection. Hum Genomics 13:34. https://doi.org/10.1186/s40246-019-0220-8

Chiappin S, Antonelli G, Gatti R, De Palo EF (2007) Saliva specimen: a new laboratory tool for diagnostic and basic investigation. Clin Chim Acta 383(1–2):30–40. https://doi.org/10.1016/j.cca.2007.04.011

Chowdhury F, Williams A, Johnson P (2009) Validation and comparison of two multiplex technologies, Luminex (R) and mesoscale discovery, for human cytokine profiling. J Immunol Methods 340:55–64. https://doi.org/10.1016/j.jim.2008.10.002

Claver JB, Mirón MV, Capitán-Vallvey LF (2009) Disposable electrochemiluminescent biosensor for lactate determination in saliva. Analyst 134:1423–1432. https://doi.org/10.1039/b821922b

Contreras-Aguilar MD, Escribano D, Martínez-Subiela S, Martínez-Miró S, Rubio M, Tvarijonaviciute A, Tecles F, Cerón JJ (2017) Influence of the way of reporting alpha- amylase values in saliva in different naturalistic situations : a pilot study. PLoS One 12:1–13. https://doi.org/10.1371/journal.pone.0180100

Costalonga M, Herzberg MC (2014) The oral microbiome and the immunobiology of periodontal disease and caries. Immunol Lett 162(2):22–38. https://doi.org/10.1016/j.imlet.2014.08.017

Denny P, Hagen FK, Hardt M, Liao L, Yan W, Arellanno M, Bassilian S, Bedi GS, Boontheung P, Cociorva D, Delahunty CM, Denny T, Dunsmore J, Faull KF, Gilligan J, Gonzalez-Begne M, Halgand F, Hall SC, Han X, Henson B, Hewel J, Hu S, Jeffrey S, Jiang J, Loo JA, Ogorzalek Loo RR, Malamud D, Melvin JE, Miroshnychenko O, Navazesh M, Niles R, Park SK, Prakobphol A, Ramachandran P, Richert M, Robinson S, Sondej M, Souda P, Sullivan MA, Takashima J, Than S, Wang J, Whitelegge JP, Witkowska HE, Wolinsky L, Xie Y, Xu T, Yu W, Ytterberg J, Wong DT, Yates JR, Fisher SJ (2008) The proteomes of human parotid and submandibular/sublingual gland Salivas collected as the ductal secretions. J Proteome Res 7:1994–2006. https://doi.org/10.1021/pr700764j

Di Lenardo D, Silva FRPD, De Carvalho França LF, Carvalho JDS, Alves EHP, Vasconcelos DFP (2019) Evaluation of biochemical parameters present in the saliva of patients with chronic periodontitis: results from a Meta-analysis. Genet Test Mol Biomarkers 23:255–263. https://doi.org/10.1089/gtmb.2017.0272

Diab-Ladki R, Pellat B, Chahine R (2003) Decrease in the total antioxidant activity of saliva in patients with periodontal diseases. Clin Oral Investig 7:103–107. https://doi.org/10.1007/s00784-003-0208-5

Dipaola C, Mandel ID (1980) Lactoferrin concentration in human-parotid saliva as measured by an enzyme-linked immunosorbent-assay (ELISA). J Dent Res 59:1463–1465. https://doi.org/10.1177/00220345800590090101

Drake RR, Cazares LH, Semmes OJ, Wadsworth JT (2005) Serum, salivary and tissue proteomics for discovery of biomarkers for head and neck cancers. Expert Rev Mol Diagn 5:93–100. https://doi.org/10.1586/14737159.5.1.93

Escribano D, Cerón JJ (2014) Salivary testosterone measurements in growing pigs : validation of an automated chemiluminescent immunoassay and its possible use as an acute stress marker. Res Vet Sci 97:20–25. https://doi.org/10.1016/j.rvsc.2014.04.001

Escribano D, Fuentes-Rubio M, Ceron JJ (2012a) Validation of an automated chemiluminescent immunoassay for salivary cortisol measurements in pigs. J Vet Diagnostic Investig 24:918–923. https://doi.org/10.1177/1040638712455171

Escribano D, Gutierrez AM, Martinez Subiela S, Tecles F, Ceron JJ (2012b) Validation of three commercially available immunoassays for quantification of IgA, IgG, and IgM in porcine saliva samples. Res Vet Sci 93:682–687. https://doi.org/10.1016/j.rvsc.2011.09.018

Escribano D, Soler L, Gutiérrez AM, Martínez-Subiela S, Cerón JJ (2013) Measurement of chromogranin a in porcine saliva: validation of a time-resolved immunofluorometric assay and evaluation of its application as a marker of acute stress. Animal 7:640–647. https://doi.org/10.1017/S1751731112002005

European Medicine Agency (2012) Guideline on bioanalytical method validation. EMA, London. EMEA/CHMP/EWP/192217/2009

Evenson MA (2001) Spectrophotometric techniques. In: Burtis CA, Ashwood ER (eds) Tietz fundamentals of clinical chemistry. W.B. Saunders, Philadelphia, pp 56–73

Farrell JJ, Zhang L, Zhou H, Chia D, Elashoff D, Akin D, Paster BJ, Joshipura K, Wong DTW (2012) Variations of oral microbiota are associated with pancreatic diseases including pancreatic cancer. Gut 64(4):582–588. https://doi.org/10.1136/gutjnl-2011-300784

Fernández B, Lobo L, Pereiro R, (2018) Atomic absorption spectrometry: fundamentals, instrumentation and capabilities, 3rd ed. Reference Module in Chemistry, Molecular Sciences and Chemical Engineering Elsevier Inc https://doi.org/10.1016/b978-0-12-409547-2.14116-2

Fernie R, Trethewey R, Krotzky A, Willmitzer L (2004) Metabolite profiling: from diagnostics to systems biology. Nat Rev Mol Cell 5:763–769

Floriano PN, Christodoulides N, Miller CS, Ebersole JL, Spertus J, Rose BG, Kinane DF, Novak MJ, Steinhubl S, Acosta S, Mohanty S, Dharshan P, Yeh C, Redding S, Furmaga W, McDevitt JT (2009) Use of saliva-based Nano-biochip tests for acute myocardial infarction at the point of care: a feasibility study. Clin Chem 55:1530–1538. https://doi.org/10.1373/clinchem.2008.117713

Franco-Martínez L, Romero D, García-Navarro JA, Tecles F, Teles M, Tvarijonaviciute A (2016) Measurement of p-nitrophenyl acetate esterase activity (EA), total antioxidant capacity (TAC), total oxidant status (TOS) and acetylcholinesterase (AChE) in gills and digestive gland of Mytilus galloprovincialis exposed to binary mixtures of Pb, Cd and Cu. Environ Sci Pollut Res 23:25385–25392. https://doi.org/10.1007/s11356-016-7677-y

Franco-Martínez L, Tvarijonaviciute A, Horvatić A, Guillemin N, Cerón JJ, Escribano D, Eckersall D, Kocatürk M, Yilmaz Z, Lamy E, Martínez-Subiela S, Mrljak V (2018) Changes in salivary analytes in canine parvovirus: A high-resolution quantitative proteomic study. Comp Immunol Microbiol Infect Dis 60:1–10

Franco-Martínez L, Tvarijonaviciute A, Horvatić A, Guillemin N, Bernal LJ, Barić Rafaj R, Cerón JJ, Thomas M d C, López MC, Tecles F, Martínez-Subiela S, Mrljak V (2019) Changes in saliva of dogs with canine leishmaniosis: A proteomic approach. Vet Parasitol 272:44–52

Gallo A, Tandon M, Alevizos I, Illei GG (2012) The majority of microRNAs detectable in serum and saliva is concentrated in exosomes. PLoS One 7(3):e30679. https://doi.org/10.1371/journal.pone.0030679

Govindaraju, S., Lee, M.H., Yun, K., 2019. Recent trends in the development of paper-based diagnostic chips for the detection of human viruses. In: Recent developments in applied microbiology and biochemistry. Elsevier Inc., pp 349–361. https://doi.org/10.1016/b978-0-12-816328-3.00025-8

Guillemin N, Horvatić A, Kuleš J, Galan A, Mrljak V, Bhide M (2016) Omics approaches to probe markers of disease resistance in animal sciences. Mol BioSyst 12:2036–2046. https://doi.org/10.1039/C6MB00220J

Gutiérrez AM, Martínez-Subiela S, Cerón J (2009a) Evaluation of an immunoassay for determination of haptoglobin concentration in various biological specimens from swine. Am J Vet Res 70(6):691–696

Gutiérrez AM, Martinez-Subiela S, Eckersall PD, Ceron JJ (2009b) C-reactive protein quantification in porcine saliva: a minimally invasive test for pig health monitoring. Vet J 181:261–265. https://doi.org/10.1016/j.tvjl.2008.03.021

Hansen AM, Garde AH, Christensen JM, Eller NH, Netterstrom B (2003) Evaluation of a radioimmunoassay and establishment of a reference interval for salivary cortisol in healthy subjects in Denmark. Scand J Clin Lab Invest 63:303–310. https://doi.org/10.1080/00365510310001942

Hansen TVO, Simonsen MK, Nielsen FC, Hundrup YA (2007) Collection of blood, saliva, and buccal cell samples in a pilot study on the Danish nurse cohort: comparison of the response rate and quality of genomic DNA. Cancer Epidemiol Biomark Prev 16(10):2072–2076. https://doi.org/10.1158/1055-9965.EPI-07-0611

Hemmila I (1985) Fluoroimmunoassays and immunofluorometric assays. Clin Chem 31:359–370

Holcombe JA, Hassell DC (1990) Atomic absorption, atomic emission, and Flame Emission Spectrometry. Anal Chem 62:78–84. https://doi.org/10.1021/ac00211a007

Horvatić A, Kuleš J, Guillemin N, Galan A, Mrljak V, Bhide M (2016) High-throughput proteomics and the fight against pathogens. Mol BioSyst 3:2373. https://doi.org/10.1039/C6MB00223D

Hu L, Xu G (2010) Applications and trends in electrochemiluminescence. Chem Soc Rev 39:3275–3304. https://doi.org/10.1039/b923679c

Hu S, Gao K, Pollard R, Arellano-Garcia M, Zhou H, Zhang L, Elashoff D, Kallenberg C, Vissink A, Wong D (2010) Preclinical validation of salivary biomarkers for primary Sjögren's syndrome. Arthritis Care Res (Hoboken)

Hu W, Dang T, Li Z, Lei L, Wang G, Li Y, Xu H, Zhou Z, Liu GL (2019) C-reaction protein detection in human saliva by nanoplasmonic color imaging. J Biomed Nanotechnol 15:1724–1733. https://doi.org/10.1166/jbn.2019.2769

Iavarone F, Melis M, Platania G, Cabras T, Manconi B, Petruzzelli R, Cordaro M, Siracusano A, Faa G, Messana I, Zanasi M, Castagnola M (2014) Characterization of salivary proteins of schizophrenic and bipolar disorder patients by top-down proteomics. J Proteome 103:15–22. https://doi.org/10.1016/j.jprot.2014.03.020

Ioannidis JPA (2018) Why most published research findings are false. In: Getting to good: research integrity in the biomedical sciences. Springer International Publishing, pp 2–8. https://doi.org/10.1371/journal.pmed.0020124

Ioannidis JPA, Khoury MJ (2011) Improving validation practices in "omics" research. Science (80-.). https://doi.org/10.1126/science.1211811

Ishihara K, Miura T, Ebihara Y, Hirayama T, Kamiya S, Okuda K (2001) Shared antigenicity between helicobacter pylori and periodontopathic campylobacter rectus strains. FEMS Microbiol Lett 197(1):23. https://doi.org/10.1016/S0378-1097(01)00078-7

Ishikawa S, Sugimoto M, Kitabatake K, Tu M, Sugano A, Yamamori I, Iba A, Yusa K, Kaneko M, Ota S, Hiwatari K, Enomoto A, Masaru T, Iino M (2017) Effect of timing of collection of salivary metabolomic biomarkers on oral cancer detection. Amino Acids 49(4):761–770. https://doi.org/10.1007/s00726-017-2378-5

Jensen AL, Kjelgaard-Hansen M (2010) Diagnostic test validation. In: Weiss DJ, Wardrop KJ, Ames I (eds) Schalms veterinary hematology. Wiley-Blackwell Publishing, Ames, pp 1027–1033

Jiang J, Lee EJ, Gusev Y, Schmittgen TD (2005) Real-time expression profiling of microRNA precursors in human cancer cell lines. Nucleic Acids Res 33(17):5394–5403. https://doi.org/10.1093/nar/gki863

Ju H, Lai G, Yan F (2017a) Introduction. In: Ju H, Lai G, Yan F (eds) Immunosensing for detection of protein biomarkers. Elsevier, pp 1–30. https://doi.org/10.1016/b978-0-08-101999-3.00001-3

Ju H, Lai G, Yan F (2017b) Chemiluminescent immunoassay. In: Ju H, Lai G, Yan F (eds) Immunosensing for detection of protein biomarkers. Elsevier, Amsterdam, pp 143–169. https://doi.org/10.1016/b978-0-08-101999-3.00005-0

Ju H, Lai G, Yan F (2017c) Electrochemiluminescent immunosensing. In: Ju H, Lai G, Yan F (eds) Immunosensing for detection of protein biomarkers. Elsevier, Amsterdam, pp 171–206. https://doi.org/10.1016/b978-0-08-101999-3.00006-2

Ju H, Lai G, Yan F (2017) Multianalyte immunoassay. In: Immunosensing for detection of protein biomarkers. Ju, Huangxian Lai, Guosong Yan, Feng, pp 207–237. https://doi.org/10.1016/0731-7085(89)80079-2

Kaczor-Urbanowicz KE, Martin Carreras-Presas C, Aro K, Tu M, Garcia-Godoy F, Wong DTW (2017) Saliva diagnostics – current views and directions. Exp Biol Med 242(5):459–472. https://doi.org/10.1177/1535370216681550

Kamarainen S, Maki M, Tolonen T, Palleschi G, Virtanen V, Micheli L, Sesay AM (2018) Disposable electrochemical immunosensor for cortisol determination in human saliva. Talanta 188:50–57. https://doi.org/10.1016/j.talanta.2018.05.039

Karnes HT, Shiu G, Shah VP (1991) Validation of bianalytical methods. Pharm Res 8:421

Katsiougiannis S, Wong DTW (2016) The proteomics of saliva in Sjögren's syndrome. Rheum Dis Clin N Am. https://doi.org/10.1016/j.rdc.2016.03.004

Khan MS, Dighe K, Wang Z, Srivastava I, Daza E, Schwartz-Dual AS, Ghannam J, Misra SK, Pan D (2018) Detection of prostate specific antigen (PSA) in human saliva using an ultra-sensitive nanocomposite of graphene nanoplatelets with diblock-: Co -polymers and Au electrodes. Analyst 143:1094–1103. https://doi.org/10.1039/c7an01932g

Khan R, Khurshid Z, Yahya Ibrahim Asiri F (2017) Advancing point-of-care (PoC) testing using human saliva as liquid biopsy. Diagnostics 7:39. https://doi.org/10.3390/diagnostics7030039

Khaustova S, Shkurnikov M, Grebenyuk E, Artyushenko V, Tonevitsky A (2009) Assessment of biochemical characteristics of the saliva using fourier transform mid-infrared spectroscopy. Bull Exp Biol Med 148:841–844. https://doi.org/10.1007/s10517-010-0831-5

Khaustova S, Shkurnikov M, Tonevitsky E, Artyushenko V, Tonevitsky A (2010) Noninvasive biochemical monitoring of physiological stress by Fourier transform infrared saliva spectroscopy. Analyst 135:3183–3192. https://doi.org/10.1039/c0an00529k

Kossowska B, Dudka I, Gancarz R, Antonowicz-Juchniewicz J (2009) Proteomic analysis of protein profiles in some pathological stages of the human organism. Postepy Hig Med Dosw. (Online)

Lakschevitz FS, Aboodi GM, Glogauer M (2013) Oral neutrophil transcriptome changes result in a pro-survival phenotype in periodontal diseases. PLoS One 8(7):e68983. https://doi.org/10.1371/journal.pone.0068983

Lamy E, Mau M (2012) Saliva proteomics as an emerging, non-invasive tool to study livestock physiology, nutrition and diseases. J Proteome 75:4251–4258. https://doi.org/10.1016/J.JPROT.2012.05.007

Lee YH, Kim JH, Zhou H, Kim BW, Wong DT (2012) Salivary transcriptomic biomarkers for detection of ovarian cancer: for serous papillary adenocarcinoma. J Mol Med 90(4):427–434. https://doi.org/10.1007/s00109-011-0829-0

Li Y, Elashoff D, Oh M, Sinha U, St John MAR, Zhou X, Abemayor E, Wong DT (2006) Serum circulating human mRNA profiling and its utility for oral cancer detection. J Clin Oncol 24(11):1754–1760. https://doi.org/10.1200/JCO.2005.03.7598

Li Y, St. John MAR, Zhou X, Kim Y, Sinha U, Jordan RCK, Eisele D, Abemayor E, Elashoff D, Park NH, Wong DT (2004) Salivary transcriptome diagnostics for oral cancer detection. Clin Cancer Res 10(24):8442–8450. https://doi.org/10.1158/1078-0432.CCR-04-1167

Lim Y, Totsika M, Morrison M, Punyadeera C (2017) Oral microbiome: a new biomarker reservoir for oral and oropharyngeal cancers. Theranostics 7(17):4313–4321. https://doi.org/10.7150/thno.21804

Liu R, Zhang S, Wei C, Xing Z, Zhang S, Zhang X (2016) Metal stable isotope tagging: renaissance of radioimmunoassay for multiplex and absolute quantification of biomolecules. Acc Chem Res 49:775–783. https://doi.org/10.1021/acs.accounts.5b00509

Łobiński R, Marczenko Z (1992) Recent advances in ultraviolet-visible spectrophotometry. Crit Rev Anal Chem 23:55–111. https://doi.org/10.1080/10408349208051647

Loo JA, Yan W, Ramachandran P, Wong DT (2010) Comparative human salivary and plasma proteomes. J Dent Res 89(10):1016–1023. https://doi.org/10.1177/0022034510380414

Looi ML, Zakaria H, Osman J, Jamal R (2012) Quantity and quality assessment of DNA extracted from saliva and blood. Clin Lab 58(3/4):307–312

Lousada-Fernandez F, Rapado-Gonzalez O, Lopez-Cedrun J-L, Lopez-Lopez R, Muinelo-Romay L, Mercedes Suarez-Cunqueiro M (2018) Liquid biopsy in oral cancer. Int J Mol Sci 19. https://doi.org/10.3390/ijms19061704

Luconi M, Silva MF, Olsina RA, Fernandez L (2001) Flow injection spectrophotometric analysis of lead in human saliva for monitoring environmental pollution. Talanta 54:45–52. https://doi.org/10.1016/S0039-9140(00)00624-X

Maclntyre I (1961) Flame photometry. Adv Clin Chem 4:1–28. https://doi.org/10.1017/CBO9781107415324.004

Mahato K, Kumar S, Srivastava A, Maurya PK, Singh R, Chandra P (2018) Electrochemical Immunosensors: fundamentals and applications in clinical diagnostics. In: Ju H, Lai G, Yan

F (Eds) Handbook of immunoassay technologies. Elsevier Inc., pp 359–414. https://doi.org/10.1016/B978-0-12-811762-0.00014-1

Martin J, Kagerbauer SM, Gempt J, Podtschaske A, Hapfelmeier A, Schneider G (2018) Oxytocin levels in saliva correlate better than plasma levels with concentrations in the cerebrospinal fluid of patients in neurocritical care. J Neuroendocrinol 30:e12596. https://doi.org/10.1111/jne.12596

Meghann P, Heather D, Scott T, Julie D, Jeff B, Arti S, Mahesh V, Vincent J (2014) Expression Profiling Smackdown: Human Transcriptome Array HTA 2.0 vs. RNA-Seq. J Biomol Tech

Messana I, Cabras T, Iavarone F, Vincenzoni F, Urbani A, Castagnola M (2013) Unraveling the different proteomic platforms. J Sep Sci 36(1):128–139. https://doi.org/10.1002/jssc.201200830

Metgud R, Patel S (2014) Serum and salivary levels of albumin as diagnostic tools for oral premalignancy and oral malignancy. Biotech Histochem 89:8–13. https://doi.org/10.3109/10520295.2013.793394

Mikkonen JJW, Singh SP, Herrala M, Lappalainen R, Myllymaa S, Kullaa AM (2016) Salivary metabolomics in the diagnosis of oral cancer and periodontal diseases. J Periodontal Res 51(4):431–437. https://doi.org/10.1111/jre.12327

Miocevic O, Cole CR, Laughlin MJ, Buck RL, Slowey PD, Shirtcliff EA (2017) Quantitative lateral flow assays for salivary biomarker assessment: a review. Front Public Heal 5:133. https://doi.org/10.3389/fpubh.2017.00133

Mischak H, Critselis E, Hanash S, Gallagher WM, Vlahou A, Ioannidis JPA (2015) Epidemiologic design and analysis for proteomic studies: a primer on -omic technologies. Am J Epidemiol 181:635–647. https://doi.org/10.1093/aje/kwu462

Mitsui S, Yamamoto M, Nagasawa M, Mogi K, Kikusui T, Ohtani N, Ohta M (2011) Urinary oxytocin as a noninvasive biomarker of positive emotion in dogs. Horm Behav 60:239–243. https://doi.org/10.1016/j.yhbeh.2011.05.012

Mohamed R, Campbell J-L, Cooper-White J, Dimeski G, Punyadeera C (2012) The impact of saliva collection and processing methods on CRP, IgE, and myoglobin immunoassays. Clin Transl Med 1:19. https://doi.org/10.1186/2001-1326-1-19

Monticelli S, Ansel KM, Xiao C, Socci ND, Krichevsky AM, Thai TH, Rajewsky N, Marks DS, Sander C, Rajewsky K, Rao A, Kosik KS (2005) MicroRNA profiling of the murine hematopoietic system. Genome Biol 6(8):1–15

Munoz-Prieto A, Martinez-Subiela S, Ceron JJ, Tvarijonaviciute A (2019) A new highly sensitive immunoassay for the detection of adiponectin in serum and saliva of dogs and its application in obesity and canine leishmaniosis. Res Vet Sci 125:374–381. https://doi.org/10.1016/j.rvsc.2019.07.019

Muñoz-Prieto A, Tvarijonaviciute A, Escribano D, Martínez-Subiela S, Cerón JJ, Tvarijonaviciute A (2017) Use of heterologous immunoassays for quantification of serum proteins: the case of canine C-reactive protein. PLoS One 12:e0172188. https://doi.org/10.1371/journal.pone.0172188

Musso D, Roche C, Nhan TX, Robin E, Teissier A, Cao-Lormeau VM (2015) Detection of Zika virus in saliva. J Clin Virol 68:53–55. https://doi.org/10.1016/j.jcv.2015.04.021

Nasidze I, Li J, Quinque D, Tang K, Stoneking M (2009) Global diversity in the human salivary microbiome. Genome Res 19(4):636. https://doi.org/10.1101/gr.084616.108

Nguyen TT, Ngo LQ, Promsudthi A, Surarit R (2017) Salivary oxidative stress biomarkers in chronic periodontitis and acute coronary syndrome. Clin Oral Investig 21:2345–2353. https://doi.org/10.1007/s00784-016-2029-3

Ohshima M, Sugahara K, Kasahara K, Katakura A (2017) Metabolomic analysis of the saliva of Japanese patients with oral squamous cell carcinoma. Oncol Rep 37(5):2727–2734. https://doi.org/10.3892/or.2017.5561

Ohtomo T, Igarashi S, Takagai Y (2013) Flow injection spectrophotometric analysis of human salivary α-Amylase activity using an enzyme degradation of starch-iodine complexes in flow channel and its application to human stress testing. Biol Pharm Bull 36:1857–1861. https://doi.org/10.1248/bpb.b13-00214

Olmedo P, Pla A, Hernández AF, López-Guarnido O, Rodrigo L, Gil F (2010) Validation of a method to quantify chromium, cadmium, manganese, nickel and lead in human whole blood, urine, saliva and hair samples by electrothermal atomic absorption spectrometry. Anal Chim Acta 659:60–67. https://doi.org/10.1016/j.aca.2009.11.056

Özmeriç N, Elgün S, Uraz A (2000) Salivary arginase in patients with adult periodontitis. Clin Oral Investig 4:21–24

Pan D, Khan MS, Misra SK, Wang Z, Daza E, Schwartz-Duval AS, Kus JM, Pan D (2017) Paper-based analytical biosensor Chip designed from graphene-Nanoplatelet-Amphiphilic-diblock-co-polymer composite for cortisol detection in human saliva. Anal Chem 89:2107–2115. https://doi.org/10.1021/acs.analchem.6b04769

Pappa E, Vastardis H, Mermelekas G, Gerasimidi-Vazeou A, Zoidakis J, Vougas K (2018) Saliva proteomics analysis offers insights on type 1 diabetes pathology in a pediatric population. Front Physiol. https://doi.org/10.3389/fphys.2018.00444

Park NJ, Li Y, Yu T, Brinkman BMN, Wong DT (2006) Characterization of RNA in saliva. Clin Chem 52(6):988–994. https://doi.org/10.1373/clinchem.2005.063206

Parra M, Tecles F, Martinez-Subiela S, Ceron J (2005) C-reactive protein measurement in canine saliva. J Vet Diag Investig 17:139–144

Parra MD, Martínez-Subiela S, Cerón JJ (2004) Fluorometría En Tiempo Retardado : Conceptos Generales y Aplicaciones en Medicina Veterinaria. An Vet Murcia 20:35–47

Pasquini C (2003) Near infrared spectroscopy: fundamentals, practical aspects and analytical applications. J Braz Chem Soc 14:198–219. https://doi.org/10.1590/S0103-50532003000200006

Patidar KA, Parwani RN, Wanjari SP (2011) Correlation of salivary and serum IgG, IgA levels with total protein in oral submucous fibrosis. J Oral Sci 53:97–102. https://doi.org/10.2334/josnusd.53.97

Pedersen AM, Bardow A, Jensen SB, Nauntofte B (2002) Saliva and gastrointestinal functions of taste, mastication, swallowing and digestion. Oral Dis 8:117–129. https://doi.org/10.1034/j.1601-0825.2002.02851.x

Perkampus H-H (1992) UV-VIS spectroscopy and its applications. Br J Psychiatry. https://doi.org/10.1192/bjp.112.483.211-a

PerkinElmer Inc. (2013) From small molecules to live cells – the assay that measures it all

Pulli T, Höyhtyä M, Söderlund H, Takkinen K (2005) One-step homogeneous immunoassay for small analytes. Anal Chem 77:2637–2642. https://doi.org/10.1021/ac0483791

Puurunen J, Tiira K, Lehtonen M, Hanhineva K, Lohi H (2016) Non-targeted metabolite profiling reveals changes in oxidative stress, tryptophan and lipid metabolisms in fearful dogs. Behav Brain Funct 12:1–12. https://doi.org/10.1186/s12993-016-0091-2

Qiao Y, Wu M, Feng Y, Zhou Z, Chen L, Chen F (2018) Alterations of oral microbiota distinguish children with autism spectrum disorders from healthy controls. Sci Rep 8(1):1–12. https://doi.org/10.1038/s41598-018-19982-y

Rai B, Kaur J, Jacobs R, Anand SC (2011) Adenosine deaminase in saliva as a diagnostic marker of squamous cell carcinoma of tongue. Clin Oral Investig 15:347–349. https://doi.org/10.1007/s00784-010-0404-z

Ramaswamy S, Perou CM (2003) DNA microarrays in breast cancer: the promise of personalised medicine. Lancet 361:1576–1577. https://doi.org/10.1016/S0140-6736(03)13322-3

Ransohoff DF (2005) Bias as a threat to the validity of cancer molecular-marker research. Nat Rev Cancer 5(2):142–149. https://doi.org/10.1038/nrc1550

Rao PV, Reddy AP, Lu X, Dasari S, Krishnaprasad A, Biggs E, Roberts CT, Nagalla SR (2009) Proteomic identification of salivary biomarkers of Type-2 diabetes. J Proteome Res 8:239–245. https://doi.org/10.1021/pr8003776

Rifai N, Gillette MA, Carr SA (2006) Protein biomarker discovery and validation: the long and uncertain path to clinical utility. Nat Biotechnol 24:971–983. https://doi.org/10.1038/nbt1235

Rosenfeld CS (2015) Microbiome disturbances and autism Spectrum disorders. Drug Metab Dispos 43(10):1557–1571. https://doi.org/10.1124/dmd.115.063826

Rylander-Rudqvist T, Håkansson N, Tybring G, Wolk A (2006) Quality and quantity of saliva DNA obtained from the self-administrated oragene method – a pilot study on the cohort of Swedish men. Cancer Epidemiol Biomark Prev 15(9):1742–1745. https://doi.org/10.1158/1055-9965.EPI-05-0706

Saiyudthong S, Suwannarat P, Trongwongsa T, Srisurapanon S (2010) Comparison between ECL and ELISA for the detection of salivary cortisol and determination of the relationship between cortisol in saliva and serum measured by ECL. ScienceAsia 36:169–171. https://doi.org/10.2306/scienceasia1513-1874.2010.36.169

Schmidt BL, Kuczynski J, Bhattacharya A, Huey B, Corby PM, Queiroz ELS, Nightingale K, Kerr AR, DeLacure MD, Veeramachaneni R, Olshen AB, Albertson DG (2014) Changes in abundance of oral microbiota associated with oral cancer. PLoS One 9(6):e98741. https://doi.org/10.1371/journal.pone.0098741

Schultz CP, Ahmed MK, Dawes C, Mantsch HH (1996) Thiocyanate levels in human saliva: quantitation by Fourier transform infrared spectroscopy. Anal Biochem 240:7–12. https://doi.org/10.1006/ABIO.1996.0323

Schulz BL, Cooper-White J, Punyadeera CK (2013) Saliva proteome research: current status and future outlook. Crit Rev Biotechnol 33:246–259. https://doi.org/10.3109/07388551.2012.687361

Shah FD, Begum R, Vajaria BN, Patel KR, Patel JB, Shukla SN, Patel PS (2011) A review on salivary genomics and proteomics biomarkers in oral cancer. Indian J Clin Biochem 26(4):326–334. https://doi.org/10.1007/s12291-011-0149-8

Shaw RA, Mantsch HH (2008) Infrared spectroscopy of biological fluids in clinical and diagnostic analysis. Encycl Anal Chem. https://doi.org/10.1002/9780470027318.a0106.pub2

Shirtcliff EA, Granger DA, Schwartz EB, Curran MJ, Booth A, Overman WH (2000) Assessing estradiol in biobehavioral studies using saliva and blood spots: simple radioimmunoassay protocols, reliability, and comparative validity. Horm Behav 38:137–147. https://doi.org/10.1006/hbeh.2000.1614

Shpitzer T, Bahar G, Feinmesser R, Nagler RM (2007) A comprehensive salivary analysis for oral cancer diagnosis. J Cancer Res Clin Oncol 133:613–617. https://doi.org/10.1007/s00432-007-0207-z

Siawaya JFD, Roberts T, Babb C, Black G, Golakai HJ, Stanley K, Bapela NB, Hoal E, Parida S, van Helden P, Walzl G (2008) An evaluation of commercial fluorescent bead-based Luminex cytokine assays. PLoS One 3(7):e2535. https://doi.org/10.1371/journal.pone.0002535

Spielmann N, Ilsley D, Gu J, Lea K, Brockman J, Heater S, Setterquist R, Wong DTW (2012) The human salivary RNA transcriptome revealed by massively parallel sequencing. Clin Chem 1(1):25–28. https://doi.org/10.1373/clinchem.2011.176941

Stoeppler M (1992) Analytical methods and instrumentation — a summarizing overview. Tech Instrum Anal Chem 12:97–132. https://doi.org/10.1016/S0167-9244(08)70105-7

Sugimoto M, Wong DT, Hirayama A, Soga T, Tomita M (2010) Capillary electrophoresis mass spectrometry-based saliva metabolomics identified oral, breast and pancreatic cancer-specific profiles. Metabolomics 6:78–95. https://doi.org/10.1007/s11306-009-0178-y

Swinehart DF (1962) The Beer-Lambert law. J Chem Educ 39:333. https://doi.org/10.1021/ed039p333

Tiwari G, Tiwari R (2010) Bioanalytical method validation: an updated review. Pharm Methods 1(1):25–38

Tlili C, Myung NV, Shetty V, Mulchandani A (2011) Label-free, chemiresistor immunosensor for stress biomarker cortisol in saliva. Biosens Bioelectrinics 26:4382–4386. https://doi.org/10.1016/j.bios.2011.04.045

Totan A, Miricescu D, Parlatescu I, Mohora M, Greabu M (2015) Possible salivary and serum biomarkers for oral lichen planus. Biotech Histochem 90:552–558. https://doi.org/10.3109/10520295.2015.1016115

Tvarijonaviciute A, Aznar-Cayuela C, Rubio CP, Ceron JJ, López-Jornet P (2017) Evaluation of salivary oxidate stress biomarkers, nitric oxide and C-reactive protein in patients with oral lichen planus and burning mouth syndrome. J Oral Pathol Med 46:387–392. https://doi.org/10.1111/jop.12522

Tvarijonaviciute A, Aznar-Cayuela C, Rubio CP, Ceron JJ, López-Jornet P (2016) Evaluation of salivary oxidate stress biomarkers, nitric oxide and C-reactive protein in patients with oral lichen planus and burning mouth syndrome. J Oral Pathol Med 46(5):387–392. https://doi.org/10.1111/jop.12522

Tvarijonaviciute A, Carrillo-Sanchez JD, García-Martinez JD, Tecles F, Martinez-Subiela S, German AJ, Ceron JJ (2014) Measurement of salivary adiponectin concentrations in dogs. Vet Clin Pathol 43:416–421. https://doi.org/10.1111/vcp.12169

van Amerongen A, Veen J, Arends HA, Koets M (2018) Lateral flow immunoassays. In: Vashist SK, Luong JHT (eds) Handbook of immunoassay technologies. Elsevier Inc., pp 157–182. https://doi.org/10.1016/B978-0-12-811762-0.00007-4

Vashist SK, Luong JHT (2018a) Immunoassays: an Overwiew. In: Vashist SK, Luong JHT (eds) Handbook of immunoassay technologies. Elsevier Inc., pp 1–18. https://doi.org/10.1016/B978-0-12-811762-0.00001-3

Vashist SK, Luong JHT (2018b) Immunoassays: future prospects and possibilities. In: Vashist SK, Luong JHT (eds) Handbook of immunoassay technologies. Elsevier Inc., pp 455–466. https://doi.org/10.1016/B978-0-12-811762-0.00017-7

Vashist SK, Luong JHT (2018c) Enzyme-linked immunoassays. In: Vashist SK, Luong JHT (eds) Handbook of immunoassay technologies. Elsevier Inc., pp 97–127. https://doi.org/10.1016/B978-0-12-811762-0.00005-0

Vashist SK, Luppa PB, Yeo LY, Ozcan A, Luong JHT (2015) Emerging technologies for next-generation point-of-care testing. Trends Biotechnol 33:692–705. https://doi.org/10.1016/j.tibtech.2015.09.001

Viet CT, Schmidt BL (2008) Methylation array analysis of preoperative and postoperative saliva DNA in oral cancer patients. Cancer Epidemiol Biomark Prev 17(12):3603–3611. https://doi.org/10.1158/1055-9965.EPI-08-0507

Wagner PD, Verma M, Srivastava S (2004) Challenges for biomarkers in cancer detection. Ann N Y Acad Sci 1022:9–16. https://doi.org/10.1196/annals.1318.003

Wang Q, Gao P, Wang X, Duan Y (2014) The early diagnosis and monitoring of squamous cell carcinoma via saliva metabolomics. Sci Rep 4(1):6802. https://doi.org/10.1038/srep06802

Wang Z, Gerstein M, Snyder M (2009) RNA-Seq: a revolutionary tool for transcriptomics. Nat Rev Genet 10(1):57–63. https://doi.org/10.1038/nrg2484

Wei D, Zhang X-L, Wang Y-Z, Yang C-X, Chen G (2010) Lipid peroxidation levels, total oxidant status and superoxide dismutase in serum, saliva and gingival crevicular fluid in chronic periodontitis patients before and after periodontal therapy. Aust Dent J 55:70–78. https://doi.org/10.1111/j.1834-7819.2009.01123.x

Wei J, Xie G, Zhou Z, Shi P, Qiu Y, Zheng X, Chen T, Su M, Zhao A, Jia W (2011) Salivary metabolite signatures of oral cancer and leukoplakia. Int J Cancer 129(9):2207–2217. https://doi.org/10.1002/ijc.25881

Whetsel KB (1968) Near-infrared spectrophotometry. Appl Spectrosc Rev 2:1–67. https://doi.org/10.1080/05704926808050886

Wong DT (2006a) Salivary diagnostics powered by nanotechnologies, proteomics and genomics. J Am Dent Assoc 137(3):313–321. https://doi.org/10.14219/jada.archive.2006.0180

Wong DT (2006b) Towards a simple, saliva-based test for the detection of oral cancer. "Oral fluid (saliva), which is the mirror of the body, is a perfect medium to be explored for health and disease surveillance". Expert Rev Mol Diagn 6(3):267–272. https://doi.org/10.1586/14737159.6.3.267

Wong DTW (2007) Salivaomics. J Am Dent Assoc 433. https://doi.org/10.1126/science.1137711

Wu Y, Liu S (2009) Immunosensing system for alpha-fetoprotein through boronate immunoaffinity column in combination with flow injection chemiluminescence. Analyst 134:230–235. https://doi.org/10.1039/b812642a

Xiao H, Zhang Y, Kim Y, Kim S, Kim JJ, Kim KM, Yoshizawa J, Fan L-Y, Cao C-X, Wong DTW (2016) Differential proteomic analysis of human saliva using tandem mass tags quantification for gastric Cancer detection. Sci Rep 6:22165. https://doi.org/10.1038/srep22165

Yalow RS (1980) Radioimmunoassay. Annu Rev Biophys Bioeng 9:327–345. https://doi.org/10.1146/annurev.bb.09.060180.001551

Yan W, Apweiler R, Balgley BM, Boontheung P, Bundy JL, Cargile BJ, Cole S, Fang X, Gonzalez-Begne M, Griffin TJ, Hagen F, Hu S, Wolinsky LE, Lee CS, Malamud D, Melvin JE, Menon R, Mueller M, Oiao R, Rhodus NL, Sevinsky JR, States D, Stephenson JL Jr, Than S, Yates JR III, Yu W, Xie H, Xie Y, Omenn GS, Loo JA, Wong DT (2009) Systematic comparison of the human saliva and plasma proteomes. Proteomics Clin Appl 3:116–134. https://doi.org/10.1002/prca.200800140

Yan W, Yu W, Than S, Hu Z, Zhou H, Wong D (2008) Salivaomics knowledge base (SKB). In: 37th Annual Meeting and Exhibition of the American Association for Dental Research. p. abstract 1179

Yang CY, Brooks E, Li Y, Denny P, Ho CM, Qi FX, Shi WY, Wolinsky L, Wu B, Wong DTW, Montemagno CD (2005) Detection of picomolar levels of interleukin-8 in human saliva by SPR. Lab Chip 5:1017–1023. https://doi.org/10.1039/b504737d

Yang K, Peretz-Soroka H, Liu Y, Lin F (2016) Novel developments in mobile sensing based on the integration of microfluidic devices and smartphones. Lab Chip 16(6):943–958. https://doi.org/10.1039/c5lc01524c

Yoshizawa JM, Schafer CA, Schafer JJ, Farrell JJ, Paster BJ, Wong DTW (2013) Salivary biomarkers: toward future clinical and diagnostic utilities. Clin Microbiol Rev 26(4):781–791. https://doi.org/10.1128/CMR.00021-13

Young SH, Luo JC (2016) Will saliva test be a good method to detect Helicobacter pylori in H. pylori-infected patients? J Chinese Med Assoc 79(7):351–352. https://doi.org/10.1016/j.jcma.2016.03.005

Zappacosta B, Manni A, Persichilli S, Boari A, Scribano D, Minucci A, Raffaelli L, Giardina B, De Sole P (2007) Salivary thiols and enzyme markers of cell damage in periodontal disease. Clin Biochem 40:661–665. https://doi.org/10.1016/j.clinbiochem.2007.01.011

Zeidán-Chuliá F, Rybarczyk-Filho JL, Gursoy M, Könönen E, Uitto V-J, Gursoy OV, Cakmakci L, Moreira JCF, Gursoy UK (2012) Bioinformatical and *in vitro* approaches to essential oil-induced matrix metalloproteinase inhibition. Pharm Biol 50:675–686. https://doi.org/10.3109/13880209.2012.677847

Zeidán-Chuliá F, Rybarczyk-Filho JL, Salmina AB, de Oliveira B-HN, Noda M, Moreira JCF (2013) Exploring the multifactorial nature of autism through computational systems biology: calcium and the Rho GTPase RAC1 under the spotlight. NeuroMolecular Med 15:364–383. https://doi.org/10.1007/s12017-013-8224-3

Zhang L, Farrell JJ, Zhou H, Elashoff D, Akin D, Park NH, Chia D, Wong DT (2010a) Salivary Transcriptomic biomarkers for detection of Resectable pancreatic Cancer. Gastroenterology 138(3):949–957. https://doi.org/10.1053/j.gastro.2009.11.010

Zhang L, Xiao H, Karlan S, Zhou H, Gross J, Elashoff D, Akin D, Yan X, Chia D, Karlan B, Wong DT (2010b) Discovery and preclinical validation of salivary transcriptomic and proteomic biomarkers for the non- invasive detection of breast cancer. PLoS One 5(12):e15573. https://doi.org/10.1371/journal.pone.0015573

Zhang L, Xiao H, Zhou H, Santiago S, Lee JM, Garon EB, Yang J, Brinkmann O, Yan X, Akin D, Chia D, Elashoff D, Park NH, Wong DTW (2012) Development of transcriptomic biomarker signature in human saliva to detect lung cancer. Cell Mol Life Sci 69(19):3341–3350. https://doi.org/10.1007/s00018-012-1027-0

Zhang X, Wan Y, Cooper-White J, Dimeski G, Atherton J, Punyadeera C (2013) Quantification of D-dimer levels in human saliva. Bioanalysis 5:2249–2256. https://doi.org/10.4155/bio.13.190

Part II
Salivary Biomarkers in Oral and Systemic Pathologies

Chapter 5
The Role of Saliva in Dental Practice

Eduardo Pons-Fuster and Pia López-Jornet

Objectives
Saliva is a valuable source of clinical information, since it contains biomarkers that may be useful for identifying patients at an increased risk of suffering oral disease. Current technology offers a valuable opportunity for developing such biological molecules as biomarkers (Giannobile et al. 2011; Schafer et al. 2014; Malamud 2011; Jiménez 2010; Chiappin et al. 2007; Baum et al. 2011; Kaufman and Lamster 2002; Llena-Puy 2006; Yoshizawa et al. 2013; Lorenzo-Pouso et al. 2018). The present chapter offers a review of saliva as a potential and effective tool for the diagnosis and monitoring of diseases of the oral cavity.

5.1 Dental Caries

Dental caries is a very common transmissible, chronic infectious disease characterized by local destruction of the hard-dental tissues as a result of the action of acids produced by the microorganisms that conform the biofilm adhered to the teeth.

From the etiological perspective there are a number of intervening factors, such as saliva, the characteristics of teeth, and a diet containing food that can be fermented by the oral microorganisms, thereby giving rise to caries. The protective function of saliva can be attributed to four effects:

E. Pons-Fuster
Biomedical Research Institute (IMIB-Arrixaca), Faculty of Medicine, Regional Campus of International Excellence Campus Mare Nostrum, University of Murcia, Murcia, Spain

P. López-Jornet (✉)
Department of Oral Medicine, Faculty of Medicine and Dentistry, Regional Campus of International Excellence Campus Mare Nostrum, University of Murcia, Murcia, Spain
e-mail: majornet@um.es

A. Tvarijonaviciute et al. (eds.), *Saliva in Health and Disease*,
https://doi.org/10.1007/978-3-030-37681-9_5

1. the dilution and elimination of sugars and other components
2. buffering capacity
3. demineralization/remineralization effects
4. antimicrobial action.

One of the most important functions of saliva is the clearing of microorganisms and food components from the oral cavity. A large salivary volume accelerates the elimination of sugars – a fact that explains the increased caries risk seen in patients with a low salivary flow (Jiménez 2010; Lorenzo-Pouso et al. 2018; Nordlund et al. 2009; Vitorino et al. 2006) (Fig. 5.1).

Caries is diagnosed from the clinical findings, since to date no saliva test has been found to be adequate for diagnosing the disease. Although combinations of factors have been suggested to predict the risk of caries, the fact is that none have demonstrated sufficient validity – probably because of the intervention of many risk factors at different levels in the development of the disease.

The caries susceptibility tests that can be made in a microbiology laboratory include:

1. Determination of the presence of caryogenic flora. It is known that bacterial growth is favored if incubation takes place under conditions of anaerobiosis, which plays a relevant role in the development of caries.
2. Isolation of acid-producing microorganisms in saliva, such as *Lactobacillus spp.*, which contribute to lower the oral pH value and thus favor tooth demineralization and increase dental susceptibility to the action of caryogenic bacteria. The Zinder test or Alban test can be used to quantitatively assess the presence of such acid-producing microorganisms in saliva. Depending on the intensity of the color shift generated in these tests, we can quantify the *Lactobacillus spp.* colonies.
3. Determination of the buffering capacity of saliva and of the salivary secretion rate. The buffering capacity of saliva is explained by the presence of different buffering systems in this body fluid, such as the phosphate system and the carbonic acid / carbonate system.
4. Quantification of fungal colonies. A number of studies have examined the interactions between the bacterial flora and *Candida albicans* in the formation of dental plaque, and it has been postulated that the concentration of this fungal species might act as an indicator of caries risk.

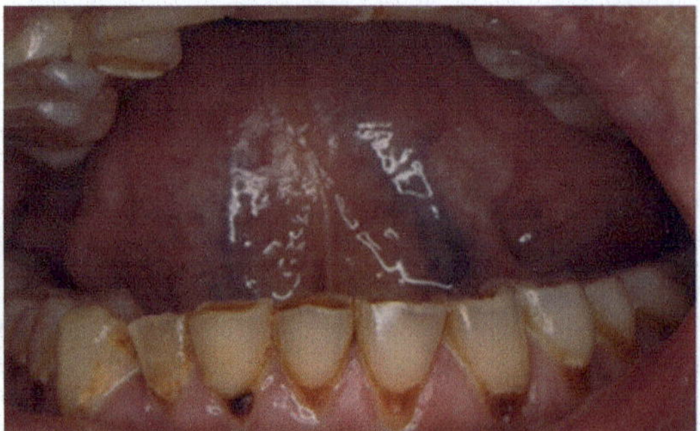

Fig. 5.1 Dental caries as a result of hyposalivation

Many studies have attempted to relate the prevalence of caries to the salivary phenotype, but the results have been contradictory. A recent field of interest in the investigation of caries risk has been the measurement of oligosaccharides, the concentrations of which have been found to be correlated to caries in young adults. In turn, significant associations have been observed between patient age and the salivary (submandibular/sublingual glands) concentrations of lactoferrin, albumin, lysozyme, mucin, cystatin, potassium, calcium, sodium and chloride. Some studies have adopted a proteomic approach to examine whether salivary proteins can act as biomarkers in the evaluation of caries risk. Their data suggest that statherin and cystatin are the best predictors of occlusal caries in saliva, though supragingival plaque and the total oral bacterial count must also be taken into account (Vitorino et al. 2006; Wang et al. 2018; Bhalla et al. 2010).

5.2 Periodontal Diseases

Periodontal disease is a chronic inflammatory disorder of bacterial origin that affects the hard and soft tissues supporting the teeth. Over 300 pathogens possibly implicated in periodontal destruction have been described in the literature. The term "periodontal disease" includes plaque-related gingivitis and periodontitis. Gingivitis is characterized by inflammation of the gums only, while in the case of periodontitis the inflammatory process extends to the alveolar bone and periodontal ligament – with progressive destruction of these elements (Slots and Slots 2011; Prakasam and Srinivasan 2014; de Lima et al. 2016; Ebersole et al. 2013; Kaufman and Lamster 2000; Miller et al. 2006; Grigoriadou et al. 2010; Canakci et al. 2009; AL-abbagh et al. 2012; Schenck et al. 1993; Ulker et al. 2008; Olayanju et al. 2012; Gonçalves Lda et al. 2010; Lee et al. 2018; Wilczynska-Borawska et al. 2006; Kaushik et al. 2011; Meschiari et al. 2013).

Periodontal disease is highly prevalent 20–50% of global population and exhibits a worldwide distribution. In most cases, periodontitis is preceded by gingivitis, and its progression towards bone and attachment loss is modulated by microbiological and immunological factors. In fact, periodontitis is regarded as a chronic inflammatory disease caused by microorganisms that form part of the subgingival biofilm, and which require a susceptible host in order to initiate the chronic inflammatory reaction that gives rise to periodontal destruction (Ebersole et al. 2013; Kaufman and Lamster 2000; Miller et al. 2006; Grigoriadou et al. 2010; Canakci et al. 2009; AL-abbagh et al. 2012; Schenck et al. 1993; Ulker et al. 2008) (Figs. 5.2 and 5.3).

In the healthy individual, the oral flora is in ecological equilibrium with the host, and this allows periodontal health to be maintained. However, this stable relationship can become altered as a result of different factors such as antimicrobial therapy or changes in host susceptibility secondary to alterations in the defense mechanisms.

Saliva contains over 1×10^8 bacteria/ml and is the usual environment of microorganisms from the supra- and subgingival biofilm, the dorsal surface of the tongue, and other surfaces of the oral mucosa. In this regard, the tongue is one of the main sources of most of the salivary bacteria. Periodontal disease can be described as one

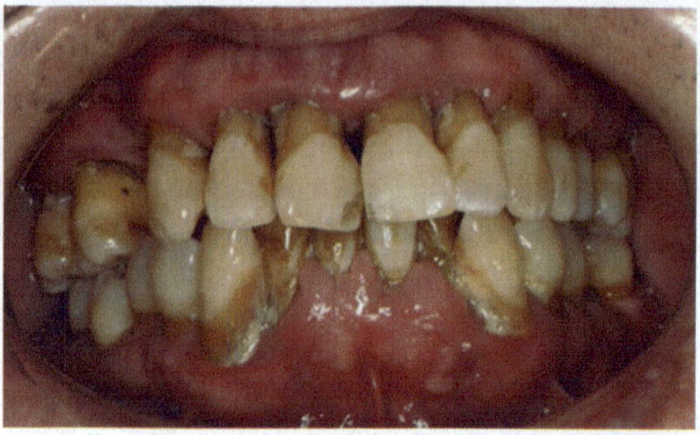

Fig. 5.2 Patient with generalized periodontitis

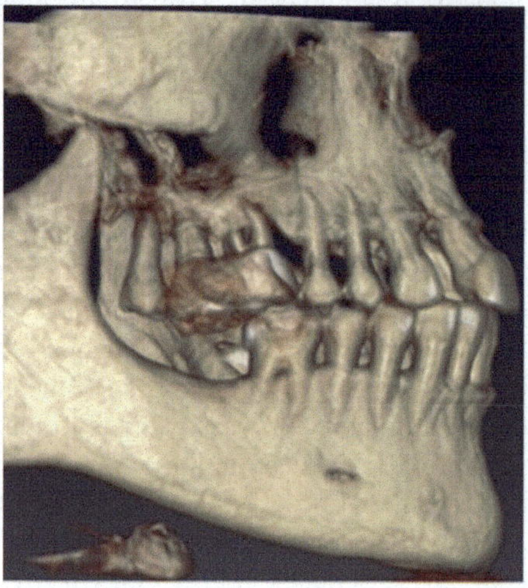

Fig. 5.3 Cone beam radiological image in a periodontal patient

of the predominant polymicrobial infections in humans. A range of organisms are implicated in the origin of periodontitis, such as *Aggregatibacter actinomycetemcomitans, Porphyromonas gingivalis, Fusobacterium nucleatum spp., Prevotella intermedia, Prevotella nigrescens, Tannerella forsythia, Treponema spp. Capnocytophaga spp., Campylobacter rectus, Eikenella corrodens* and many other bacteria (Slots and Slots 2011).

The mucins (MG1 and MG2) in turn exert cytoprotective, lubricating and protective actions. Mucin MG2 influences bacterial aggregation and adherence, and in this respect a decrease in salivary MG2 levels can increase colonization by *Aggregatibacter actinomycetemcomitans*.

Different studies have pointed to periodontal infection as a risk factor for certain systemic disorders such as cardiovascular diseases, certain respiratory conditions, obesity and metabolic syndrome. In addition, it has been postulated to alter adequate diabetes control and give rise to premature delivery and/or infants with low birth weight (de Lima et al. 2016; Ebersole et al. 2013; Kaufman and Lamster 2000; Miller et al. 2006).

Increased oxidative damage has been observed in the more advanced stages of periodontal disease, since reactive oxygen species (ROS) are found to be increased. Salivary markers of oxidative stress and antioxidant status constitute a promising tool for the investigation of oral diseases, in view of the importance of ROS in the pathogenesis of periodontal disease. An increase in ROS gives rise to periodontal tissue destruction, and this is one of the leading causes of periodontal disease. There is evidence of significant changes in oxidative stress, with increased levels of different oxidative stress markers (8-OHdG, malondialdehyde [MDA], glutathione peroxidase [GPx], superoxide dismutase [SOD] and total antioxidant capacity [TAOC]) in individuals with worsened periodontal health (Kaufman and Lamster 2000; AL-abbagh et al. 2012).

Oxidative stress increases with the number of different bacteria present in the environment. Reactive oxygen species are related to polymorphonuclear cell (PMN) destruction of periodontal pathogens, and in this regard the increase in ROS levels as a consequence of PMN action would lead to tissue degeneration and worsened periodontal disease.

In sum, the saliva of patients with periodontitis has been found to contain gingival tissue cells, crevicular fluid and a series of inflammatory mediators and molecules implicated in tissue destruction. Consequently, saliva can be expected to contain biomarkers specific of all three key aspects of periodontitis (inflammation, collagen degradation and bone turnover) and which are correlated to the clinical features of the disease. The application of salivary proteomic biomarkers to the diagnosis of periodontal disease is still in the experimental phase, and is based on the changes in profile of molecules implicated in inflammation, collagen degradation and bone loss. Some of these salivary biomarkers are enzymes such as alkaline phosphatase, esterase, metalloproteinases (MMP), β-glucuronidase, aminopeptidase, cystatin, prolactin and α-amylase, while others are immunoglobulins such as IgA and IgG, cytokines such as interleukin-1β (IL-1β), interleukin-8 (IL-8) and steroid hormones (Prakasam and Srinivasan 2014).

Recently, macrophage inflammatory protein-1α (MIP-1α) has been identified as the biomarker of greatest diagnostic capacity, followed by IL-1β and IL-6 (Grigoriadou et al. 2010; Canakci et al. 2009; AL-abbagh et al. 2012; Schenck et al. 1993; Ulker et al. 2008; Olayanju et al. 2012).

5.3 Peri-implantitis

Dental implants are currently one of the safest options for replacing lost teeth. Osseointegration is the direct structural and functional connection between the bone and the rough surface of the dental implant, needed to resist occlusal loading. The term peri-implant disease is currently a general reference to the inflammatory reactions occurring in tissues surrounding dental implants (Gomes et al. 2018; Liskmann et al. 2007) (Fig. 5.4).

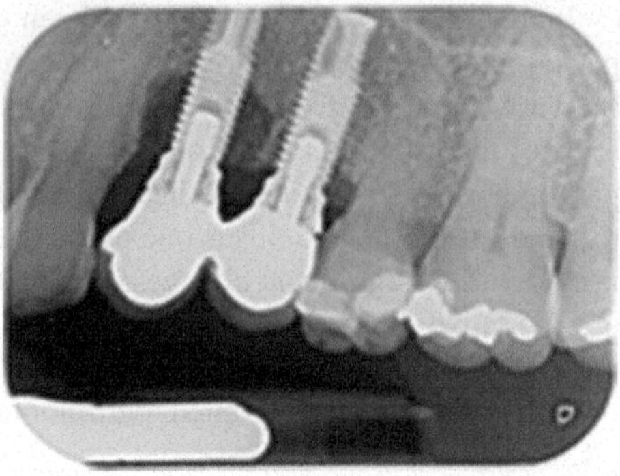

Fig. 5.4 Peri-implantitis

The causes of early implant failure include surgical trauma, bacterial contamination of the implant bed, and smoking. In turn, late implant failure can be caused by bacterial infections, while the contributing role of occlusal overload or biomechanical imbalances is more controversial. Mucositis is defined as the presence of inflammation of the peri-implant mucosa, without the loss of bone support, while peri-implantitis is characterized by inflammation with the loss of supporting bone.

The salivary interleukin-1β levels are seen to be lower in patients with healthy implants than in those with inflamed dental implants. In turn, a significant positive correlation has been observed between salivary IL-6 concentration and peri-implant inflammatory disorders.

Salivary total antioxidant capacity (TAOC) and the salivary levels of urate and ascorbate have been shown to be higher in individuals with healthy implants than in those with peri-implantitis.

5.4 Viral Diseases (Balamane et al. 2010)

The diagnosis of viral diseases of the oral cavity is based on the clinical findings and laboratory test results. In this respect, saliva can be used to detect and evaluate host immunization to measles, mumps and rubella, for example (see Chap. 12).

5.5 Fungal Diseases

Infections due to *Candida spp.* are among the most frequent mycoses. The great majority of Candida species (90–95%) isolated in patients with candidiasis correspond to *Candida albicans, Candida glabrata, Candida parapsilosis, Candida tropicalis* and *Candida krusei* (Salvatori et al. 2016; Thein et al. 2006; Canabarro et al. 2013).

The development of infection due to Candida depends on a number of factors:

(a) Damage to the skin or mucous membrane barriers.
(b) The dose or magnitude of the fungal inoculum.
(c) The defensive immune condition of the host.
(d) Fungal virulence (crucial role of dimorphism). Colonization, adhesion, invasion and damage are the successive stages involved in the pathogenesis of Candida infection.

There are a number of clinical presentations of oral candidiasis: pseudomembranous candidiasis, erythematous candidiasis, hyperplasic candidiasis, associated lesions (prosthetic stomatitis, angle cheilitis, rhomboid glossitis, exfoliativa cheilitis), and mucocutaneous candidiasis (chronic). When two or more of these clinical presentations coexist, we speak of multifocal oral candidiasis. The diagnosis of any of the forms of oral candidiasis is essentially based on the clinical findings. The observed clinical lesions in turn need to be confirmed by the microscopic identification of *Candida* in the oral samples and/or its isolation in culture. It should be noted that in the case of Candida, mere identification of the organism in the oral cavity is not indicative of infection, since it is a common component of the normal oral microflora. The definitive diagnosis of candidiasis requires the confirmation of tissue invasion by *Candida* (Fig. 5.5).

Saliva can also be used for the detection of fungal infections. Specifically, the salivary fungal count can reflect colonization of the oral mucosa.

A decrease in salivary secretion can cause dysbiosis, favoring the excessive growth of *C. albicans*. In this regard, there is an inverse correlation between the salivary flow rate and the oral Candida burden. Furthermore, it has been shown that patients with Sjögren's syndrome and oral candidiasis have high salivary levels of calprotectin – an antimicrobial peptide – possibly due to transudation of the inflamed mucosa. Therefore, salivary dysfunction may cancel the host defenses in the oral cavity secondary to changes in the levels of salivary proteins or due to loss of the protective salivary layer on the mucosal surfaces – thereby reducing epithelial barrier function (Salvatori et al. 2016; Thein et al. 2006).

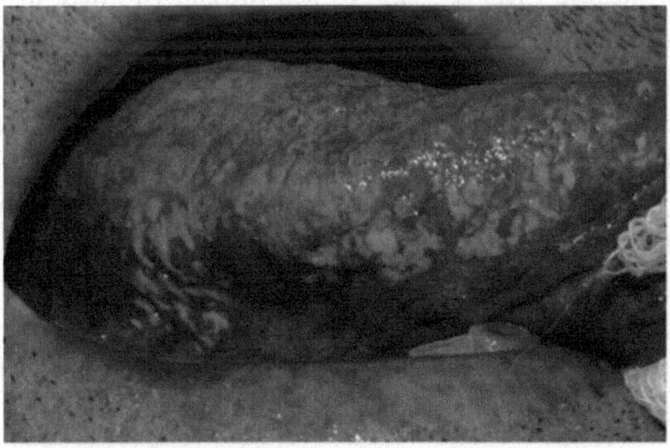

Fig. 5.5 Oral candidiasis

The development of diagnostic techniques independent of culture procedures appears to be a promising option for the early diagnosis of candidiasis. There are many limitations in this context, however, related mainly to the low yield of such techniques when used on an individual basis and to the high incidence of false-positive results, which in turn may contribute to an excessive use of antifungal agents.

Alterations of the salivary proteome related to proteins with antifungal activity, such as immunoglobulins, histatin, mucins, peroxidases and proline-rich proteins, may be of diagnostic utility, especially in recurrent cases.

Although largely regarded as a disorder of immunocompromised individuals, oral *C. albicans* infections are common among people wearing dentures and in cases of hyposalivation. Consequently, many immunocompetent individuals are susceptible to oral fungal infections. In this regard, it would be particularly interesting to know how the oral microbiome is altered in immunocompetent subjects with salivary dysfunction.

5.6 Orthodontics

Orthodontics has become one of the most popular dental treatments, though the application of appliances such as brackets can cause changes in saliva. The integrity and balance of the oral mucosa depend on the quality of saliva, the pH value and the concentration of proteins, which are factors that allow saliva to protect the hard and soft tissues of the oral cavity. The pH value and concentration of proteins may become altered in orthodontic patients, since the appliances used modify the oral environment, making saliva more acidic – though the total protein concentration is not affected (Koch et al. 2010; Eliades et al. 2003).

One of the main concerns among investigators in the last decade has been the performance of alloys in the environments in which they are intended to operate. The corrosion of orthodontic appliances raises two main concerns: (a) whether corrosion products are really produced and are absorbed in the body, causing local or systemic effects; and (b) the impact of corrosion upon the physical properties and clinical performance of the orthodontic appliances. Corrosion is characterized by deterioration of the surface of metals or alloys. In the concrete case of the metal appliances used in orthodontics, corrosion is a consequence of oxidation – reduction mechanisms.

A number of characteristics of the oral medium influence metal corrosion, such as temperature, humidity, pH changes, food, etc. Electrolytic corrosion is explained by the difference in potential between different metals, resulting in genuine degradation of the metal structure. Most studies on the release of metals from orthodontic appliances have focused on the biocompatibility of these materials, with assessment of the concentrations of Co, Cr, Ni and Fe in saliva.

Eliades et al. (Eliades et al. 2003) used inductively coupled plasma atomic emission spectroscopy (ICP-AES) to analyze the saliva of patients wearing orthodontic appliances with the aim of assessing the presence of metal traces (Ni, Cr, Fe). The authors concluded that there are no significant differences in the salivary levels of these metals between the patients and controls. In some cases the salivary metal concentrations were below the limit of detection of the measurement system used.

5.7 Conclusions

Diagnosis based on saliva can be greatly facilitated by the new technologies – becoming a powerful future tool for the diagnosis of oral and systemic diseases.

References

AL-abbagh M, Alladah A, Lin Y, Kryscio RJ, Thomas MV, Ebersole J et al (2012) Bone remodeling-associated salivary biomarker MIP-1a distinguished periodontal disease from health. J Periodontal Res 47:389–395

Balamane M, Winters MA, Dalai SC et al (2010) Detection of HIV-1 in saliva: implications for case-identification, clinical monitoring and surveillance for drug resistance. Open Virol J 4:88–93

Baum BJ, Yates JR, Srivastava S, Wong DT, Melvin JE (2011) Scientific frontiers : emerging technologies for salivary diagnostics. Adv Dent Res 23:360–368

Bhalla S, Tandon S, Satyamoorthy K (2010) Salivary proteins and early childhood caries A gel electrophoretic analysis. Contemp Clin Dent 1:17–22

Canabarro A, Valle C, Farias MR, Santos FB, Lazera M, Wanke B (2013) Association of subgingival colonization of Candida albicans and other yeasts with severity of chronic periodontitis. J Periodontal Res 48:428–432

Canakci CF, Cicek Y, Yildirim A, Sezer U, Canakci V (2009) Increased levels of 8-hydroxydeoxyguanosine and malondialdehyde and its relationship with antioxidant enzymes in saliva of periodontitis patients. Eur J Dent 3:100–106

Chiappin S, Antonelli G, Gatti R, De Palo EF (2007) Saliva specimen: a new laboratory tool for diagnostic and basic investigation. Clin Chim Acta 383:30–40

de Lima CL, Acevedo AC, Grisi DC, Taba M Jr, Guerra E, Canto GL (2016) Host-derived salivary biomarkers in diagnosing periodontal disease: systematic review and meta-analysis. J Clin Periodontol 43:492–502

Ebersole JL, Schuster JL, Stevens J, Dawson D 3rd, Kryscio RJ, Lin Y et al (2013) Patterns of salivary analytes provide diagnostic capacity for distinguishing chronic adult periodontitis from health. J Clin Immunol 33:271–279

Eliades T, Trapalis C, Eliades G, Katsavrias E (2003) Salivary metal levels of orthodontic patients: a novel methodological and analytical approach. Eur J Orthod 25:103–106

Giannobile WV, McDevitt JT, Niedbala RS, Malamud D (2011) Translational and clinical applications of salivary diagnostics. Adv Dent Res 23:375–380

Gomes AM, Douglas-de-Oliveira DW, Oliveira Costa F (2018) Could the biomarker levels in saliva help distinguish between healthy implants and implants with peri-implant disease? A systematic review. Arch Oral Biol 96:216–222

Gonçalves Lda R, Soares MR, Nogueira FC, Garcia C, Camisasca DR, Domont G, Feitosa AC, Pereira Dde A, Zingali RB, Alves G (2010) Comparative proteomic analysis of whole saliva from chronic periodontitis patients. J Proteome 73:1334–1341

Grigoriadou ME, Koutayas SO, Madianos PN, Strub JR (2010) Interleukin-1 as a genetic marker for periodontitis: review of the literature. Quintessence Int 41:517–525

Jiménez Y. La saliva como fluido diagnostico (2010) En: Bagán JV, Jiménez Y. Fisiopatología de las glándulas salivales. Valencia : eds Medicina Oral SL, pp 347–355

Kaufman E, Lamster IB (2000) Analysis of saliva for periodontal diagnosis – a review. J Clin Periodontol 27:453–465

Kaufman E, Lamster IB (2002) The diagnostic applications of saliva – a review. Crit Rev Oral Biol Med 13:197–212

Kaushik R, Yeltiwar RK, Pushpanshu KR (2011) Salivary interleukin-1beta levels in patients with chronic periodontitis before and after periodontal phase I therapy and healthy controls: a case-control study. J Periodontol 82:1353–1359

Koch KE, Collantes CL, Lewintre MI, Latyn K (2010) Influencia de la aparatología ortodóntica fija en la viscosidad y flujo salival. Rev Fac Odontol UNNE 3:48–51

Lee CH, Chen YW, Tu YK, Wu YC, Chang PC (2018) The potential of salivary biomarkers for predicting the sensitivity and monitoring the response to nonsurgical periodontal therapy: a preliminary assessment. J Periodontal Res 53:545–554

Liskmann S, Vihalemm T, Salum O, Zilmer K, Fischer K, Zilmer M (2007) Characterization of the antioxidant profile of human saliva in peri-implant health and disease. Clin Oral Implants Res 18:27–33

Llena-Puy C (2006) The rôle of saliva in maintaining oral health and as an aid to diagnosis. Med Oral Patol Oral Cir Bucal 11:E449–E455

Lorenzo-Pouso AI, Pérez-Sayáns M, Bravo SB, López-Jornet P, García-Vence M, Alonso-Sampedro M, Carballo J, García-García A (2018) Protein-based salivary profiles as novel biomarkers for oral diseases. Dis Markers 7:6141845

Malamud D (2011) Saliva as a diagnostic fluid. Dent Clin N Am 55:159–178

Meschiari CA, Marcaccini AE, Moura S, Zuardi LR, Tanus-Santos JE, Gerlach RF (2013) Salivary MMPs, TIMPs, and MPO levels in periodontal disease patients and controls. Clin Chem Acta 421:140–146

Miller CS, King CP Jr, Langub MC, Kryscio RJ, Thomas MV (2006) Salivary biomarkers of existing periodontal disease: across- sectional study. J Am Dent Assoc 137:322–329

Nordlund A, Johansson I, Källestål C, Ericson T, Sjöström M, Strömberg N (2009) Improved ability of biological and previous caries multimarkers to predict caries disease as revealed by multivariate PLS modelling. BMC Oral Health 9:28

Olayanju OA, Rahamon SK, Joseph IO, Arinola OG (2012) Salivary immunoglobulin classes in Nigerians with periodontitis. J Contemp Dent Pract 13:163–166

Prakasam S, Srinivasan M (2014) Evaluation of salivary biomarker profiles following non-surgical management of chronic periodontitis. Oral Dis 20:171–177

Salvatori O, Puri S, Tati S, Edgerton M (2016) Innate immunity and saliva in Candida albicans–mediated oral diseases. J Dent Res 95:365–371

Schafer CA, Schafer JJ, Yakob M, Lima P, Camargo P, Wong DT (2014) Saliva diagnostics: utilizing oral fluids to determine health status. Monogr Oral Sci 24:88–98

Schenck K, Poppelsdorf D, Denis C, Tollefsen T (1993) Levels of salivary IgA antibodies reactive with bacteria from dental plaque are associated with susceptibility to experimental gingivitis. J Clin Periodontol 20:411–417

Slots J, Slots H (2011) Bacterial and viral pathogens in saliva: disease relationship and infectious risk. Periodontol 55:48–69

Thein ZM, Samaranayake YH, Samaranayake LP (2006) Effect of oral bacteria on growth and survival of Candida albicans biofilms. Arch Oral Biol 51:672–680

Ulker AE, Tulunoglu O, Ozmeric N, Can M, Demirtas S (2008) The evaluation of cystatin C, IL-1beta, and TNF-alpha levels in total saliva and gingival crevicular fluid from 11- to 16-year-old children, J. Periodontology 79:854–860

Vitorino R, de Morais Guedes S, Ferreira R, Lobo MJ, Duarte J, Ferrer-Correia AJ, Tomer KB, Domingues PM, Amado FM (2006) Two-dimensional electrophoresis study of in vitro pellicle formation and dental caries susceptibility. Eur J Oral Sci 114:147–153

Wang K, Wang Y, Wang X, Ren Q, Han S, Ding L, Li Z, Zhou X, Li W, Zhang LK (2018) Comparative salivary proteomics analysis of children with and without dental caries using the iTRAQ/MRM approach. J Transl Med 16:018-1388-8

Wilczynska-Borawska M, Borawski J, Kovalchuk O, Chyczewski L, Stokowska W (2006) Hepatocyte growth factor in saliva is a potential marker of symptomatic periodontal disease. J Oral Sci 48:47–50

Yoshizawa JM, Schafer CA, Schafer JJ, Farrell JJ, Paster BJ, Wong DT (2013) Salivary biomarkers: toward future clinical and diagnostic utilities. Clin Microbiol Rev 26:781–791

Chapter 6
Salivary Biomarkers in Respiratory Diseases

Jovita Mazeikiene

6.1 Introduction

Respiratory diseases englobe diseases that affect any part of the respiratory system impeding adequate gas exchange by narrowing or blocking the airways. Respiratory diseases are among the most frequently occurring diseases in humans and animals (WHO 2018). Due to their high variability in terms of origin (e.g., inflammatory, infectious or neoplastic), intensity (acute, chronic), severity (self-limiting to life-threatening), and anatomical localization (trachea, bronchi, bronchioles, alveoli, pleurae, pleural cavity, and the nerves and muscles of respiration) the diagnostics of respiratory tract disorders is complex (Table 6.1).

In the present chapter, the existing information about the possible utility of salivary biomarkers for diagnosis and monitoring of the most commonly studied respiratory diseases are reviewed. Furthermore, since sputum (a coughed-up mucus from the lower airways, trachea and bronchi) is a sample already used in clinics and has a close relationship with saliva, data related to sputum analysis was also included. Nevertheless, animal salivary biomarkers are not described in this chapter as they were mainly studied in respiratory diseases of infectious origin and, though, are reviewed in Chap. 12.

J. Mazeikiene (✉)
InMedica Vilnius-Alfa Clinic, Vilnius, Lithuania

A. Tvarijonaviciute et al. (eds.), *Saliva in Health and Disease*,
https://doi.org/10.1007/978-3-030-37681-9_6

Table 6.1 Respiratory disease diagnosis

Steps to follow
(a) Complaints of the patient and medical history;
(b) Physical examination findings (inspection, auscultation and percussion);
(c) Radiological tests (X-ray, computer tomography, scintigraphy, angiography, ultrasound and magnetic resonance test);
(d) Functional tests (spirometry, blood gas analysis, lung diffusion test, lung capacity measurement);
(e) Salivary sputum tests (cytological and microbiological tests, biomarkers).
(f) Bronchoscopy (including tests of bronchoalveolar lavage);
(g) Biopsies (including tests of pleural fluid);

6.2 Bronchial Asthma

The heterogeneity of asthma phenotypes represents a challenge for adequate assessment and treatment of the disease (Wenzel 2012). The molecular analysis of highly characterized cohorts of asthma patients was suggested to help to identify biomarkers of asthma subtypes, which may lead to more efficient and personalized therapies (Wenzel 2012; Wagener et al. 2013). So far, asthma has been mainly divided into two subtypes based on airway and systemic eosinophilia and response to glucocorticoids: T-helper cell type 2 (Th2)-high asthma (presence of eosinophilia, respond to glucocorticoids) and Th2-low asthma (without airway or systemic eosinophilia; do not respond to glucocorticoids) (Fahy 2015).

Sputum Patients with poorly controlled asthma present higher sputum eosinophils, leukotriene E4 (LTE4), eosinophil cationic protein (ECP) and regulated upon activation normal T-cell expressed and secreted (RANTES) levels, and interleukin (IL) 13, but not IL-8, compared to patients with controlled asthma (Romagnoli et al. 2002; Truyen et al. 2006). However, sputum eosinophilia and sputum inflammatory biomarkers were associated with poorly controlled asthma rather than with the severity (Romagnoli et al. 2002). On the other hand, sputum IL-6 was suggested as a biomarker for asthma monitoring since strongly correlated with forced expiratory volume in 1 s (FEV_1) and FEV_1/FVC (forced vital capacity) (Neveu et al. 2010; Poynter and Irvin 2016). Both clinical and functional responsiveness of patients with asthma to treatment with steroids was related to sputum eosinophilia indicating its utility for disease monitoring (Romagnoli et al. 2002).

Gene expression analysis of induced sputum revealed the tryptase *TPSAB1*, *CPA3* (a mast cell granule marker) *and CLC* (an eosinophilic granule protein) genes to be mostly upregulated in patients with asthma *versus* control subjects (Foresi et al. 1997; Saha et al. 2008; Wang et al. 2016; Reber and Fahy 2016). According to tryptase *TPSAB1* and *CPA3* gene expression in sputum, patients with asthma could be grouped into 3 groups: (1) low tryptase *TPSAB1* and *CPA3* expression; (2) detectable tryptase *TPSAB1* (MC_T); (3) detectable both tryptase and *CPA3* genes ($MC_{T/CPA3}$). The $MC_{T/CPA3}$ group was characterized by elevated exhaled NO, sputum eosinophilia, bronchial sensitivity and reactivity compared with the MC_T group and responded better to the therapy with corticosteroids (Dougherty et al. 2010). Furthermore, the close relation between mRNA of *CLC* and *CPA3* in induced sputum and response to glucocorticoid treatment was observed reinforcing their clinical utility (Saha et al. 2008). Other gene expression profiles have been also identified in induced sputum including IL-4, IL-5 and IL-13 being useful in phenotyping asthma as Th2-low and Th2-high (Naseer et al. 1997). Overall, although technically more complex than blood analysis, sputum gene expression signatures were reported to have a great potential in detecting underlying mechanisms and guiding personalized treatment and management strategies in patients suffering from asthma (Wang et al. 2016).

Saliva Patients with asthma were shown to present alterations in salivary stress markers. Just for instance, salivary alpha-amylase (sAA) and cortisol were lower in patients with asthma than in saliva of healthy controls (Tan and Chew 2009; Bakkeheim et al. 2010). Furthermore, lower morning salivary cortisol levels were observed in patients with uncontrolled asthma as compared to patients with controlled asthma (Shin et al. 2014). For this, the authors recommended measuring the morning salivary cortisol levels to assess asthma control status. However, cautions should be taken since a big overlap in salivary cortisol levels between groups existed (Shin et al. 2014).

Inflammatory biomarkers, such as C-reactive protein (CRP), haptoglobin (Hp), leukotriene and Eosinophil cationic protein (ECP), were increased in patients with asthma (Schmekel et al. 2001; Gaber et al. 2008; Rao et al. 2011). Interestingly, salivary leukotriene contributed to the differentiation of aspirin-intolerant asthmatic patients from tolerant counterparts (Gaber et al. 2008). ECP, a protein located in the eosinophil primary matrix and released during degranulation of eosinophils, was associated with asthma activity (Schmekel et al. 2001).

6.3 Bronchiectasis

Bronchiectasis is a chronic destructive disease of respiratory tract characterized by persistent bacterial colonization, purulent inflammation and chronic irreversible dilatation of bronchi.

Sputum Sputum neutrophil elastase (NE), a 29-kDa serine protease released from the azurophilic granules of neutrophils, is considered one of the best biomarkers of bronchiectasis being related to disease activity, severity (including clinical data, antibiotic responsiveness, quality of life), and inflammatory biomarkers (leukocytes, IL-1b, and TNF-a). Patognostic features of NE include degradation of elastin present in the airways and impairment of the immune response (Gramegna et al. 2017; Polverino et al. 2017). Some authors suggest sputum NE be able to predict future exacerbations in bronchiectasis patients (Chalmers et al. 2017; Oscullo Yepez et al. 2019), although their levels did not increase in all cases (Sibila et al. 2019). Therefore, NE could not be an ideal marker for diagnosing bronchiectasis exacerbations. Furthermore, it should be noted that sputum NE is not a pathognomonic bio-

Table 6.2 Behavior antimicrobial peptides in the sputum of patients with bronchiectasis according to Sibila et al. (2019)

Antimicrobial peptide and its behavior	Associated with	Clinical utility
At diagnosis		
↑ LL-37 ↓ SLPI	Frequent exacerbator phenotype of bronchiectasis Bronchiectasis severity index Lower FEV_1 (forced expiratory volume in 1 s) *Pseudomonas aeruginosa* infection	Could predict future risk of exacerbations in bronchiectasis
At follow-up		
↑ LL-37 ↓ SLPI	Shorter time to the next exacerbation	Prediction of future risk exacerbations in bronchiectasis and frequency
↑ LL-37	Exacerbation frequency over the next 12 months	

LL-37 cathelicidin, *SLPI* secretory leukocyte protease inhibitor

marker of bronchiectasis and can also increase in other respiratory diseases that present inflammatory status such as COPD or cystic fibrosis (Gramegna et al. 2017; Polverino et al. 2017). Nevertheless, scientific evidences suggest NE be a key molecule in the pathophysiology of bronchiectasis in most patients, and, therefore, its use in clinical practice would greatly contribute to the accurate disease management, although point-of-care devices facilitating its determination in situ are needed (Oscullo Yepez et al. 2019).

Altered sputum antimicrobial peptides (AMP) was reported in patients with bronchiectasis (Table 6.2) (Sibila et al. 2019).

6.4 Chronic Obstructive Pulmonary Disease (COPD)

COPD is characterized by airway inflammation, chronic airflow limitation, progressive tissue destruction, extra-pulmonary manifestations and systemic inflammation (Agustí et al. 2012; Thomsen et al. 2013).

Sputum The inflammatory response in COPD is dominated by neutrophils and chemokines/cytokines such as tumor necrosis factor-α (TNF-α) and interleukin-8 (IL-8), which are of importance for neutrophils recruitment (Keatings et al. 1996; Lipińska-Ojrzanowska et al. 2017). In consequence, sputum from patients with COPD contains increased levels of eosinophils, IL-5, IL-8, and TNF-α being related to the severity of COPD and its exacerbation (Papi et al. 2006). Furthermore, IL-8 and IL-6 were able to differentiate patients with COPD with and without coexisting bronchiectasis (Patel et al. 2004).

Sputum mucin concentration, microstructure and biochemical properties of main airway mucins, MUC5AC and MUC5B, were also suggested to contribute to clinical management and follow-up of COPD (Kesimer et al. 2017; Garudadri and Woodruff 2018; Chisholm et al. 2019). Sputum mucin concentrations were higher in patients with two or more respiratory exacerbations per year than in those with zero exacerbations (Kesimer et al. 2017; Garudadri and Woodruff 2018). Furthermore, both MUC5AC and MUC5B sputum concentrations increase with increasing airflow obstruction in COPD. In addition, since MUC5AC increases to a greater degree than MUC5B, ratio MUC5AC:MUC5B was related to the severity of the obstruction (Kesimer et al. 2017). Nevertheless, smoking (current or former) was also related to increment concentrations of mucins, MUC5B and MUC5AC, in sputum, thus, smoking should be taken into account when interpreting results (Kesimer et al. 2017). Recently, Chisholm et al. (2019) reported that microscopic mucus structure could serve as a risk factor for COPD progression and severity, although further studies are needed to collaborate these findings.

Saliva Salivary CRP, procalcitonin (PCT) and neutrophil elastase (NE) increase during COPD exacerbations (Ilumets et al. 2008; Özçaka et al. 2011; Patel et al. 2015). Furthermore, CRP and PCT concentrations in saliva were correlated with their concentrations in serum, while salivary NE did not correlate significantly with its levels in serum, but did reflect sputum NE levels (Patel et al. 2015). Therefore, salivary CRP, PCT and NE could have clinical value for COPD management.

6.5 Pneumonia

Pneumonia is an inflammation of the lung tissue affecting alveoli. Due to infection microorganisms, edemic liquid and inflammatory cells accumulate in alveoli and cause damage to normal gas exchange. Death rate from pneumonia takes 5-6th places in industrial countries (WHO 2018) being the greatest risk in the neonatal period (Nissen 2007). Thus, accurate early diagnosis is of high importance to enable efficient treatment and improve the prognosis.

Sputum Alpkvist et al. (2018) reported sputum *lytA* load being associated with respiratory failure and severity of disease. While, although sputum HMGB1 (High Mobility Group Box 1) levels were not related to severity (pneumonia severity index or presence of sepsis), its high levels were associated with pneumococcal etiology, indicating a potential role for HMGB1 in bacterial dissemination (Alpkvist et al. 2018).

Saliva A number of studies reported significant increments of salivary concentrations of CRP in neonates with late-onset pneumonia in comparison with healthy controls (Xiao et al. 2015; Omran et al. 2018). Therefore, salivary CRP was indicated to be one of the best single tests for the early non-invasive detection of pneumonia in children (Hansen et al. 2000; Omran et al. 2018)

6.6 Tuberculosis (TB)

TB is an infectious disease usually affecting the lungs but also may pleura and lymph nodes among others (Ketata et al. 2015). In 2017, globally 10.0 million people developed TB disease and 1.3 million people died from this disease (WHO 2019). Furthermore, a number of patients present drug-resistant TB constituting so denominated a public health crisis (WHO 2019). For these reasons, TB is considered to be one of the most important and dangerous diseases worldwide (Jacobs et al. 2016).

Sputum Sputum mRNA, especially isocitrate lyase (*icl*) mRNA, quantified by RT-PCR was reported to be a promising marker of *M. tuberculosis* viability and, therefore, useful to monitor the response to short and long-term TB treatments. Furthermore, *icl* mRNA was stated to potentially replace the quantitative culture (CFU counts) (Li et al. 2010). On the other hand, the utility of a commercially available real-time PCR assay, Xpert MTB/RIF, to detect Mycobacterium-tuberculosis-specific DNA sequences in sputum samples was evaluated (Friedrich et al. 2013). And, although high sensitivity was observed (97·0%, 95% CI 95·8-97·9), specificity was poor (48·6%, 45·0-52·2). Therefore, the authors concluded that the method was not clinically useful for monitoring tuberculosis treatment and should not replace standard analyses (culture, smear microscopy) (Friedrich et al. 2013). Nevertheless, in any case, longitudinal studies are needed in order to confirm the reliability of the sputum as a sample for TB detection and monitoring.

Saliva Salivary pro-inflammatory proteins including ILs (5, 6, 15), TNFα and CRP were showed to present potential in the diagnosis of TB disease (Phalane et al. 2013; Jacobs et al. 2016). However, only CRP and Hp were shown to present clinical utility independently presenting AUC >0.8 when data of patients with TB and other respiratory diseases (ORD) were studied by ROC curve analysis (Jacobs et al. 2016). Nevertheless, ideally, the use of a combination of salivary biomarkers was proposed in order to increase the specificity and sensitivity of the diagnosis of TB disease. The proposed panel would ideally include CRP, ferritin, serum amyloid P (SAP), monocyte chemotactic protein (MCP)-1, alpha2 macroglobulin (A2M), fibrinogen and haptoglobin (Jacobs et al. 2016). Finally, a number of salivary biomarkers, namely IL-15, granzyme A, MCP-1, IL-1β, IL-9, MIP-1β, IL-10, SAA and ferritin, significantly changed in a course of treatment suggesting their potential as biomarkers of TB disease monitoring (Jacobs et al. 2016). Further large scale studies are now required in order to confirm these promising findings and to deepen the subject since the use of saliva could greatly improve the screening of the potential patents permitting early diagnosis and initiation of treatment and follow-up.

6.7 Lung Cancer

Lung cancer was the most frequently diagnosed cancer type (11.6%) and a reason for death (18.4%) in 2018 as compared to other types of cancers (International Agency for Research on Cancer and WHO). Both sputum and saliva were studied as a possible source of lung cancer biomarkers such as different proteins, RNA and DNA (I and Cho 2015). Furthermore, since saliva can be noninvasively collected and contains a large array of biomarkers, it was suggested to be a good biofluid for early detection of lung cancer (I and Cho 2015). For more information, related to salivary biomarkers in cancer, see Chap. 13.

6.8 Conclusion

Overall, saliva presents the potential for respiratory diseases diagnosis and treatment monitoring. However, further studies are now required in order to confirm these evidences and increase the knowledge related to a broader range of diseases, therapies, biomarkers among others, what would permit the use of this non-invasive biofluid in clinics.

References

Agustí A, Edwards LD, Rennard SI et al (2012) Persistent systemic inflammation is associated with poor clinical outcomes in COPD: a novel phenotype. PLoS One 7:e37483. https://doi.org/10.1371/journal.pone.0037483

Alpkvist H, Athlin S, Mölling P et al (2018) High HMGB1 levels in sputum are related to pneumococcal bacteraemia but not to disease severity in community-acquired pneumonia. Sci Rep 8:13428. https://doi.org/10.1038/s41598-018-31504-4

Bakkeheim E, Mowinckel P, Carlsen KH et al (2010) Reduced basal salivary cortisol in children with asthma and allergic rhinitis. Acta Paediatr 99:1705–1711. https://doi.org/10.1111/j.1651-2227.2009.01598.x

Chalmers JD, Moffitt KL, Suarez-Cuartin G et al (2017) Neutrophil elastase activity is associated with exacerbations and lung function decline in bronchiectasis. Am J Respir Crit Care Med 195:1384–1393. https://doi.org/10.1164/rccm.201605-1027OC

Chisholm JF, Shenoy SK, Shade JK et al (2019) Nanoparticle diffusion in spontaneously expectorated sputum as a biophysical tool to probe disease severity in COPD. Eur Respir J 54:1900088. https://doi.org/10.1183/13993003.00088-2019

Dougherty RH, Sidhu SS, Raman K et al (2010) Accumulation of intraepithelial mast cells with a unique protease phenotype in T(H)2-high asthma. J Allergy Clin Immunol 125:1046–1053.e8. https://doi.org/10.1016/j.jaci.2010.03.003

Fahy JV (2015) Type 2 inflammation in asthma – present in most, absent in many. Nat Rev Immunol 15:57–65. https://doi.org/10.1038/nri3786

Foresi A, Leone C, Pelucchi A et al (1997) Eosinophils, mast cells, and basophils in induced sputum from patients with seasonal allergic rhinitis and perennial asthma: relationship to methacholine responsiveness. J Allergy Clin Immunol 100:58–64. https://doi.org/10.1016/S0091-6749(97)70195-7

Friedrich SO, Rachow A, Saathoff E et al (2013) Assessment of the sensitivity and specificity of Xpert MTB/RIF assay as an early sputum biomarker of response to tuberculosis treatment. Lancet Respir Med 1:462–470. https://doi.org/10.1016/S2213-2600(13)70119-X

Gaber F, Daham K, Higashi A et al (2008) Increased levels of cysteinyl-leukotrienes in saliva, induced sputum, urine and blood from patients with aspirin-intolerant asthma. Thorax 63:1076–1082. https://doi.org/10.1136/thx.2008.101196

Garudadri S, Woodruff PG (2018) Targeting chronic obstructive pulmonary disease phenotypes, endotypes, and biomarkers. Ann Am Thorac Soc 15:S234–S238. https://doi.org/10.1513/AnnalsATS.201808-533MG

Gramegna A, Amati F, Terranova L et al (2017) Neutrophil elastase in bronchiectasis. Respir Res 18:211. https://doi.org/10.1186/s12931-017-0691-x

Hansen AB, Verder H, Staun-Olsen P (2000) Soluble intercellular adhesion molecule and C-reactive protein as early markers of infection in newborns. J Perinat Med 28:97–103. https://doi.org/10.1515/JPM.2000.012

I H, Cho J-Y (2015) Lung cancer biomarkers. Adv Clin Chem 72:107–170. https://doi.org/10.1016/BS.ACC.2015.07.003

Ilumets H, Rytilä PH, Sovijärvi AR et al (2008) Transient elevation of neutrophil proteinases in induced sputum during COPD exacerbation. Scand J Clin Lab Invest 68:618–623. https://doi.org/10.1080/00365510801983773

International Agency for Research on Cancer, WHO Cancer Today. Data visualization for exploring the global cancer burden in 2018

Jacobs R, Tshehla E, Malherbe S et al (2016) Host biomarkers detected in saliva show promise as markers for the diagnosis of pulmonary tuberculosis disease and monitoring of the response to tuberculosis treatment. Cytokine 81:50–56. https://doi.org/10.1016/j.cyto.2016.02.004

Keatings VM, Collins PD, Scott DM, Barnes PJ (1996) Differences in interleukin-8 and tumor necrosis factor-alpha in induced sputum from patients with chronic obstructive pulmonary disease or asthma. Am J Respir Crit Care Med 153:530–534. https://doi.org/10.1164/ajrccm.153.2.8564092

Kesimer M, Ford AA, Ceppe A et al (2017) Airway mucin concentration as a marker of chronic bronchitis. N Engl J Med 377:911–922. https://doi.org/10.1056/NEJMoa1701632

Ketata W, Rekik WK, Ayadi H, Kammoun S (2015) Les tuberculoses extrapulmonaires. Rev Pneumol Clin 71:83–92. https://doi.org/10.1016/j.pneumo.2014.04.001

Li L, Mahan CS, Palaci M et al (2010) Sputum Mycobacterium tuberculosis mRNA as a marker of bacteriologic clearance in response to antituberculosis therapy. J Clin Microbiol 48:46–51. https://doi.org/10.1128/JCM.01526-09

Lipińska-Ojrzanowska A, Marcinkiewicz A, Walusiak-Skorupa J (2017) Usefulness of biomarkers in work-related airway disease. Curr Treat Options Allergy 4:181–190. https://doi.org/10.1007/s40521-017-0121-9

Naseer T, Minshall EM, Leung DY et al (1997) Expression of IL-12 and IL-13 mRNA in asthma and their modulation in response to steroid therapy. Am J Respir Crit Care Med 155:845–851. https://doi.org/10.1164/ajrccm.155.3.9117015

Neveu WA, Allard JL, Raymond DM et al (2010) Elevation of IL-6 in the allergic asthmatic airway is independent of inflammation but associates with loss of central airway function. Respir Res 11:28. https://doi.org/10.1186/1465-9921-11-28

Nissen MD (2007) Congenital and neonatal pneumonia. Paediatr Respir Rev 8:195–203. https://doi.org/10.1016/j.prrv.2007.07.001

Omran A, Ali M, Saleh MH, Zekry O (2018) Salivary C-reactive protein and mean platelet volume in diagnosis of late-onset neonatal pneumonia. Clin Respir J 12:1644–1650. https://doi.org/10.1111/crj.12723

Oscullo Yepez G, Sierra-Párraga JM, Posadas Blazquez T, Martínez-García MÁ (2019) Towards precision medicine in bronchiectasis: what is the role of neutrophilic elastase determination? Eur Respir J 53:1900765. https://doi.org/10.1183/13993003.00765-2019

Özçaka Ö, Bıçakcı N, Pussinen P et al (2011) Smoking and matrix metalloproteinases, neutrophil elastase and myeloperoxidase in chronic periodontitis. Oral Dis 17:68–76. https://doi.org/10.1111/j.1601-0825.2010.01705.x

Papi A, Bellettato CM, Braccioni F et al (2006) Infections and airway inflammation in chronic obstructive pulmonary disease severe exacerbations. Am J Respir Crit Care Med 173:1114–1121. https://doi.org/10.1164/rccm.200506-859OC

Patel IS, Vlahos I, Wilkinson TMA et al (2004) Bronchiectasis, exacerbation indices, and inflammation in chronic obstructive pulmonary disease. Am J Respir Crit Care Med 170:400–407. https://doi.org/10.1164/rccm.200305-648OC

Patel N, Belcher J, Thorpe G et al (2015) Measurement of C-reactive protein, procalcitonin and neutrophil elastase in saliva of COPD patients and healthy controls: correlation to self-reported wellbeing parameters. Respir Res 16:62. https://doi.org/10.1186/s12931-015-0219-1

Phalane KG, Kriel M, Loxton AG et al (2013) Differential expression of host biomarkers in saliva and serum samples from individuals with suspected pulmonary tuberculosis. Mediat Inflamm 2013:981984–981910. https://doi.org/10.1155/2013/981984

Polverino E, Rosales-Mayor E, Dale GE et al (2017) The role of neutrophil elastase inhibitors in lung diseases. Chest 152:249–262. https://doi.org/10.1016/j.chest.2017.03.056

Poynter ME, Irvin CG (2016) Interleukin-6 as a biomarker for asthma: hype or is there something else? Eur Respir J 48:979–981. https://doi.org/10.1183/13993003.01597-2016

Rao NL, Shetty S, D'Souza J et al (2011) Saliva C- reactive protein levels in patients with acute urticaria. Biomed Res 22(1):90–92

Reber LL, Fahy JV (2016) Mast cells in asthma: biomarker and therapeutic target. Eur Respir J 47:1040–1042. https://doi.org/10.1183/13993003.00065-2016

Romagnoli M, Vachier I, Tarodo de la Fuente P et al (2002) Eosinophilic inflammation in sputum of poorly controlled asthmatics. Eur Respir J 20:1370–1377. https://doi.org/10.1183/09031936.02.00029202

Saha SK, Berry MA, Parker D et al (2008) Increased sputum and bronchial biopsy IL-13 expression in severe asthma. J Allergy Clin Immunol 121:685–691. https://doi.org/10.1016/j.jaci.2008.01.005

Schmekel B, Ahlner J, Malmström M, Venge P (2001) Eosinophil cationic protein (ECP) in saliva: a new marker of disease activity in bronchial asthma. Respir Med 95:670–675. https://doi.org/10.1053/rmed.2001.1123

Shin YS, Liu JN, Kim J-H et al (2014) The impact of asthma control on salivary cortisol level in adult asthmatics. Allergy, Asthma Immunol Res 6:463–466. https://doi.org/10.4168/aair.2014.6.5.463

Sibila O, Perea L, Cantó E et al (2019) Antimicrobial peptides, disease severity and exacerbations in bronchiectasis. Thorax 74:835–842. https://doi.org/10.1136/thoraxjnl-2018-212895

Tan M, Chew F (2009) Identification of saliva biomarkers for asthma by proteomic analysis, pp 1–4

Thomsen M, Ingebrigtsen TS, Marott JL et al (2013) Inflammatory biomarkers and exacerbations in chronic obstructive pulmonary disease. JAMA 309:2353–2361. https://doi.org/10.1001/jama.2013.5732

Truyen E, Coteur L, Dilissen E et al (2006) Evaluation of airway inflammation by quantitative Th1/Th2 cytokine mRNA measurement in sputum of asthma patients. Thorax 61:202–208. https://doi.org/10.1136/thx.2005.052399

Wagener AH, Yick CY, Brinkman P et al (2013) Toward composite molecular signatures in the phenotyping of asthma. Ann Am Thorac Soc 10(Suppl):S197–S205. https://doi.org/10.1513/AnnalsATS.201302-035AW

Wang G, Baines KJ, Fu JJ et al (2016) Sputum mast cell subtypes relate to eosinophilia and corticosteroid response in asthma. Eur Respir J 47:1123–1133. https://doi.org/10.1183/13993003.01098-2015

Wenzel SE (2012) Asthma phenotypes: the evolution from clinical to molecular approaches. Nat Med 18:716–725. https://doi.org/10.1038/nm.2678

WHO (2018) WHO|World Health Statistics 2018: monitoring health for the SDGs. World Health Organization

WHO (2019) WHO|Global tuberculosis report 2018. World Health Organization

Xiao X, Xue L, Sheng H-L, Xiao Z-H (2015) Correlation between serum levels of C-reactive protein and infant pneumonia: a meta-analysis. Exp Ther Med 9:2331–2338. https://doi.org/10.3892/etm.2015.2417

Chapter 7
Salivary Biomarkers in Neurologic Diseases

Zeineb Zian and Assia Bouhoudan

Objectives
Neurologic diseases (ND) are disorders of the brain affecting the central and peripheral nervous system, and characterized by a progressive loss of memory and cognition. Because of their complicated causes and manifestations, early diagnosis of these neurological conditions remains a great challenge and may involve invasive and painful tests.

Seeking biomarkers from biological fluids has revealed several indicators of ND. Unlike blood and other invasive fluids used to diagnose ND, saliva presented a non-invasive physiological fluid characterized by simple access and collection and can be evaluated using different analytical tests.

In this chapter, we will review some studies revealing salivary biomarkers of most known ND in humans and animals in order to establish an overview about the advances made in this field. Our goal is to provide some information about the salivary biomarkers used in diagnosis, treatment or monitoring of the ND progression.

Authors Zeineb Zian and Assia Bouhoudan have equally contributed to this chapter.

Z. Zian (✉)
Biomedical Genomics and Oncogenetics Research Laboratory, Faculty of Sciences and Techniques of Tangier, Abdelmalek Essaadi University, Tétouan, Morocco
e-mail: z.zian@uae.ac.ma

A. Bouhoudan
Faculty of Sciences of Tetouan, Abdelmalek Essaadi University, Tétouan, Morocco
e-mail: assia.bouhoudan@uae.ac.ma

A. Tvarijonaviciute et al. (eds.), *Saliva in Health and Disease*,
https://doi.org/10.1007/978-3-030-37681-9_7

7.1 In Humans

7.1.1 Introduction

Neurologic diseases (ND) constitute a family of neurologic disorders affecting the central and peripheral nervous system, and ranging from neurodegenerative to neurodevelopmental and neuropsychiatric. They include epilepsy, bacterial and viral neuroinfections, brain tumors, cerebrovascular diseases, dementias, headache disorders, multiple sclerosis, Parkinson's disease, and traumatic disorders due to head trauma. The neurodegeneration occurring during ND results from a progressive loss of some classes of neurons and can affect memory, cognition and/or motor skill (Farah et al. 2018). Several combined factors including genetic and environmental factors are involved in these conditions (Bertram 2005; Landrigan et al. 2005). According to Khan et al. the screening examination of ND is based on six areas including: Mental status, Motor, Sensory perception, Reflexes, Coordination and Cranial nerves (Khan et al. 2018). Since their presentation and clinical manifestations are mimicked by other connective tissue diseases, inflammatory and autoimmune diseases, early diagnosis is an essential key for patients with these neurological conditions in order to their treatment be effective. We summarized in Table 7.1 some most known ND in humans including; Alzheimer's, Parkinson's, Huntington's, Amyotrophic lateral sclerosis, Multiple sclerosis, Autism spectrum disorders and neuropsychiatric disorders (schizophrenia, bipolar disorder and attention deficit hyperactivity disorder).

7.1.2 Currently Used Biomarkers

Several biomarkers associated to ND have been identified in different biological media such as cerebrospinal fluid (CSF), blood and saliva (Table 7.2).

7.1.2.1 Alzheimer's Biomarkers

Amyloid beta (Aβ) peptide and tau protein are both suggested to be the major factors that contribute to the occurrence of Alzheimer's disease (AD) (Ballatore et al. 2007; Bloom 2014; Buee et al. 2000; Ittner and Götz 2011; Mudher and Lovestone 2002). Aβ42 is a peptide fraction of the large Aβ protein precursor (AβPP), which plays normal physiological roles in memory formation, lipid homeostasis, regulation of neuron activity and neurite growth (Lee et al. 2016; Garcia-Osta and Alberini 2009; Grimm et al. 2007; Kontush et al. 2001; Nicolas and Hassan 2014). The second hallmark of AD is tau protein, also called the Microtubule Associated Protein T(MAPT). Total tau and phosphorylated tau (t-tau and p-tau) both have been identified in CSF

Table 7.1 Description of most known neurologic diseases in humans

Neurologic diseases		Description	References
Neuropsychiatric disorders	SZ	Complex neuropsychiatric disorder with genetic and epigenetic factors contributing to its pathophysiology. It is characterized by hallucinations, delusions and cognitive deficits. Genetic alterations affecting some candidate genes were considered as the primary cause of its pathogenesis.	Abdolmaleky et al. (2015), Mellios et al. (2009) and Straub et al. (2002)
	BD	Severe mood disorder characterized by periodic manic episodes alternating with depression. It shares phenotypic similarities with other psychiatric disorders and results in chronic disease with moderate to severe impairment.	Psychiatric GWAS Consortium Bipolar Disorder Working Group (2011)
	ADHD	Complex condition and one of the most common childhood mental health disorders, characterized by developmentally inappropriate levels of inattention, hyperactivity and impulsivity. Environmental and genetic factors might influence this pathology.	Lasky-Su et al. (2008) and Wilmot et al. (2016)
Cognitive disorders	AD	Severe neurodegenerative disorder of the central nervous system (CNS) mainly manifested by dementia, confusion and cognitive injury. It is the single largest cause of dementia and characterized by a progressive loss of memory and cognitive function.	Farah et al., (2018) and Lee et al. (2016)
Motor disorders	PD	Progressive neurodegenerative disorder with unknown cause which results in multiple motor and cognitive deficits. It causes loss of two types of neurons in the brain; the dopaminergic and the serotoninergic. There are two groups of PD symptoms; motor and non-motor. Based on the dominant symptom, PD has been classified into 3 main types; tremor dominant type, akinetic-rigid dominant type and the mixed type.	Chaudhuri et al. (2006), Lee et al. (2016), Ruiz et al. (2011) and Thenganatt and Jankovic (2014)

(continued)

Table 7.1 (continued)

Neurologic diseases		Description	References
	HD	Genetically inherited and autosomal dominant neurodegenerative disorder caused by an expansion of trinucleotide cytosine-adenine-guanine (CAG) repeat in the huntingtin (Htt) gene. It is characterized by psychiatric troubles and progressive deterioration in cognitive and motor function leading to early death. It is characterized by the dysfunction and death of various neuron classes in the neocortex, striatum, cerebellum, hippocampus, substantia nigra, and brainstem nuclei. Mutant Htt proteins have been shown to form cytoplasmic aggregates, whose some reports suggest their role as neuroprotective, but others find them toxic	Killoran, (2016), Miller et al. (2010), Ross and Tabrizi (2011), Sapp et al. (1997) and Saudou et al. (1998)
	MS	Neurodegenerative, autoimmune and inflammatory disorder of the CNS in which infiltration of focal lymphocytic leads to damage of CNS neurons. It is characterized by a heterogeneous pathophysiological manifestations and a very complicated etiology. Its main symptoms are sensory, motor and cognitive.	Ahlgren et al. (2011), Betts et al. (1993), Chiaravalloti and DeLuca (2008), Chwastiak et al. (2002), Compston and Coles (2008), Krupp et al. (1988), Lassmann et al. (2007), Steinman MD (1996), Rizzo et al. (2004) and Stevens et al. (2013)
	ALS	Motor neuron disorder characterized by a progressive weakness, muscular wasting, dysphagia, dysarthria, and respiratory impairment leading to death. Familial and sporadic form of ALS have shown genetic mutations in different genes which are tested for in case of there exist a family history of the disease. Early diagnosis is an essential key for ALS patients in order to their treatment be effective.	Hayward C et al. (1996), Lomen-Hoerth et al. (2003), Riviere et al. (1998), Saberi et al. (2015) and Sreedharan et al. (2008)
Neurodevelopmental disorders	ASD	Neurodevelopmental disorder characterized by dysfunction in communicative skills, social deficits, and repetitive behaviors, and by stereotyped motor and mental process appearing 3 years after birth.	Castagnola et al. (2008) and Hollander et al. (2007)

AD Alzheimer's disease, *ADHD* Attention deficit hyperactivity disorder, *ALS* Amyotrophic lateral sclerosis, *ASD* Autism spectrum disorder, *BD* Bipolar disorder, *HD* Huntington's disease, *MS* Multiple sclerosis, *PD* Parkinson's disease, *SZ* Schizophrenia

Table 7.2 Summary of neurologic diseases biomarkers identified in CSF, blood, and saliva

Neurologic diseases	CSF and/or blood biomarkers	Salivary biomarkers	References
SZ	DTNBP1, HTR2A, MB-COMT,	DTNBP1, HTR2A, MB-COMT, α-defensins 1–4, S100A12, cystatin A S-glutathionylated of cystatin B and S-cysteinylated of cystatin B	Abdolmaleky et al. (2015), Ghadirivasfi et al. (2011), Iavarone et al. (2014), Nohesara et al. (2011) and Wilmot et al. (2016)
BD	DTNBP1, HTR2A	DTNBP1, HTR2A	Abdolmaleky et al. (2015), Ghadirivasfi et al. (2011), Nohesara et al. (2011) and Wilmot et al. (2016)
ADHD	VIPR2	VIPR2	Wilmot et al. (2016)
AD	Aβ 42, T-tau and p-tau,	Aβ42, p-tau and/or the p-tau/t-tau, AChE and Lactoferrin	Bermejo-Pareja et al. (2010), Carro et al. (2017), Farah et al. (2018), Sayer et al. (2004) and Shi et al. (2011)
PD	α-Syn, DJ-1, AChE	AChE andα-Syn_{total}, α-Syn_{olig},α-Syn_{olig}/ α-Syn_{total}, DJ-1	Bermejo-Pareja et al. (2010), Devic et al. (2011), Fedorova et al. (2015), Kang et al. (2014), Masters et al. (2015), Moreau et al. (2016) and Shi et al. (2010, 2011)
HD	mHtt protein	mHtt protein	Corey-Bloom et al. (2016), Massai et al. (2013) and Wild et al. (2015)
MS	sHLA class II, TBARS, AGEs, AOPP, Fructosamine	sHLA class II, TBARS, AGEs	Adamashvili et al. (2005) and Karlík et al. (2015)
ALS	CgA	CgA	Kaiserova et al. (2017) and Obayashi et al. (2008)
ASD	miRNAs, OT	miRNAs and OT, LTF, PIP, ANXA1, DEFA1/ DEFA1B,IGKC, IGHG1, Ig lambda-2 chain C regions, ELANE, LPO, PIGR,DMBT1 and MPO	Hicks et al. (2016), Ngounou Wetie et al. (2015) and Rutigliano et al. (2016)

AChE Acetylcholinesterase, *AD* Alzheimer's disease, *ADHD* Attention deficit hyperactivity disorder, *AGEs* advanced glycation end, *ALS* Amyotrophic lateral sclerosis, *ANXA1* Annexin A1, *AOPP* Advanced Oxidation Protein, Products, *ASD* Autism spectrum disorder, *Aβ* Amyloid β, *BD* Bipolar disease, *CgA* Chromogranins A, *CSF* Cerebrospinal fluid, *DEFA1/DEFA1B* Neutrophil-defensin 1, *DMBT1* Deleted in malignant brain tumors 1, *DTNBP1* Dystrobrevin binding protein, *ELANE* Neutrophil elastase, *HD* Huntingtin's disease, *HTR2A* 5-hydroxytryptamine receptor 2A, *IGHG1* Ig gamma-1 chain C region, *IGKC* Ig kappa chain C region, *LPO* Lactoperoxidase, *LTF* Lactotransferrin, *MB-COMT* catechol-O-methyltransferase, *mHtt* mutant Huntingtin protein, *miRNAs* MicroRNAs, *MPO* Myeloperoxidase, *MS* Multiple Sclerosis, *OT* Oxytocin, *PD* Parkinson's disease, *PIGR* Polymeric immunoglobulin receptor, *PIP* Prolactin-inducible protein, *p-tau* phosphorylated tau, *sHLA* Human Leukocyte Antigen, *SZ* Schizophrenia, *TBARS* Thiobarbituric Acid Reacting Substances, *T-tau* Total tau, *VIPR2* vasoactive intestinal peptide receptor 2, *α-Syn* α-synuclein of patients with AD (Parnetti et al. 2013). The p-tau can work synergistically with Aβ to increase its cytotoxic functions (Amadoro et al. 2011; Bloom 2014; Hyman et al. 2005). Thus, the high concentration of p-tau regarding to the t-tau may be used as a potential biomarker for evaluation of AD status (Farah et al. 2018).

7.1.2.2 Parkinson's Biomarkers

The proteins α-synuclein (α-Syn) andDJ-1 are both markers which decrease in the cerebrospinal fluid (CSF) of patients with Parkinson's disease (PD).The α-Syn is a member of the synuclein family, which is found in multiple brain regions (Iwai et al. 1995). In familial PD, α-Syn pathology has been linked to its increased concentrations in neurons (Chartier-harlin et al. 2004; Maraganore et al. 2006; Pals et al. 2004; Singleton 2003). DJ-1 is a protein located in mitochondria and nuclei of dopaminergic neurons, but mainly in the cytoplasm. It inhibits the formation of α-Syn fibrils which is necessary for the formation of α-syn oligomers, a key element in PD pathology (Shendelman et al. 2004). In contrast to its concentrations in CSF, urinary DJ-1 concentrations exhibited a significantly elevation in Korean PD males compared to non-PD males (Ho et al. 2014).

7.1.2.3 Huntingtin's Biomarkers

Huntingtin (Htt) protein, encoded by the Htt gene, is considered the most important molecular target for disease modifying therapies and several therapeutic approaches. Normal Htt protein has several physiological roles such as embryonic development, cell survival, tissue maintenance, and cell morphology (Saudou and Humbert 2016). CSF and blood Htt protein levels have been measured for diagnostic confirmation of Huntingtin's disease (HD). However, this measure remains not easy because of invasiveness of the method and due to the low Htt concentrations in the samples (Massai et al. 2013; Wild et al. 2015).

7.1.2.4 Multiple Sclerosis Biomarkers

The Human Leukocyte Antigen (HLA) has been demonstrated to be a biomarker for Multiple sclerosis (MS). Acorrelation between HLA mutations and levels of HLA class II molecules with MS has been reported (Martin 1991; Oksenberg et al. 2004; Olerup and Hillert 1991). Free radicals such as reactive oxygen or nitrogen species (ROS, RNS) are produced, in MS, by activated macrophages, microglia and mitochondrial dysfunction (van Horssen et al. 2011). Many studies reported that ROS contribute to the formation of pathological plaque in MS (Ortiz et al. 2013; van Horssen et al. 2011; Wang et al. 2014). Moreover, four oxidative stress markers; Advanced Oxidation Protein Products (AOPP) a marker of protein oxidation, Thiobarbituric Acid Reacting Substances (TBARS) a marker of lipoperoxidation and advanced glycation end products (AGEs) and fructosamine which are markers of carbonyl stress have been determined by Karlik et al. in blood samples of MS patients (Karlík et al. 2015). Their concentrations in plasma of MS patients were higher than controls. Previous studies reported higher TBARS levels in CSF of MS patients as compared to healthy controls (Ghabaee et al. 2010; Hunter et al. 1985; Naidoo and Knapp 1992; Wang et al. 2014). On the other hand, Karlik et al. have measured the Total Antioxidant Capacity (TAC) and Ferric Ion Reducing Ability (FRA) in blood of patients suffering from MS (Karlík et al. 2015). They found that concentration of plasma TAC exhibited a significantly decrease in MS patients when compared with controls group, while plasma FRA remained unchanged.

7.1.2.5 Amyotrophic Lateral Sclerosis Biomarkers

Chromogranins, namely Chromogranins A (CgA) and Chromogranins B (CgB) are soluble acidic glycophosphoproteins that found in large dense core vesicles of neurons and endocrine cells containing neuropeptides and hormones respectively (Ezzi et al. 2010). Chromogranin A (CgA) has been reported to have several physiological functions, and a major role as modulator of the neuroendocrine system (Taupenot 2003). It was shown that CgA interacts with mutant forms of superoxide dismutase that are linked to Amyotrophic lateral sclerosis (ALS). Its CSF level was shown increased in ALS patients compared to the controls (Kaiserova et al. 2017) and decreased in PD patients (O'Connor et al. 1993).

7.1.2.6 Autism Spectrum Disorder Biomarkers

Most of MicroRNAs (miRNA) expression profiles found in Autism spectrum disorder (ASD) are involved in nervous system development and function. The miR-132 is induced by endotoxins (Lagos et al. 2010; Shaked et al. 2009; Taganov et al. 2006), and affected by cytomegalovirus infection (Wang et al. 2008), that contribute to the pathogenesis of autism (Onore et al. 2012; Stubbs et al. 1984). A large number of miRNA expression profiles known to be altered in schizophrenia, showed changes in ASD samples (Abu-Elneel et al. 2008; Ghahramani Seno et al. 2011; Sarachana et al. 2010; Talebizadeh et al. 2008). The miR-195 and miR-128b are known to target brain-derived neurotrophic factor (BDNF) and inhibit cAMP-response-element-binding protein(CREB), respectively (Lin et al. 2011; Mellios et al. 2009), while miR-132 and miR-212 are induced by BDNF and positively regulate CREB signaling (Hollander et al. 2010; Remenyi et al. 2010). Oxytocin (OT) is a neuropeptide that play a key role in the brain as well as in social behavior. OT was suggested to contribute to the development of the repetitive behaviors found in ASD patients and their social deficit (Guastella et al. 2010). Furthermore, some reports suggested that its intervention as a treatment contribute to attenuate social impairment symptoms (Guastella et al. 2010; Hollander et al. 2003, 2007; Yrigollen et al. 2008).

7.1.2.7 Neuropsychiatric's Biomarkers

Four genes have been identified involved in schizophrenia (SZ), bipolar disorder (BD) and/or attention deficit hyperactivity disorder (ADHD), with epigenetic implication (Craddock 2005; López-Figueroa et al. 2004). Dystrobrevin binding protein 1 (DTNBP1), has been first reported by Straub et al. (2002) as a susceptibility gene in SZ. Raybould et al. have reported a modestly significant evidence for the involvement of this gene in a group of bipolar patients with predominantly psychotic episodes of mood disturbance with a similar pattern of findings found in this group in SZ (Green et al. 2005). The catechol-O-methyltransferase (COMT), one of the most studied genes in SZ, degrades catecholamines including dopamine, epinephrine, and norepinephrine in the synapse. In addition, it lies at 22q11, a region implicated in both SZ and BD (Craddock 2005). The gene encoding the type 2A serotonin receptor, 5-hydroxytryptamine receptor 2A (HTR2A), contains a number of polymorphous regions associating with SZ (Alfimova et al. 2010). In fact, genetic variation in DTNBP1, COMT and HTR2A affects memory, hallucinations, glutamate signalling, and symptom severity in both schizophrenic and bipolar patients (Alfimova et al. 2010; Cheah et al. 2015; Goghari and Sponheim 2008; Karlsgodt et al. 2011; Walton et al. 2014; Wright et al. 2012). The vasoactive intestinal peptide receptor 2 (VIPR2) gene is known to have a role in neuronal function and neuropsychiatric diseases such as ADHD (Peter et al. 2016). Hamza et al. have shown a decrease level of C-phosphate-G (CpG) methylation in VIPR2 in ADHD patients compared to healthy control (Hamza et al. 2017). However, further researches are warranted to study the function of this gene in ADHD.

7.1.3 Salivary Biomarkers

Since CSF and blood analysis is not always practical measure and is invasive, saliva has received special attention since this body fluid is readily accessible and easily obtained. Several salivary biomarkers have been identified indifferent ND (Table 7.2).

7.1.3.1 Alzheimer and Parkinson Salivary Biomarkers

Aβ deposits are found in brain and peripheral regions, including skin, nasal mucosa, and the lacrimal and lingual glands (Attems et al. 2005; Frederikse et al. 1996; Fukuchi et al. 2000; Gasparini et al. 1998). Due to the expression of both APP and Aβ in human salivary epithelial cells, human salivary gland biopsies have been recently used as a tool for research on AD (Oh and Turner 2006). Using ELISA, Bermejo Pareja et al. compared levels of Aβ42 between AD patients and two group of controls: healthy controls and patients with PD (Bermejo-Pareja et al. 2010).

They found that Aβ42 levels increased in patients with mild and moderate AD, while their levels in severe AD remained comparable to healthy controls. Moreover, it was suggested by Farah et al. that salivary Aβ42 may be more insightful of familial AD genotype rather than sporadic AD (Farah et al. 2018). Shi et al. were the only who examined p-tau as a salivary biomarker for AD using Mass Spectrometry analysis (Shi et al. 2011). They also suggested that salivary tau proteins could be excellent biomarkers for AD diagnosis, in particular for asymptomatic patients or in early stages of the disease.

Acetylcholinesterase (AChE) is the enzyme involved in the hydrolysis of the neurotransmitter Acetylcholine in the CNS of mammals. It was demonstrated that AChE accumulates inamyloid plaques and neurofibrillary tangles of AD brains. Moreover, AChE was implicated in Aβ pathology and enhancing neurotoxicity (García-Ayllón 2011; Inestrosa et al. 1996). It has been suggested to be a useful salivary biomarker of central cholinergic activity which is a key event in AD (Sayer et al. 2004). The AChE salivary activity was reported firstly by Sayer et al. using the Ellman colorimetric method (Sayer et al. 2004). Assessment of this activity showed a decrease with age among healthy subjects, and a significant decrease in AD patients who did not respond to AChE inhibitors (AChE-I) treatment. In contrast, patients with AD who respond to AChE-I treatment did not show any variation compared to controls (Sayer et al. 2004). Recent study showed that AChE-I was effective to alleviate some PD symptoms (Moreau et al. 2016). Interestingly, Fedorova et al. revealed an increase in enzymatic activity of salivary AChE in PD patients when compared to controls (Fedorova et al. 2015). They also calculated the ratio AChE activity/Total protein concentration in the saliva of PD patient and showed that this increased catalytic activity of salivary AChE can be related to an alteration in enzymatic function and not enzymatic concentration.

Recently, lactoferrin was presented by Carro et al. as one of the first salivary biomarkers for AD early detection and diagnosis (Carro et al. 2017). Lactoferrin is an antimicrobial peptide characterized by a known Aβ-binding ability and contributes to the modulation of immune reactions and inflammation (Beljaars et al. 2004; Berlutti et al. 2006; Valenti et al. 2011; van der Strate et al. 2001; Wang et al. 2010). The study conducted by Carro et al. showed that salivary lactoferrin concentration has been found significantly reduced in AD patients compared to PD patients and healthy controls. Moreover, controls who presented with reduced salivary lactoferrin concentrations comparable to AD associated concentrations developed either mild cognitive impairment or AD. Whereas, controls who presented with normal or high lactoferrin salivary concentrations did not develop any form of cognitive impairment (Carro et al. 2017).

Since α-Syn and DJ-1 cannot be detected in a blood test (Shi et al. 2010), these biomarkers have also been identified in human saliva (Bermejo-Pareja et al. 2010; Shi et al. 2010, 2011). Using western blotting and mass spectrometry, Devic et al. detected the presence of α-Syn in human saliva and measured differences in α-Syn and DJ-1 levels between PD patients and controls (Devic et al. 2011). No statistical significance has been obtained in these differences. Concentrations of α-Syn and

DJ-1 significantly decrease and increase, respectively, in the saliva of PD patients when compared to controls. Another study performed by Vivacqua et al., measured in saliva the concentrations of α-Syn oligomer (α-Syn_{olig}) and totalα-Syn_{total} (Vivacqua et al. 2016). α-Syn_{olig} and α-Syn_{olig}/α-Syn_{total} ratio showed a significant increase in saliva of PD patients as compared to healthy controls. A significant decrease was detected in α-Syn_{total} in saliva of PD patients when compared to healthy controls. Farah et al. suggested that α-Syn_{total} is an important marker of PD patients, since its salivary levels showed correlations with disease severity, progression, stages and cognitive impairments (Farah et al. 2018). Using Luminex assay, Kang et al. identified a significant increase in DJ-1 concentration in saliva of PD patients classified as stage 4 compared to those who were classified as stages 1–3 (Kang et al. 2014). They reported that DJ-1 concentration showed a significant decrease in saliva of patients with mixed type PD when compared to tremor dominant and akinetic-rigid dominant type. Recent study using quantitative western blotting showed a significant increase in DJ-1 salivary concentration and a positive correlation between its concentration and clinical test scores for the UPDRS motor test (Masters et al. 2015). But, these findings are in disagreement with those reported by Devic et al. and Kang et al. who showed that DJ-1 levels did not correlate with UPDRS motor score (Devic et al. 2011; Kang et al. 2014).

7.1.3.2 Huntington's Salivary Biomarkers

Similar to these previous salivary biomarkers, Htt protein, the protein implicated in HD, is detected in saliva and has been proposed for the diagnosis of this disease. Bloom et al. detected a significant increase in salivary concentration of total Htt protein in HD patients when compared to controls (Corey-Bloom et al. 2016). They also reported a significant increase in mHtt concentration in saliva of pre-symptomatic HD patients compared to healthy controls. Knowing that non-invasive methods to measure Htt in the CNS do not exist, and CSF and blood Htt protein concentrations are low, salivary Htt could be used as biomarker for early detection of HD.

7.1.3.3 Multiple Sclerosis Salivary Biomarkers

Looking for easily accessible early diagnostic markers indicative of MS, Adamashvili et al. measured, by ELISA, the concentration of soluble HLA in saliva of patients with MS (Adamashvili et al. 2005). They found a significant increase in soluble HLA class II in the saliva of patients with Relapsing Remitting MS (RRMS) compared to healthy controls and a low concentration of salivary soluble HLA class I. Using the same technique, Minagar et al. found that salivary soluble HLA class II was elevated in RRMS patients when compared to healthy subjects (Minagar et al. 2007). However, the concentration of salivary soluble HLA class I was undetectable. In addition, Minagar et al. demonstrated that the concentration of salivary soluble HLA class II showed an increase in response to interferon β1-a treatment (Minagar et al. 2007). In the other hand, Karlik et al. assessed four oxidative stress markers; AOPP, TBARS, AGEs and fructosamine, in saliva samples obtained from patients suffering from MS (Karlík et al. 2015). Consequently, they found a significant increase in salivary TBARS and AGEs of MS patients as compared to healthy controls, whereas AOPP remained unchanged. Karlik et al. also evaluated TAC and FRA in MS patient saliva samples, and they reported lower salivary FRA levels in MS patient compared to healthy subjects. TAC levels were lower without reaching significance. They explained this difference by the fact that FRAS measures the antioxidative effects of non-protein molecules, while TAC is affected by multiple variables since it measures the total antioxidative capacity of the fluid.

7.1.3.4 Amyotrophic Lateral Sclerosis Salivary Biomarkers

Salivary CgA might be a useful marker to reflective of ALS disease severity and the affective state of patients. This has been demonstrated by Obayashi et al. (2008) that measured the levels of salivary CgA in patients with ALS using a YK070 chromogranin A EIA kit. Interestingly, they found a significant increase in salivary CgA concentrations in terminal ALS patients as compared to healthy controls, moderate ALS patients and patients with vascular dementia. On the other hand, the authors reported a positive correlation between the CgA concentration and the E1 Escorial score of emotional function compared to that of physical mobility, communication and alimentation.

7.1.3.5 Autism Spectrum Disorder Salivary Biomarkers

In the search for diagnostic salivary biomarkers for ASD, Hicks et al. identified 14 differentially expressed miRNAs in saliva samples of ASD patients compared to healthy subjects (Hicks et al. 2016). Four out of the 14 miRNAs were down-regulated in saliva of ASD patients (miR-23a-3p, miR-27a-3p, miR-30e-5p and miR-32-5p), while 10 were up-regulated (miR-140-3p, miR2467-5p, miR-218-5p, miR- 28-5p, miR-335-3p, miR-628-5p, miR-7-5p, miR-191-5p, miR-127-3p and miR-3529-3p). Except for miR-140-3p that did not show significant correlations, all miRNA levels exhibited significant correlations with Vineland neurodevelopmental scores. In a pilot study performed by Ngounou Wetie et al. the salivary proteome of ASD patients and healthy controls has been analyzed using Mass Spectrometry-based proteomics (Ngounou Wetie et al. 2015). They identified 12 salivary protein biomarkers (Lactotransferrin (LTF), Prolactin-inducible protein (PIP), Annexin A1 (ANXA1),Neutrophil-defensin 1(DEFA1/DEFA1B), Ig kappa chain C region (IGKC), Ig gamma-1 chain C region(IGHG1), Ig lambda-2 chain C regions, Neutrophil elastase (ELANE), Lactoperoxidase (LPO),Polymeric immunoglobulin receptor(PIGR), Deleted in malignant brain tumors 1 (DMBT1) and Myeloperoxidase

(MPO)) with elevated concentrations and 4 (Submaxillary gland androgen-regulated protein 3B (SMR3B), Salivary acidic proline rich phosphoprotein (PRH1/2), Statherin (STATH) and Histatin (HTN1)) with reduced concentrations in ASD patients compared to healthy subjects. Moreover, all these proteins are implicated in immune reactions, which validate that the deregulation of the immune system may be etiological or consequential to ASD (Estes and McAllister 2015; Gottfried et al. 2015; Pardo et al. 2005). Farah et al. have reported these proteins with elevated and reduced concentrations in saliva of ASD patients along with their physiological functions (Farah et al. 2018). Castagnola et al. have demonstrated hypo-phosphorylation of four of these salivary proteins (Statherin, Histatin and Proline rich proteins 1 and 3) in ASD patients compared to healthy subjects (Castagnola et al. 2008). Thus, they considered hypo-phosphorylation as an explanation for failed protein mechanisms which could explain the essence of protein-protein interaction deficits. These observations can lead to conclude that biochemical, physiological and molecular changes in salivary proteins could affect their functions in ASD. Several researchers have measured the concentrations of the OT neuropeptide in saliva of patients with ASD and healthy subjects (Feldman et al. 2014; Fujisawa et al. 2014). Findings showed that salivary OT levels in ASD patients were significantly lower than those of healthy controls. However, a meta-analysis conducted by Grazia et al. reported no significant differences in salivary OT concentrations between ASD patients and healthy individuals (Rutigliano et al. 2016).

7.1.3.6 Salivary Neuropsychiatric's Biomarkers

Research of methylation profiles of salivary DNA inpatients with SZ, BD and ADHD has shown the involvement of the 4genes mentioned above (HTR2A, DTNBP1, MB-COMT and VIPR2) (Abdolmaleky et al. 2015; Ghadirivasfi et al. 2011; Nohesara et al. 2011; Wilmot et al. 2016). In fact, Abdolmaleky et al. reported lower levels of DTNBP1 promoter methylation in the saliva of schizophrenic and bipolar patients who underwent and responded to treatment compared to those who were untreated (Abdolmaleky et al. 2015). Furthermore, they found an association between elevated levels of DNA methylation and early disease onset. These findings suggest that DTNBP1 may contribute to the disease prediction and its level of gene methylation may indicate the treatment responsiveness for these patients. Regarding the HTR2A, its hypomethylation was detected in first degree relatives of patients with SZ and BD which suggest the heritability of the epigenetic alterations (Ghadirivasfi et al. 2011). In general, the salivary methylation status of DTNBP1, HTR2A and MB-COMT was reported to reflect the brain's methylation status for these genes. Similar to these 3 genes, Wilmot et al. have performed a genome wide DNA methylation analysis for patients with ADHD and they detected VIPR2 CpG methylation in their saliva (Wilmot et al. 2016). On the other hand, a salivary proteome analysis of patients with SZ and BD has revealed the presence of 8 salivary proteins and peptides (α-defensins 1–4, S100A12, cystatin A and the S-derivatives of cystatin B (S-glutathionylated and S-cysteinylated)) significantly elevated in these patients when compared to healthy controls (Iavarone et al. 2014). All these proteins and peptides are involved in the innate immunity which suggest an immunologic imbalance in patients with SZ and BD (Ganz 2003; Magister and Kos 2013; Pietzsch and Hoppmann 2009).

7.2 In Animals

7.2.1 Short Description of the Disease and Importance

Nervous system (NS) is the most complex part of the body not only for humans but also for animals. This last, depends on the NS for survival, reproduction and production of food and fiber (for livestock). Therefore, a neurological disease in animals can make great damage. These affections include diseases of the brain, spinal cord or the peripheral and muscular nervous system which cause various clinical signs and which can correspond to various diseases.

In animals, it is difficult to explain all the symptoms produced by the nervous system diseases. CNS disorders have a wide variety of clinical signs varies with the location and severity of lesions. The common point of these affections is the alteration of behavior. It is therefore important to know the normal behavior of animals. Unusual behavior is suspect, especially if it extends over time. These alterations can range from a subtle change of expression or attitude to obvious disorders such as aggression, a change of voice, drowsiness or the irrepressible need to consume or drink unusual substances, absolute blindness, moving in a circle, abnormal gait, with exaggerated elevation members for example, support the head against objects, abnormal movements of the head, in coordination and seizures.

CNS diseases are likely to cause spasms, often triggered, as in the case of tetanus, by brutal nerve stimulation such as a sudden loud noise. In extreme cases, CNS disease can lead to loss of control of skeletal muscles, leading to an inability to move or stand (paralysis). Rabies is a fatal disease responsible for a serious inflammation of the brain (encephalitis). Aggressiveness or paralyses are well-known signs. However, the first signs may be limited to a slight change in behavior or facial expression. Encephalomyelitis is diseases that cause inflammation of the brain and spinal cord (Hunter 1996). It has been proven that some congenital neurological diseases can lead to disability or death of the animal. Some of these, such as idiopathic epilepsy and hydrocephalus, have racial or familial predispositions (Passantino and Masucci 2016). Dystocia is very common in cattle. Difficult cases of calving paralysis may present a variable pathology caused by pelvic fractures or muscular contusions causing paralysis of the obturator and other nerves. The motor functions are performed in the disorder that blocks this neurotransmission, as in the ascending placid paralysis of Ixodes tick poisoning (Finnie et al. 2011).

The investigation of any disease suspected of involving the nervous system requires a neurological examination to determine the cause and therefore find the treatment. The following table lists the most common diseases in domestic animals and food and fiber producing animals (Table 7.3):

Table 7.3 Most common neurological diseases in domestic animals, food and fiber producing animals

Disease	Animal	References
Babesia (Babesia bovis)	Cattle	Adjou et al. (2012) and Hunter (2006)
Sodium and copper deficiency		
Theileriosis		
Paresis, Ataxia, Dysmetria		
Nervous form of coccidiosis		
Hypovitaminosis A		
Aujesky's disease		
Bovine herpesvirus meningoencephalitis (BoHV1 and BoHV5)		
Gangrene coryza		
Rare viral encephalitis (European Sporadic Bovine Encephalitis (ESBE))		
Nerve ketosis		
Lead poisoning, salt and ammonia (NH3)		
Poisoning by plants and tremorine toxins		
Cerebrospinal encephalitis	Cattle, sheep and goats	Brugère-Picoux (2004)
Necrosis of the cerebral cortex (NCC)		
Coenurus cerebralis		
Meningitis and brain abscess		
Heartwater		
Acaricide poisoning (chlorinated hydrocarbons, etc.)	Cattle, sheep, goats, equines and pigs	Peter Constable et al. (2016)
Organophosphorus poisoning	Cattle, equines and pigs	
Hypomagnesemia and hypocalcemia		
Rabies	Cattle, sheep, goats, equines, pigs and dogs	
Poisoning with fern large eagle	Sheep and goats	Brugère-Picoux (2004)
Pregnancy toxemia		
Scrapie		
Visna		
Dourine	Equines	Hennequin et al. (1998)
Amitraz poisoning		
Castor seed cake poisoning		
Equine meningoencephalomyelitis		
Tetanus		
Trypanosomiasis		

(continued)

Table 7.3 (continued)

Disease	Animal	References
Talfan disease (porcine encephalomyelitis)	Pigs	Hunter (2006)
Teschen's disease		
African swine fever		
Classical swine fever		
Ticks	Dogs and cats	Passantino and Masucci (2016)
Distemper Leptospirosis		
Typhus		
Prion diseases, or Transmissible Spongiform Encephalopathies (TSEs)	Sheep (Scrapie), cattle (bovine spongiform encephalopathy) and mule, deer, elk (Chronic wasting disease)	Wang et al. (2014)

7.2.2 Currently Used Biomarkers in Veterinary Medicine

The role of biomarkers is not limited to understanding the cause, diagnosis and severity of nervous system disorders, but also has implications for the interpretation of treatment outcomes. Frequently used biomarkers for information about the nervous system are blood, brain, CSF, muscles, nerves, skin, and urine (Mayeux 2004). The sampling of these biomarkers is not identical, there are some that are easy to collect as are blood and urine. On the other hand, tissue samples from the CNS and CSF are more difficult and can lead to a significant risk of morbidity (Nishida 2014).

There are several types of biomarkers of nervous system disorder namely: pigments, metabolic biomarkers, cell-specific biomarkers, CNS-specific biomarkers, free radicals, inflammatory biomarkers, and immunological biomarkers (Divya et al. 2018).

The cell-specific biomarkers are the glial fibrillary acidic protein, the phosphorylated axonal form of the NFH neurofilament, and the Myelin core protein (Oji et al. 2007; Weiss et al. 2009). While the CNS-specific biomarkers are S100β, neuron-specific enolase (NSE), Tau protein, spectrin Alpha II, ubiquitin carboxyterminal hydrolase L1, and creatine kinase BB (Płonek et al. 2016) (Table 7.4) .

7.2.2.1 Central Nervous System Specific Biomarkers

Neuron-Specific Enolase

Neuron-specific enolase (NSE) was initially found in the cytoplasm of neurons, but is also localized in neuroendocrine cells, oligodendrocytes, thrombocytes, and erythrocytes. The biological half-life of NSE in serum is 24 h (Marangos and Schmechel 1987). NSE has been used as a marker for brain damage. It is released by damaged neurons because of ischemia, multiple sclerosis, spinal cord injury and

Table 7.4 Biomarkers frequently used for animal neurological diseases

Biomarker	CSF/ plasma	Diagnosis and/or progression	References
Tau	CSF	Disease progression	Süssmuth et al. (2003)
S100b	CSF	Disease progression	Süssmuth et al. (2003)
Neurofilament heavy chain	CSF	Diagnosis and progression	Brettschneider et al. (2006)
Glial fibrillary acidic protein	CSF	Diagnosis	Grundström et al. (2006)
Myelin basic protein	CSF	Diagnosis	Lamers et al. (2003)
Neuron-specific enolase	CSF	Diagnosis	Lamers et al. (2003)
Matrix metalloprotease-9	Plasma	Diagnosis	Beuche et al. (2000)
α-II-spectrin	Plasma	Diagnosis and progression	Jain et al. (2014)
Ubiquitin carboxy terminal hydrolase L1	CSF, Plasma	Diagnosis	Sekiguchi et al. (2006)
Creatine kinase	Plasma	Disease progression	Süssmuth et al. (2003)
MMPs	CSF, Plasma	Diagnosis and progression	Ayvazova et al. (2016)

brain trauma in rodents (Kari et al. 2006). Therefore, an increase of NSE in CSF could be useful for detecting neuronal damage in the CNS. Concentrations of NSE in CSF were increased in dogs with Gangliosidosis and in dogs with meningoencephalitis. However, there is not currently enough evidence to use as a diagnostic test because it does not differentiate the various causes of neuronal damage (Pratamastuti et al. 2017).

S100β

S100b is a neurotrophic protein mainly located in glial cells of the CNS and Schwann cells in the peripheral nervous system (PNS) (Donato 1986). It is known to promote the growth of neurites and promote the activation of astrocytes (Winningham-Major et al. 1989). These in vitro and in vivo neurotrophic effects prompted researchers to analyze its effect on axon-mediated motor spinal neuron death in the neonatal rat (Iwasaki et al. 1997).

An increased level of S100B is associated with pathological injury or clinical severity of various CNS disorders including head trauma or neurodegenerative diseases. Increases in S100B levels in serum or CSF are a sensitive measure for determining CNS lesions at the molecular level before significant changes occur, allowing for crucial medical intervention before irreversible damage occurs (Yardan et al. 2011).

Tau Protein

Tau protein is part of the family of microtubule-associated proteins (MT), which stabilizes and regulates the assembly of TM. Tau has been given special attention because of its links with neurodegenerative diseases (Joachim et al. 1987; Bancher et al. 1993). It is found primarily in neuronal cells and was originally shown to be a regulator of tubulin assembly in vivo (Drubin 1986). Tau serves to maintain the cytoskeletal network integrity and, through interaction with the motor proteins dynein and kinesin, tau is involved in axonal transport along this network (De Vos et al. 2008; Wolfe 2012).

The c-Tau protein (cleaved Tau) can be detected in serum and CSF after neuronal injury. High levels of c-tau protein in CSF have been detected in dogs with Intervertebral disc herniation (IVDH) or dogs with paresis (Dixit et al. 2008; Lei et al. 2010).

Alpha II Spectrin

The α-II-spectrin is a major structural component of the cytoskeleton of the cortical membrane abundant in axons and presynaptic terminals (Goodman et al. 1995; Riederer et al. 1986). The α-II-spectrin is a major substrate for cysteine proteases involved in the necrotic cell death (calpain) and apoptosis (caspase-3) (Wang et al. 1998). Although α-II-spectrin is not specific for the CNS, it is present at high levels in neurons, in addition, its presence in glial cells. These properties make α-II-SBDP attractive candidates for biomarkers despite the relative lack of cerebral specificity. These unique properties highlight the potential importance of α-II-SBDP for brain injury monitoring from a variety of etiologies, including those occurring from Basic calcium phosphate (BCP) and Hepatocellular adenoma (HCA) (Weiss et al. 2009).

Ubiquitin Carboxy Terminal Hydrolase L1

Ubiquitin C-terminal hydrolase L1 (UCH-L1) is a highly abundant protein in the brain, it is estimated that it constitutes 1–5% of the total neuronal protein. UCH-L1 is not essential for neuronal development, but it is necessary to maintain axonal integrity and its dysfunction is implicated in neurodegenerative diseases (Bishop et al. 2016). The UCHL1 can play an important role in the repair of axons and neurons after injury by eliminating abnormal proteins through ubiquitin-proteasome (UPP) and autophagy (Kowalski and Juo 2012). It also regulates synaptic function and long-term potentiation (LTP) and may be involved in the memory function (Liu et al. 2019). In addition, UCHL1 activity protects primary neurons from hypoxia-induced cell death (Liu et al. 2019). A different spontaneous mutation in UCH-L1 in mice leads to gritty axonal dystrophy (gad), characterized by an accumulation of ubiquitinated protein aggregates in neurons, axonal degeneration in the spinal tract and late progressive ataxia (Kowalski and Juo 2012; Chen et al. 2010). Remarkably, this neuronal protein can be easily detected in the CSF and in the blood soon after injury, providing a valuable time window for potential neuroprotective strategies (Mondello et al. 2012).

Creatine Kinase BB

Creatine kinase (CK, EC 2.7.3.2) is an intracellular enzyme found mainly in high-energy metabolism tissues such as muscles, heart and brain. Each of these organs contains a specific CK isoenzyme: specifically, CK-BB is present in brain cells (Dawson and Fine 1967). A small amount of CK-BB can also be produced in the intestine and spleen (Paltrinieri et al. 2017). CK-BB leakage from damaged brain cells results in increased activity in the serum (Kramer and Hoffmann 1997). Another factor to consider is that CK-BB has a half-life of 1–5 h and some of the enzymes are inactivated in and around damaged cells. Indeed, only a part is released in the CSF, where the activity can be further reduced by dilution, thermal inactivation, diffusion and light (Ferreira 2016). An increase in serum CK-BB activity has been reported in dogs with CNS disorders (Paltrinieri et al. 2017). However, no studies on the possible differences in serum activity of CK iso- or macro-enzymes (Macro-CK1 and Macro-CK2) in different neurological diseases are available (Paltrinieri et al. 2017).

7.2.2.2 Cell-Specific Biomarkers

Myelin Basic Protein

Basic myelin protein (BMP) is produced by oligodendrocytes and is a major constituent of the axonal sheath of myelin. Myelin is essential for the saltatory propagation of nerve impulses in the peripheral nervous system. Persistent dysfunctions of peripheral myelin possibly result from axonal degeneration and a sensory loss in diseases such as hereditary demyelinating neuropathies like Charcot-Marie-Tooth disease (CMT) (Warner et al. 1999; Houlden and Reilly 2006). BMP has been detected in the CSF of dogs with white matter lesions (Whitaker 2016). This marker could be used to define important differences in severity of this disease, particularly in the thoraco-lumbar region. MBP concentration in CSF was higher in dogs with thoraco-lumbar intervertebral discs.

Experimental allergic encephalomyelitis (EAE) is an autoimmune disease (form of multiple sclerosis in humans) mediated by T lymphocytes (Wucherpfennig et al. 1994). It begins with inflammation and is followed by demyelination of the CNS. It thus causes in the animal a chronic disease with cycles of relapse (Zamvil and Steinman 1990). The increase in the level of myelin PPL (PLP1 accounts for about half of myelin protein and is the most abundant component of myelin) is known to be one of the main causes of this disease (Bouchard 1999).

Glial Fibrillary Acid Protein

Glial fibrillary acidic protein (GFAP) is a monomeric intermediate filament protein found in astrocytes that forms an important cellular component in the maintenance of the blood brain barrier (BBB). Serum GFAP has been studied for the potential diagnosis of progressive myelomalacia in dogs. In addition, dogs with necrotizing meningoencephalitis (NME) have been shown to have higher GFAP levels than naive controls and those with various diseases of the central nervous system, including idiopathic epilepsy, lymphoma malignant, meningioma and glioma (Salouci and Gabriel 2016).

Phosphorylated Neurofilament Heavy Chain

Neurofilaments are components of the axonal cytoskeleton and include: heavy chain (NF-H 190–210 kDa), medium chain (160 kDa), light chain (NF-L 68 kDa) and α-internexin neurofilaments (66 kDa) (Mrozek and Geeraerts 2012). NF-H is one of the most abundant protein components of neurons; its function is to maintain the axons structural integrity. This function is restricted to the intracellular compartment. As major components of the cytoskeleton, NFs are good candidates for the function of biomarkers, especially for the quantification of axonal loss. Rupture of the axonal membrane results in their release into the CSF (Petzold 2005). NFs were thus, demonstrated in numerous acute (Cerebrovascular Vascular: Stroke …) or chronic (ALS) neurologic pathologies (Brettschneider et al. 2006). Numerous studies in experimental allergic encephalomyelitis (EAE) were performed. They show a correlation between NF-H levels present in the CSF and disability. This seems to be a predictor of poor prognosis (Petzold 2005). Miyazawa et al. (2007) report a correlation between CSF concentrations of NF-H and prognosis in dogs with paraplegia with profound absence of pain perception. Plasma pNF-H values in CSF elevated were significantly elevated compared with intact controls.

Matrix Metalloproteinases

Matrix metalloproteinases (MMPs) are endopeptidase-dependent, termed metzincins, that are responsible for remodeling the extracellular matrix (ECM) and maintaining tissue homeostasis in developing and adult brains (Mukherjee and Swarnakar 2015). MMPs play a role in acute neuroinflammatory lesions as well as in chronic neurodegeneration. The critical function of individual MMPs in tissue repair is also very important. MMPs perform important functions in CNS during growth and development (Mukherjee and Swarnakar 2015). Since the CNS encounters most MMPs only in small amounts under normal conditions, the simultaneous and significant increase in the number of MMPs may result in widespread proteolysis with undue consequences. In contrast, the loss of this control may be associated with proliferation or cell death processes and tissues remodeling that characterize the malignant and degenerative diseases of many non-nervous tissues (Rosenberg 2015).

The control of MMP activity by their physiological inhibitors, tissue inhibitors of metalloproteinases (TIMPs), contributes to tissue homeostasis. The expression of MMPs and its TIMPs in the nervous system of the healthy adult animal is usually neuronal. However, during various diseases of the nervous system, reactive neuroglia, BBB cells, and infiltrating cells of the immune system become an important source of these proteins. This spatiotemporal modulation of MMP / TIMP system expression and the resulting net proteolytic activity have been implicated in excitotoxicity, neuronal death, permeabilization of the BBB, neuroinflammation and demyelination (Rivera et al. 2004). An increase in MMP-9 activity in CSF was detected in dogs with thoraco-lumbar IVDH (Yong et al. 2007).

7.2.3 *Salivary Biomarkers*

In animals, the diagnosis of ND remains a challenge as regards the methods of sampling. The invasive nature of these methods usually causes discomfort, pain and unpleasant side effects. The procedure of contention and blood collection may lead to a temporary stress causing biased measures (Redbo 1993; Negrão et al. 2004). Therefore, and in order to reduce animal discomfort during sampling, the research focuses on the development of non-invasive sampling devices, as well as media that do not immediately respond to the stress of sampling such as saliva (Schwinn et al. 2016).

Although many diseases have confirmed salivary biomarkers (Farah et al. 2018), diseases affecting the nervous system have few markers available in saliva, even rarer in animals. Literature data are limited in referring biomarkers of ND in animals. The stress is more accessible given the interest that is oriented towards the welfare of animals (domestic and companion animals) (Muneta et al. 2010). Apart from the stress associated with the collection of samples and which may interfere with the diagnosis results (Gutiérrez et al. 2014), prolonged stress conditions will likely result in ND and salivary proteins can be applied to identify such disease (Salekdeh 2016).

Salivary biomarkers used in stress assessment in animals, which will be presented in detail in the chapter of "salivary biomarkers of welfare/stress", include: cortisol (Hellhammer et al. 2009), α-amylase (Fuentes et al. 2011; Takai et al. 2004), lipase, adenosine deaminase (ADA) (Pikula et al. 1992) and chromogranin A in pigs, horses, sheep (Muneta et al. 2010) and the dog (Escribano et al. 2013).

7.3 Conclusions

For the purpose of diagnosis, treatment or monitoring of the ND progression, biomarkers can be obtained from body fluids such as blood, plasma, urine, sweat and saliva, or any other accessible material, such as hair and stool. The results of a large number of animal and human studies showed that the identification of these biomarkers could provide important information. However, the difficulty of identifying biomarkers responsible for the characterization of ND was one of the reasons to deal with easily accessible peripheral tissues such as saliva. Thus, the use of salivary biomarkers as a non-invasive tool of control has been increasingly required. Although salivary biomarkers appear to be promising for the diagnosis of ND, there are currently only a few biomarkers that can be quantified in saliva samples. Additionally, we think that salivary biomarkers, in their initial stages, will not be able to detect only neurological pathologies, but rather, they will serve to facilitate the diagnosis or replace other invasive tests.

References

Abdolmaleky HM, Pajouhanfar S, Faghankhani M, Joghataei MT, Mostafavi A, Thiagalingam S (2015) Antipsychotic drugs attenuate aberrant DNA methylation of *DTNBP1* (dysbindin) promoter in saliva and post-mortem brain of patients with schizophrenia and psychotic bipolar disorder. Am J Med Genet B Neuropsychiatr Genet 168:687–696. https://doi.org/10.1002/ajmg.b.32361

Abu-Elneel K, Liu T, Gazzaniga FS, Nishimura Y, Wall DP, Geschwind DH, Lao K, Kosik KS (2008) Heterogeneous dysregulation of microRNAs across the autism spectrum. Neurogenetics 9:153–161. https://doi.org/10.1007/s10048-008-0133-5

Adamashvili I, Minagar A, Gonzalez-Toledo E, Featherston L, Kelley RE (2005) Soluble HLA measurement in saliva and cerebrospinal fluid in Caucasian patients with multiple sclerosis: a preliminary study. J Neuroinflammation 2:13–20

Adjou KT, Comoy E, Deslys J-P, Grassi J, Ouidja MO, Brugere H, Brugere-Picoux J (2012) Méthodes de diagnostic des « maladies à prions » chez l'homme et chez l'animal. Bull Acad Vet Fr (1):429

Ahlgren C, Oden A, Lycke J (2011) High nationwide prevalence of multiple sclerosis in Sweden. Mult Scler J 17:901e8

Alfimova MV, Monakhov MV, Abramova LI, Golubev SA, Golimbet VE (2010) Polymorphism of serotonin receptor genes (5-HTR2A) and dysbindin (DTNBP1) and individual components of short-term verbal memory processes in schizophrenia. Neurosci Behav Physiol 40:934–940. https://doi.org/10.1007/s11055-010-9348-7

Amadoro G, Corsetti V, Ciotti MT, Florenzano F, Capsoni S, Amato G, Calissano P (2011) Endogenous Aβ causes cell death via early tau hyperphosphorylation. Neurobiol Aging 32:969–990. https://doi.org/10.1016/j.neurobiolaging.2009.06.005

Attems J, Lintner F, Jellinger KA (2005) Olfactory involvement in aging and Alzheimer's disease: an autopsy study. J Alzheimers Dis 7:149–157. https://doi.org/10.3233/JAD-2005-7208

Ayvazova NP, Rilcheva VS, Konova EI, Roussev RG, Rashev PI (2016) Original article. Matrix metalloproteinase Mmp-2 and Mmp-9 activities in seminal plasma. Journal of Biomedical and Clinical Research 9(2):114–120

Ballatore C, Lee VM-Y, Trojanowski JQ (2007) Tau-mediated neurodegeneration in Alzheimer's disease and related disorders. Nat Rev Neurosci 8:663–672. https://doi.org/10.1038/nrn2194

Bancher C, Braak H, Fischer P, Jellinger KA (1993) Neuropathological staging of Alzheimer lesions and intellectual status in Alzheimer's and Parkinson's disease patients. Neurosci Lett 162(1–2):179–182

Beljaars L, Vanderstrate B, Bakker H, Rekersmit C, Vanloenenweemaes A, Wiegmans F, Harmsen M, Molema G, Meijer D (2004) Inhibition of cytomegalovirus infection by lactoferrin in vitro and in vivo. Antivir Res 63:197–208. https://doi.org/10.1016/j.antiviral.2004.05.002

Berlutti F, Schippa S, Morea C, Sarli S, Perfetto B, Donnarumma G, Valenti P (2006) Lactoferrin downregulates pro-inflammatory cytokines upexpressed in intestinal epithelial cells infected with invasive or noninvasive *Escherichia coli* strains. This paper is one of a selection of papers published in this special issue, entitled 7th international conference on Lactoferrin: structure, function, and applications, and has undergone the Journal's usual peer review process. Biochem Cell Biol 84:351–357. https://doi.org/10.1139/o06-039

Bermejo-Pareja F, Antequera D, Vargas T, Molina JA, Carro E (2010) Saliva levels of Abeta1-42 as potential biomarker of Alzheimer's disease: a pilot study. BMC Neurol 10:108. https://doi.org/10.1186/1471-2377-10-108

Bertram L (2005) The genetic epidemiology of neurodegenerative disease. J Clin Invest 115:1449–1457. https://doi.org/10.1172/JCI24761

Betts CD, D'Mellow MT, Fowler CJ (1993) Urinary symptoms and the neurological features of bladder dysfunction in multiple sclerosis. J Neurol Neurosurg Psychiatry 56:245–250. https://doi.org/10.1136/jnnp.56.3.245

Beuche W, Yushchenko M, Mäder M, Maliszewska M, Felgenhauer K, Weber F (2000) Matrix metalloproteinase-9 is elevated in serum of patients with amyotrophic lateral sclerosis. Neuroreport 11(16):3419–3422

Bishop P, Rocca D, Henley JM (2016) Ubiquitin C-terminal hydrolase L1 (UCH-L1): structure, distribution and roles in brain function and dysfunction. Biochem J 473(16):2453–2462

Bloom GS (2014) Amyloid-β and Tau: the trigger and bullet in Alzheimer disease pathogenesis. JAMA Neurol 71:505–508. https://doi.org/10.1001/jamaneurol.2013.5847

Bouchard G (1999) Caractérisation De Lymphocytes T Anti-Myéline Et Coronavirus Humain Oc43 Dans La Sclérose En Plaques (masters). In: Université du Québec. Québec, Institut national de la recherche scientifique

Brettschneider J, Petzold A, Schöttle D, Claus A, Riepe M, Tumani H (2006) The Neurofilament heavy chain (NfH^{SMI35}) in the cerebrospinal fluid diagnosis of Alzheimer's disease. Dement Geriatr Cogn Disord 21(5–6):291–295

Brugère-Picoux J (2004) Maladies des moutons. France Agricole Editions

Buee L, Bussiere T, Buee-Scherrer V, Delacourte A, Hof PR (2000) Tau protein isoforms, phosphorylation and role in neurodegenerative disorders. Brain Res Rev 36(1):95–130

Carro E, Bartolomé F, Bermejo-Pareja F, Villarejo-Galende A, Molina JA, Ortiz P, Calero M, Rabano A, Cantero JL, Orive G (2017) Early diagnosis of mild cognitive impairment and Alzheimer's disease based on salivary lactoferrin. Alzheimers Dement Diagn Assess Dis Monit 8:131–138. https://doi.org/10.1016/j.dadm.2017.04.002

Castagnola M, Messana I, Inzitari R, Fanali C, Cabras T, Morelli A, Pecoraro AM, Neri G, Torrioli MG, Gurrieri F (2008) Hypo-phosphorylation of salivary peptidome as a clue to the molecular pathogenesis of autism spectrum disorders. J Proteome Res 7:5327–5332. https://doi.org/10.1021/pr8004088

Chartier-Harlin MC, Kachergus J, Roumier C, Mouroux V, Douay X, Lincoln S et al (2004) A-synuclein locus duplication as a cause of familial Parkinson's disease. Lancet 364:1167e9

Chaudhuri KR, Healy DG, Schapira AHV (2006) Non-motor symptoms of Parkinson's disease: diagnosis and management. Lancet Neurol 5(3):235–245

Cheah S-Y, Lawford BR, Young RM, Morris CP, Voisey J (2015) Dysbindin (DTNBP1) variants are associated with hallucinations in schizophrenia. Eur Psychiatry 30:486–491. https://doi.org/10.1016/j.eurpsy.2015.01.008

Chen F, Sugiura Y, Myers KG, Liu Y, Lin W (2010) Ubiquitin carboxyl-terminal hydrolase L1 is required for maintaining the structure and function of the neuromuscular junction. Proc Natl Acad Sci 107(4):1636–1641
Chiaravalloti ND, DeLuca J (2008) Cognitive impairment in. Mult Scler 7:13
Chwastiak L, Ehde DM, Gibbons LE, Sullivan M, Bowen JD, Kraft GH (2002) Depressive symptoms and severity of illness in multiple sclerosis: epidemiologic study of a large community sample. Am J Psychiatry 159:1862–1868. https://doi.org/10.1176/appi.ajp.159.11.1862
Compston A, Coles A (2008) Multiple sclerosis. Lancet 372:16
Corey-Bloom J, Aikin A, Garza M, Haque A, Park S, Herndon A, Thomas E (2016) Salivary Huntington protein as a peripheral biomarker for Huntington's disease (P1.053). Neurology 86:P1.053
Craddock N (2005) Genes for schizophrenia and bipolar disorder? Implications for psychiatric nosology. Schizophr Bull 32:9–16. https://doi.org/10.1093/schbul/sbj033
Dawson DM, Fine IH (1967) Creatine kinase in human tissues. Arch Neurol 16(2):175–180
Devic I, Hwang H, Edgar JS, Izutsu K, Presland R, Pan C, Goodlett DR, Wang Y, Armaly J, Tumas V, Zabetian CP, Leverenz JB, Shi M, Zhang J (2011) Salivary α-synuclein and DJ-1: potential biomarkers for Parkinson's disease. Brain 134:e178–e178. https://doi.org/10.1093/brain/awr015
De Vos KJ, Grierson AJ, Ackerley S, Miller CCJ (2008) Role of axonal transport in neurodegenerative diseases. Annu Rev Neurosci 31(1):151–173
Divya D, Deepty D, Rajashree M, Ashok D (2018) Review on use of biomarkers in small animal disease diagnosis
Dixit R, Ross JL, Goldman YE, Holzbaur ELF (2008) Differential regulation of dynein and Kinesin motor proteins by tau. Science 319(5866):1086–1089
Donato R (1986) S-100 proteins. Cell Calcium 7(3):123–145
Drubin DG (1986) Tau protein function in living cells. J Cell Biol 103:2739–2746. https://doi.org/10.1083/jcb.103.6.2739
Escribano D, Soler L, Gutiérrez AM, Martínez-Subiela S, Cerón JJ (2013) Measurement of chromogranin a in porcine saliva: validation of a time-resolved immunofluorometric assay and evaluation of its application as a marker of acute stress. Animal 7(4):640–647
Estes ML, McAllister AK (2015) Immune mediators in the brain and peripheral tissues in autism spectrum disorder. Nat Rev Neurosci 16:469–486. https://doi.org/10.1038/nrn3978
Ezzi SA, Larivière R, Urushitani M, Julien J-P (2010) Neuronal over-expression of chromogranin A accelerates disease onset in a mouse model of ALS: chromogranins and ALS pathogenesis. J Neurochem 115:1102–1111. https://doi.org/10.1111/j.1471-4159.2010.06979.x
Farah R, Haraty H, Salame Z, Fares Y, Ojcius DM, Said Sadier N (2018) Salivary biomarkers for the diagnosis and monitoring of neurological diseases. Biom J 41:63–87. https://doi.org/10.1016/j.bj.2018.03.004
Fedorova T, Knudsen CS, Mouridsen K, Nexo E, Borghammer P (2015) Salivary acetylcholinesterase activity is increased in Parkinson's disease: a potential marker of parasympathetic dysfunction. Park Dis 2015:1–7. https://doi.org/10.1155/2015/156479
Feldman R, Golan O, Hirschler-Guttenberg Y, Ostfeld-Etzion S, Zagoory-Sharon O (2014) Parent–child interaction and oxytocin production in pre-schoolers with autism spectrum disorder. Br J Psychiatry 205:107–112. https://doi.org/10.1192/bjp.bp.113.137513
Ferreira A (2016) Diagnostic value of creatine kinase activity in canine cerebrospinal fluid. Can Vet J 57:1081–1086
Finnie JW, Windsor PA, Kessell AE (2011) Neurological diseases of ruminant livestock in Australia. I: general neurological examination, necropsy procedures and neurological manifestations of systemic disease, trauma and neoplasia. Aust Vet J 89(7):243–246
Frederikse PH, Garland D, Zigler JS, Piatigorsky J (1996) Oxidative stress increases production of -amyloid precursor protein and -amyloid (A) in mammalian lenses, and a has toxic effects on lens epithelial cells. J Biol Chem 271:10169–10174. https://doi.org/10.1074/jbc.271.17.10169

Fuentes M, Tecles F, Gutiérrez A, Otal J, Martínez-Subiela S, Cerón JJ (2011) Validation of an automated method for salivary alpha-amylase measurements in pigs ( ) and its application as a stress biomarker. J Vet Diagn Investig 23(2):282–287

Fujisawa TX, Tanaka S, Saito DN, Kosaka H, Tomoda A (2014) Visual attention for social information and salivary oxytocin levels in preschool children with autism spectrum disorders: an eye-tracking study. Front Neurosci 8:295. https://doi.org/10.3389/fnins.2014.00295

Fukuchi K-I, Li L, Hart M, Lindsey JR (2000) Accumulation of amyloid-b protein in exocrine glands of transgenic mice overexpressing a carboxyl terminal portion of amyloid protein precursor. Int J Exp Pathol 81(4):231–239

Ganz T (2003) Defensins: antimicrobial peptides of innate immunity. Nat Rev Immunol 3:710–720. https://doi.org/10.1038/nri1180

García-Ayllón M-S (2011) Revisiting the role of acetylcholinesterase in Alzheimer's disease: cross-talk with P-tau and β-amyloid. Front Mol Neurosci 4:22. https://doi.org/10.3389/fnmol.2011.00022

Garcia-Osta A, Alberini CM (2009) Amyloid beta mediates memory formation. Learn Mem 16:267–272. https://doi.org/10.1101/lm.1310209

Gasparini L, Racchi M, Binetti G, Trabucchi M, Solerte SB, Alkon D, Etcheberrigaray R, Gibson G, Blass J, Paoletti R, Govoni S (1998) Peripheral markers in testing pathophysiological hypotheses and diagnosing Alzheimer's disease. FASEB J 12:17–34. https://doi.org/10.1096/fasebj.12.1.17

Ghabaee M, Jabedari B, Al-E-Eshagh N, Ghaffarpour M, Asadi F (2010) Serum and cerebrospinal fluid antioxidant activity and lipid peroxidation in Guillain–Barre syndrome and multiple sclerosis patients. Int J Neurosci 120:301–304. https://doi.org/10.3109/00207451003695690

Ghadirivasfi M, Nohesara S, Ahmadkhaniha HR, Eskandari MR, Mostafavi S, Thiagalingam S et al (2011) Hypomethylation of the serotonin receptor type-2A Gene (HTR2A) at T102C polymorphic site in DNA derived from the saliva of patients with schizophrenia and bipolar disorder. Am J Med Genet Part B: Neuropsychiatric Genet 156:536e45

Ghahramani Seno MM, Hu P, Gwadry FG, Pinto D, Marshall CR, Casallo G, Scherer SW (2011) Gene and miRNA expression profiles in autism spectrum disorders. Brain Res 1380:85–97. https://doi.org/10.1016/j.brainres.2010.09.046

Goghari VM, Sponheim SR (2008) Differential association of the COMT Val158Met polymorphism with clinical phenotypes in schizophrenia and bipolar disorder. Schizophr Res 103:186–191. https://doi.org/10.1016/j.schres.2008.05.015

Goodman SR, Zimmer WE, Blair Clark M, Zagon IS, Barker JE, Bloom ML (1995) Brain spectrin: of mice and men. Brain Res Bull 36(6):593–606

Gottfried C, Bambini-Junior V, Francis F, Riesgo R, Savino W (2015) The impact of neuroimmune alterations in autism spectrum disorder. Front Psychol 6:121. https://doi.org/10.3389/fpsyt.2015.00121

Green EK, Raybould R, Macgregor S, Gordon-Smith K, Heron J, Hyde S, Grozeva D, Hamshere M, Williams N, Owen MJ, O'Donovan MC, Jones L, Jones I, Kirov G, Craddock N (2005) Operation of the schizophrenia susceptibility gene, neuregulin 1, across traditional diagnostic boundaries to increase risk for bipolar disorder. Arch Gen Psychiatry 62:642–648. https://doi.org/10.1001/archpsyc.62.6.642

Grimm MOW, Grimm HS, Hartmann T (2007) Amyloid beta as a regulator of lipid homeostasis. Trends Mol Med 13:337–344. https://doi.org/10.1016/j.molmed.2007.06.004

Grundström E, Lindholm D, Johansson A, Blennow K, Askmark H (2006) GDNF but not BDNF is increased in cerebrospinal fluid in amyotrophic lateral sclerosis. Neuroreport 11(8):1781–1783

Guastella AJ, Einfeld SL, Gray KM, Rinehart NJ, Tonge BJ, Lambert TJ, Hickie IB (2010) Intranasal oxytocin improves emotion recognition for youth with autism spectrum disorders. Biol Psychiatry 67:692–694. https://doi.org/10.1016/j.biopsych.2009.09.020

Gutierrez A, Ceron J, Fuentes-Rubio M, Tecles F, Beeley J (2014) A proteomic approach to porcine saliva. Curr Protein Pept Sci 15:56–63. https://doi.org/10.2174/1389203715666140221115704

Hamza M, Halayem S, Bourgou S, Daoud M, Charfi F, Belhadj A (2017) Epigenetics and ADHD: toward an integrative approach of the disorder pathogenesis. J Atten Disord 23(7):655–664. https://doi.org/10.1177/1087054717696769

Hayward C, Swingler RJ, Simpson SA, Brock D (1996) A specific superoxide dismutase mutation is on the same genetic background in sporadic and familial cases of amyotrophic lateral sclerosis. Am J Hum Genet 59:1165e7

Hellhammer DH, Wüst S, Kudielka BM (2009) Salivary cortisol as a biomarker in stress research. Psychoneuroendocrinology 34(2):163–171

Hennequin C, Tredaniel J, Chevret S, Durdux C, Dray M, Manoux D, Perret M, Bonnaud G, Homasson J-P, Chotin G, Hirsch A, Maylin C (1998) Predictive factors for late toxicity after endobronchial brachytherapy: a multivariate analysis. Int J Radiat Oncol Biol Phys 42(1):21–27

Hicks SD, Ignacio C, Gentile K, Middleton FA (2016) Salivary miRNA profiles identify children with autism spectrum disorder, correlate with adaptive behavior, and implicate ASD candidate genes involved in neurodevelopment. BMC Pediatr 16:52. https://doi.org/10.1186/s12887-016-0586-x

Ho DH, Yi S, Seo H, Son I, Seol W (2014) Increased DJ-1 in urine exosome of Korean males with Parkinson's disease. Biomed Res Int 2014:1–8. https://doi.org/10.1155/2014/704678

Hollander E, Novotny S, Hanratty M, Yaffe R, DeCaria CM, Aronowitz BR, Mosovich S (2003) Oxytocin infusion reduces repetitive behaviors in adults with autistic and Asperger's disorders. Neuropsychopharmacology 28:193–198. https://doi.org/10.1038/sj.npp.1300021

Hollander E, Bartz J, Chaplin W, Phillips A, Sumner J, Soorya L, Anagnostou E, Wasserman S (2007) Oxytocin increases retention of social cognition in autism. Biol Psychiatry 61:498–503. https://doi.org/10.1016/j.biopsych.2006.05.030

Hollander JA, Im H-I, Amelio AL, Kocerha J, Bali P, Lu Q, Willoughby D, Wahlestedt C, Conkright MD, Kenny PJ (2010) Striatal microRNA controls cocaine intake through CREB signalling. Nature 466:197–202. https://doi.org/10.1038/nature09202

Houlden H, Reilly MM (2006) Molecular genetics of autosomal-dominant demyelinating Charcot-Marie-tooth disease. NeuroMolecular Med 8(1–2):43–62

Hunter A (1996) Animal health. In: General principles, vol Vol 1. Centre for Agricultural and Rural Cooperation, Technical

Hunter A (2006) La santé animale Volume 1. Généralités, Éditions Cirad. ed

Hunter MIS, Nlemadim BC, Davidson DLW (1985) Lipid peroxidation products and antioxidant proteins in plasma and cerebrospinal fluid from multiple sclerosis patients. Neurochem Res 10:1645–1652. https://doi.org/10.1007/BF00988606

Hyman BT, Augustinack JC, Ingelsson M (2005) Transcriptional and conformational changes of the tau molecule in Alzheimer's disease. Biochim Biophys Acta (BBA) - Mol Basis Dis 1739:150–157. https://doi.org/10.1016/j.bbadis.2004.06.015

Iavarone F, Melis M, Platania G, Cabras T, Manconi B, Petruzzelli R, Cordaro M, Siracusano A, Faa G, Messana I, Zanasi M, Castagnola M (2014) Characterization of salivary proteins of schizophrenic and bipolar disorder patients by top-down proteomics. J Proteome 103:15–22. https://doi.org/10.1016/j.jprot.2014.03.020

Inestrosa NC, Alvarez A, Perez CA, Moreno RD, Vicente M, Linker C et al (1996) Acetylcholinesterase accelerates assembly of amyloid-b-peptides into Alzheimer's fibrils: possible role of the peripheral site of the enzyme. Neuron 16:881e91

Ittner LM, Götz J (2011) Amyloid-β and tau – a toxic pas de deux in Alzheimer's disease. Nat Rev Neurosci 12:67–72. https://doi.org/10.1038/nrn2967

Iwai A, Masliah E, Yoshimoto M, Ge N, Flanagan L, De Silva HR (1995) et al. The precursor protein of non-Ab component of Alzheimer's disease amyloid is a presynaptic protein of the central nervous system. Neuron 14:467e75

Iwasaki Y, Shiojima T, Kinoshita M (1997) S100β prevents the death of motor neurons in newborn rats after sciatic nerve section. J Neurol Sci 151(1):7–12

Jain P, Spaeder MC, Donofrio MT, Sinha P, Jonas RA, Levy RJ (2014) Detection of alpha II-Spectrin breakdown products in the serum of neonates with congenital heart disease*. Pediatr Crit Care Med 15(3):229–235

Joachim CL, Morris JH, Kosik KS, Selkoe DJ (1987) Tau antisera recognize neurofibrillary tangles in a range of neurodegenerative disorders. Ann Neurol 22(4):514–520

Kaiserova M, Grambalova Z, Otruba P, Stejskal D, Prikrylova Vranova H, Mares J, Mensikova K, Kanovsky P (2017) Cerebrospinal fluid levels of chromogranin A and phosphorylated neurofilament heavy chain are elevated in amyotrophic lateral sclerosis. Acta Neurol Scand 136:360–364. https://doi.org/10.1111/ane.12735

Kang W-Y, Yang Q, Jiang X-F, Chen W, Zhang L-Y, Wang X-Y, Zhang L-N, Quinn TJ, Liu J, Chen S-D (2014) Salivary DJ-1 could be an indicator of Parkinson's disease progression. Front Aging Neurosci 6:102. https://doi.org/10.3389/fnagi.2014.00102

Kari K, Liu W, Gautama K, Mammen MP, Clemens JD, Nisalak A, Subrata K, Kim HK, Xu Z-Y (2006) A hospital-based surveillance for Japanese encephalitis in Bali, Indonesia. BMC Med 4(1)

Karlík M, Valkovič P, Hančinová V, Krížová L, Tóthová Ľ, Celec P (2015) Markers of oxidative stress in plasma and saliva in patients with multiple sclerosis. Clin Biochem 48:24–28. https://doi.org/10.1016/j.clinbiochem.2014.09.023

Karlsgodt KH, Robleto K, Trantham-Davidson H, Jairl C, Cannon TD, Lavin A, Jentsch JD (2011) Reduced dysbindin expression mediates N-methyl-D-aspartate receptor hypofunction and impaired working memory performance. Biol Psychiatry 69:28–34. https://doi.org/10.1016/j.biopsych.2010.09.012

Khan RS, Ahmed MR, Khalid B, Mahmood A, Hassan R (2018) Biomarker detection of neurological disorders through spectroscopy analysis. Int Dent Med J Adv Res 2015(4):1–9. https://doi.org/10.15713/ins.idmjar.86

Killoran A (2016) Biomarkers for Huntington's disease: a brief overview. J Rare Dis Res Treat 1:46–50. https://doi.org/10.29245/2572-9411/2016/2.1029

Kontush A, Berndt C, Weber W, Akopyan V, Arlt S, Schippling S et al (2001) Amyloid-b is an antioxidant for lipoproteins in cerebrospinal fluid and plasma. Free Radic Biol Med 30:119e28

Kowalski JR, Juo P (2012) The role of Deubiquitinating enzymes in synaptic function and nervous system diseases. Neural Plast 2012:1–13

Kramer JW, Hoffmann WE (1997) Chapter 12 - clinical enzymology. In: Kaneko JJ, Harvey JW, Bruss ML (eds) Clinical biochemistry of domestic animals, Fifth edn. Academic, San Diego, pp 303–325. https://doi.org/10.1016/B978-012396305-5/50013-0

Krupp LB, Alvarez LA, LaRocca NG, Scheinberg LC (1988) Fatigue in multiple sclerosis. Arch Neurol 45:435e7

Lagos D, Pollara G, Henderson S, Gratrix F, Fabani M, Milne RSB, Gotch F, Boshoff C (2010) miR-132 regulates antiviral innate immunity through suppression of the p300 transcriptional co-activator. Nat Cell Biol 12:513–519. https://doi.org/10.1038/ncb2054

Lamers KJB, Vos P, Verbeek MM, Rosmalen F, van Geel WJA, van Engelen BGM (2003) Protein S-100B, neuron-specific enolase (NSE), myelin basic protein (MBP) and glial fibrillary acidic protein (GFAP) in cerebrospinal fluid (CSF) and blood of neurological patients. Brain Res Bull 61(3):261–264

Landrigan PJ, Sonawane B, Butler RN, Trasande L, Callan R, Droller D (2005) Early environmental origins of neurodegenerative disease in later life. Environ Health Perspect 113:1230–1233. https://doi.org/10.1289/ehp.7571

Lasky-Su J, Neale BM, Franke B, Anney RJL, Zhou K, Maller JB, Vasquez AA, Chen W, Asherson P, Buitelaar J, Banaschewski T, Ebstein R, Gill M, Miranda A, Mulas F, Oades RD, Roeyers H, Rothenberger A, Sergeant J, Sonuga-Barke E, Steinhausen HC, Taylor E, Daly M, Laird N, Lange C, Faraone SV (2008) Genome-wide association scan of quantitative traits for attention deficit hyperactivity disorder identifies novel associations and confirms candidate gene associations. Am J Med Genet B Neuropsychiatr Genet 147B:1345–1354. https://doi.org/10.1002/ajmg.b.30867

Lassmann H, Brück W, Lucchinetti CF (2007) The immunopathology of multiple sclerosis: an overview. Brain Pathol 17:210–218. https://doi.org/10.1111/j.1750-3639.2007.00064.x

Lee M, Guo J-P, Kennedy K, McGeer EG, McGeer PL (2016) A method for diagnosing Alzheimer's disease based on salivary amyloid-β protein 42 levels. J Alzheimers Dis 55:1175–1182. https://doi.org/10.3233/JAD-160748

Lei P, Ayton S, Finkelstein DI, Adlard PA, Masters CL, Bush AI (2010) Tau protein: relevance to Parkinson's disease. Int J Biochem Cell Biol 42(11):1775–1778

Lin Q, Wei W, Coelho CM, Li X, Baker-Andresen D, Dudley K, Ratnu VS, Boskovic Z, Kobor MS, Sun YE, Bredy TW (2011) The brain-specific microRNA miR-128b regulates the formation of fear-extinction memory. Nat Neurosci 14:1115–1117. https://doi.org/10.1038/nn.2891

Liu H, Povysheva N, Rose ME, Mi Z, Banton JS, Li W, Chen F, Reay DP, Barrionuevo G, Zhang F, Graham SH (2019) Role of UCHL1 in axonal injury and functional recovery after cerebral ischemia. Proc Natl Acad Sci 116(10):4643–4650

Lomen-Hoerth C, Murphy J, Langmore S, Kramer J, Olney R, Miller B (2003) Are amyotrophic lateral sclerosis patients cognitively normal? Neurology 60:1094e7

López-Figueroa AL, Norton CS, López-Figueroa MO, Armellini-Dodel D, Burke S, Akil H, López JF, Watson SJ (2004) Serotonin 5-HT1A, 5-HT1B, and 5-HT2A receptor mRNA expression in subjects with major depression, bipolar disorder, and schizophrenia. Biol Psychiatry 55:225–233. https://doi.org/10.1016/j.biopsych.2003.09.017

Magister Š, Kos J (2013) Cystatins in immune system. J Cancer 4:45–56. https://doi.org/10.7150/jca.5044

Maraganore DM, De Andrade M, Elbaz A, Farrer MJ, Ioannidis JP, Kru¨ ger R et al (2006) Collaborative analysis of asynuclein gene promoter variability and Parkinson disease. JAMA 296:661e70

Marangos PJ, Schmechel DE (1987) Neuron specific Enolase, a clinically useful marker for neurons and neuroendocrine cells. Annu Rev Neurosci 10(1):269–295

Martin R (1991) A myelin basic protein peptide is recognized by cytotoxic T cells in the context of four HLA-DR types associated with multiple sclerosis. J Exp Med 173:19–24. https://doi.org/10.1084/jem.173.1.19

Massai L, Petricca L, Magnoni L, Rovetini L, Haider S, Andre R, Tabrizi SJ, Süssmuth SD, Landwehrmeyer BG, Caricasole A, Pollio G, Bernocco S (2013) Development of an ELISA assay for the quantification of soluble huntingtin in human blood cells. BMC Biochem 14:34. https://doi.org/10.1186/1471-2091-14-34

Masters JM, Noyce AJ, Warner TT, Giovannoni G, Proctor GB (2015) Elevated salivary protein in Parkinson's disease and salivary DJ-1 as a potential marker of disease severity. Parkinsonism Relat Disord 21:1251–1255. https://doi.org/10.1016/j.parkreldis.2015.07.021

Mayeux R (2004) Biomarkers: potential uses and limitations. NeuroRx 1(2):182–188

Mellios N, Huang H-S, Baker SP, Galdzicka M, Ginns E, Akbarian S (2009) Molecular determinants of dysregulated GABAergic gene expression in the prefrontal cortex of subjects with schizophrenia. Biol Psychiatry 65:1006–1014. https://doi.org/10.1016/j.biopsych.2008.11.019

Miller J, Arrasate M, Shaby BA, Mitra S, Masliah E, Finkbeiner S (2010) Quantitative relationships between huntingtin levels, polyglutamine length, inclusion body formation, and neuronal death provide novel insight into Huntington's disease molecular pathogenesis. J Neurosci 30:10541–10550. https://doi.org/10.1523/JNEUROSCI.0146-10.2010

Minagar A, Adamashvili I, Kelley RE, Gonzalez-Toledo E, McLarty J, Smith SJ (2007) Saliva soluble HLA as a potential marker of response to interferon-β1a in multiple sclerosis: a preliminary study. J Neuroinflammation 4:16. https://doi.org/10.1186/1742-2094-4-16

Miyazawa I, Nakashima I, Petzold A, Fujihara K, Sato S, Itoyama Y (2007) High CSF neurofilament heavy chain levels in neuromyelitis optica. Neurology 68(11):865–867

Mondello S, Palmio J, Streeter J, Hayes RL, Peltola J, Jeromin A (2012) Ubiquitin Carboxy-terminal hydrolase L1 (UCH-L1) is increased in cerebrospinal fluid and plasma of patients after epileptic seizure. BMC Neurol 12(1)

Moreau C, Devos D, Defebvre L (2016) Acetylcholinesterase inhibitors and gait: a steadying hand? Lancet Neurol 15:232–233. https://doi.org/10.1016/S1474-4422(16)00003-X

Mrozek S, Geeraerts T (2012) Biomarqueurs et cerveau. Presented at the MAPAR 2012. Centre Hospitalier Universitaire de Toulouse, France, Pôle Anesthésie Réanimation, p 624

Mudher A, Lovestone S (2002) Alzheimer's disease – do tauists and baptists finally shake hands? Trends Neurosci 25:22–26. https://doi.org/10.1016/S0166-2236(00)02031-2

Mukherjee A, Swarnakar S (2015) Implication of matrix metalloproteinases in regulating neuronal disorder. Mol Biol Rep 42(1):1–11

Muneta Y, Yoshikawa T, Minagawa Y, Shibahara T, Maeda R, Omata Y (2010) Salivary IgA as a useful non-invasive marker for restraint stress in pigs. J Vet Med Sci 72:1295–1300. https://doi.org/10.1292/jvms.10-0009

Naidoo R, Knapp ML (1992) Studies of lipid peroxidation products in cerebrospinal fluid and serum in multiple sclerosis and other conditions. Clin Chem 38:2449e54

Negrão JA, Porcionato MA, de Passillé AM, Rushen J (2004) Cortisol in saliva and plasma of cattle after ACTH administration and milking. J Dairy Sci 87:1713–1718

Ngounou Wetie AG, Wormwood KL, Russell S, Ryan JP, Darie CC, Woods AG (2015) A pilot proteomic analysis of salivary biomarkers in autism spectrum disorder: a pilot proteomic analysis of salivary biomarkers in ASD. Autism Res 8:338–350. https://doi.org/10.1002/aur.1450

Nicolas M, Hassan BA (2014) Amyloid precursor protein and neural development. Development 141:2543–2548. https://doi.org/10.1242/dev.108712

Nishida H (2014) Biomarkers for neural injury and infection in small animals. Vet Clin N Am Small Anim Pract 44(6):1187–1199

Nohesara S, Ghadirivasfi M, Mostafavi S, Eskandari M-R, Ahmadkhaniha H, Thiagalingam S, Abdolmaleky HM (2011) DNA hypomethylation of MB-COMT promoter in the DNA derived from saliva in schizophrenia and bipolar disorder. J Psychiatr Res 45:1432–1438. https://doi.org/10.1016/j.jpsychires.2011.06.013

Obayashi K, Sato K, Shimazaki R, Ishikawa T, Goto K, Ueyama H, Mori T, Ando Y, Kumamoto T (2008) Salivary chromogranin a: useful and quantitative biochemical marker of affective state in patients with amyotrophic lateral sclerosis. Intern Med 47:1875–1879. https://doi.org/10.2169/internalmedicine.47.1278

O'Connor DT, Cervenka JH, Stone RA, Parmer RJ, Franco-Bourland RE, Madrazo I, Langlais PJ (1993) Chromogranin a immunoreactivity in human cerebrospinal fluid: properties, relationship to noradrenergic neuronal activity, and variation in neurologic disease. Neuroscience 56:999–1007. https://doi.org/10.1016/0306-4522(93)90146-7

Oh YS, Turner RJ (2006) Effect of gamma-secretase inhibitors on muscarinic receptor-mediated calcium signaling in human salivary epithelial cells. Am J Phys Cell Physiol 291:8

Oji T, Kamishina H, Cheeseman JA, Clemmons RM (2007) Measurement of myelin basic protein in the cerebrospinal fluid of dogs with degenerative myelopathy. Vet Clin Pathol 36(3):281–284

Oksenberg JR, Barcellos LF, Cree BAC, Baranzini SE, Bugawan TL, Khan O, Lincoln RR, Swerdlin A, Mignot E, Lin L, Goodin D, Erlich HA, Schmidt S, Thomson G, Reich DE, Pericak-Vance MA, Haines JL, Hauser SL (2004) Mapping multiple sclerosis susceptibility to the HLA-DR locus in African Americans. Am J Hum Genet 74:160–167. https://doi.org/10.1086/380997

Olerup O, Hillert J (1991) HLA class II-associated genetic susceptibility in multiple sclerosis: a critical evaluation. Tissue Antigens 38:1–15. https://doi.org/10.1111/j.1399-0039.1991.tb02029.x

Onore C, Careaga M, Ashwood P (2012) The role of immune dysfunction in the pathophysiology of autism. Brain Behav Immun 26:383–392. https://doi.org/10.1016/j.bbi.2011.08.007

Ortiz GG, Pacheco-Moisés FP, Bitzer-Quintero OK, Ramírez-Anguiano AC, Flores-Alvarado LJ, Ramírez-Ramírez V, Macias-Islas MA, Torres-Sánchez ED (2013) Immunology and oxidative stress in multiple sclerosis: clinical and basic approach. Clin Dev Immunol 2013:1–14. https://doi.org/10.1155/2013/708659

Pals P, Lincoln S, Manning J, Heckman M, Skipper L, Hulihan M, Van den Broeck M, De Pooter T, Cras P, Crook J, Van Broeckhoven C, Farrer MJ (2004) alpha-Synuclein promoter confers susceptibility to Parkinson's disease. Ann Neurol 56:591–595. https://doi.org/10.1002/ana.20268

Paltrinieri S, Pintore L, Balducci F, Giordano A, Costabile A, Bernardini M (2017) Serum creatine kinase isoenzymes and macroenzymes in dogs with different neurologic diseases. Vet Clin Pathol 46(1):91–99

Pardo CA, Vargas DL, Zimmerman AW (2005) Immunity, neuroglia and neuroinflammation in autism. Int Rev Psychiatry 17:485–495. https://doi.org/10.1080/02646830500381930

Parnetti L, Castrioto A, Chiasserini D, Persichetti E, Tambasco N, El-Agnaf O, Calabresi P (2013) Cerebrospinal fluid biomarkers in Parkinson disease. Nat Rev Neurol 9:131–140. https://doi.org/10.1038/nrneurol.2013.10

Passantino A, Masucci M (2016) Congenital and inherited neurologic diseases in dogs and cats: legislation and its effect on purchase in Italy. Veterinary World 9(5):437–443

Peter CJ, Fischer LK, Kundakovic M, Garg P, Jakovcevski M, Dincer A, Amaral AC, Ginns EI, Galdzicka M, Bryce CP, Ratner C, Waber DP, Mokler D, Medford G, Champagne FA, Rosene DL, McGaughy JA, Sharp AJ, Galler JR, Akbarian S (2016) DNA methylation signatures of early childhood malnutrition associated with impairments in attention and cognition. Biol Psychiatry 80:765–774. https://doi.org/10.1016/j.biopsych.2016.03.2100

Peter C, Kenneth W Hinchcliff, Stanley Done, Walter Gruenberg, 2016. Veterinary Medicine - 11th Edition

Petzold A (2005) Neurofilament phosphoforms: surrogate markers for axonal injury, degeneration and loss. J Neurol Sci 233(1–2):183–198

Pietzsch J, Hoppmann S (2009) Human S100A12: a novel key player in inflammation? Amino Acids 36:381–389. https://doi.org/10.1007/s00726-008-0097-7

Pikula DL, Harris EF, Desiderio DM, Fridland GH, Lovelace JL (1992) Methionine enkephalin-like, substance P-like, and beta-endorphin-like immunoreactivity in human parotid saliva. Arch Oral Biol 37:705–709

Płonek M, Wrzosek M, Nicpoń J (2016) Biomarkers of injury to neural tissue in veterinary medicine. Journal of Veterinary Research 60(3):365–371

Pratamastuti D, Gunawan PI, Saharso D (2017) Serum neuron specific enolase is increased in pediatric acute encephalitis syndrome. Korean J Pediatr 60(9):302

Psychiatric GWAS Consortium Bipolar Disorder Working Group (2011) Large-scale genome-wide association analysis of bipolar disorder identifies a new susceptibility locus near ODZ4. Nat Genet 43:977–983. https://doi.org/10.1038/ng.943

Redbo I (1993) Stereotypies and cortisol secretion in heifers subjected to tethering. Appl Anim Behav Sci 38:213–225

Remenyi J, Hunter CJ, Cole C, Ando H, Impey S, Monk CE, Martin KJ, Barton GJ, Hutvagner G, Arthur JSC (2010) Regulation of the miR-212/132 locus by MSK1 and CREB in response to neurotrophins. Biochem J 428:281–291. https://doi.org/10.1042/BJ20100024

Riederer BM, Zagon IS, Goodman SR (1986) Brain spectrin(240/235) and brain spectrin(240/235E): two distinct spectrin subtypes with different locations within mammalian neural cells. J Cell Biol 102(6):2088–2097

Rivera S, Jourquin J, Ogier C, Bernard A, Charton G, Tremblay É, Khrestchatisky M (2004) Le système MMP/TIMP dans le système nerveux. médecine/sciences 20(1):55–60

Riviere M, Meininger V, Zeisser P, Munsat T (1998) An analysis of extended survival in patients with amyotrophic lateral sclerosis treated with riluzole. Arch Neurol 55:526–528. https://doi.org/10.1001/archneur.55.4.526

Rizzo MA, Hadjimichael OC, Preiningerova J, Vollmer TL (2004) Prevalence and treatment of spasticity reported by multiple sclerosis patients. Mult Scler J 10:589–595. https://doi.org/10.1191/1352458504ms1085oa

Rosenberg G (2015) Metalloproteinases and neurodegenerative diseases: pathophysiological and therapeutic perspectives. Metalloproteinases In Medicine:39

Ross CA, Tabrizi SJ (2011) Huntington's disease: from molecular pathogenesis to clinical treatment. Lancet Neurol 10:83–98. https://doi.org/10.1016/S1474-4422(10)70245-3
Ruiz PJG, Catalán MJ, Carril JMF (2011) Initial motor symptoms of Parkinson disease. Neurologist 17:S18–S20. https://doi.org/10.1097/NRL.0b013e31823966b4
Rutigliano G, Rocchetti M, Paloyelis Y, Gilleen J, Sardella A, Cappucciati M, Palombini E, Dell'Osso L, Caverzasi E, Politi P, McGuire P, Fusar-Poli P (2016) Peripheral oxytocin and vasopressin: biomarkers of psychiatric disorders? A comprehensive systematic review and preliminary meta-analysis. Psychiatry Res 241:207–220. https://doi.org/10.1016/j.psychres.2016.04.117
Saberi S, Stauffer JE, Schulte DJ, Ravits J (2015) Neuropathology of amyotrophic lateral sclerosis and its variants. Neurol Clin 33:855–876. https://doi.org/10.1016/j.ncl.2015.07.012
Salekdeh GH (2016) Agricultural proteomics volume 2: environmental stresses. Springer
Salouci M, Gabriel A (2016) Focus sur les astrocytes: cellules de la glie aux rôles multiples et importants dans le SNC. Le mouton est-il un excellent modèle animal? Rev. Médecine Vét
Sapp E, Schwarz C, Chase K, Bhide PG, Young AB, Penney J, Vonsattel JP, Aronin N, DiFiglia M (1997) Huntingtin localization in brains of normal and Huntington's disease patients. Ann Neurol 42:604–612. https://doi.org/10.1002/ana.410420411
Sarachana T, Zhou R, Chen G, Manji HK, Hu VW (2010) Investigation of post-transcriptional gene regulatory networks associated with autism spectrum disorders by microRNA expression profiling of lymphoblastoid cell lines. Genome Med 2:23. https://doi.org/10.1186/gm144
Saudou F, Humbert S (2016) The biology of huntingtin. Neuron 89:910–926. https://doi.org/10.1016/j.neuron.2016.02.003
Saudou F, Finkbeiner S, Devys D, Greenberg ME (1998) Huntingtin acts in the nucleus to induce apoptosis but death does not correlate with the formation of intranuclear inclusions. Cell 95:55–66. https://doi.org/10.1016/S0092-8674(00)81782-1
Sayer R, Law E, Connelly PJ, Breen KC (2004) Association of a salivary acetylcholinesterase with Alzheimer's disease and response to cholinesterase inhibitors. Clin Biochem 37:98–104. https://doi.org/10.1016/j.clinbiochem.2003.10.007
Schwinn A-C, Knight CH, Bruckmaier RM, Gross JJ (2016) Suitability of saliva cortisol as a biomarker for hypothalamic–pituitary–adrenal axis activation assessment, effects of feeding actions, and immunostimulatory challenges in dairy cows1. J Anim Sci 94(6):2357–2365
Sekiguchi S, Kwon J, Yoshida E, Hamasaki H, Ichinose S, Hideshima M, Kuraoka M, Takahashi A, Ishii Y, Kyuwa S, Wada K, Yoshikawa Y (2006) Localization of ubiquitin C-terminal hydrolase L1 in mouse ova and its function in the plasma membrane to block Polyspermy. Am J Pathol 169(5):1722–1729
Shaked I, Meerson A, Wolf Y, Avni R, Greenberg D, Gilboa-Geffen A, Soreq H (2009) MicroRNA-132 potentiates cholinergic anti-inflammatory signaling by targeting acetylcholinesterase. Immunity 31:965–973. https://doi.org/10.1016/j.immuni.2009.09.019
Shendelman S, Jonason A, Martinat C, Leete T, Abeliovich A (2004) DJ-1 is a redox-dependent molecular chaperone that inhibits a-synuclein aggregate formation. PLoS Biol 2:10
Shi M, Zabetian CP, Hancock AM, Ginghina C, Hong Z, Yearout D, Chung KA, Quinn JF, Peskind ER, Galasko D, Jankovic J, Leverenz JB, Zhang J (2010) Significance and confounders of peripheral DJ-1 and alpha-synuclein in Parkinson's disease. Neurosci Lett 480:78–82. https://doi.org/10.1016/j.neulet.2010.06.009
Shi M, Sui Y-T, Peskind ER, Li G, Hwang H, Devic I, Ginghina C, Edgar JS, Pan C, Goodlett DR, Furay AR, Gonzalez-Cuyar LF, Zhang J (2011) Salivary Tau species are potential biomarkers of Alzheimer's disease. J Alzheimers Dis 27:299–305. https://doi.org/10.3233/JAD-2011-110731
Singleton AB (2003) Synuclein locus triplication causes Parkinson's disease. Science 302:841–841. https://doi.org/10.1126/science.1090278
Sreedharan J, Blair IP, Tripathi VB, Hu X, Vance C, Rogelj B, Ackerley S, Durnall JC, Williams KL, Buratti E, Baralle F, de Belleroche J, Mitchell JD, Leigh PN, Al-Chalabi A, Miller CC, Nicholson G, Shaw CE (2008) TDP-43 mutations in familial and sporadic amyotrophic lateral sclerosis. Science 319:1668–1672. https://doi.org/10.1126/science.1154584

Steinman MDL (1996) Multiple sclerosis: a coordinated immunological attack against myelin in the central nervous system. Cell 85:299–302. https://doi.org/10.1016/S0092-8674(00)81107-1

Stevens V, Goodman K, Rough K, Kraft GH (2013) Gait impairment and optimizing mobility in multiple sclerosis. Phys Med Rehabil Clin N Am 24:573–592. https://doi.org/10.1016/j.pmr.2013.07.002

Straub RE, Jiang Y, MacLean CJ, Ma Y, Webb BT, Myakishev MV, Harris-Kerr C, Wormley B, Sadek H, Kadambi B, Cesare AJ, Gibberman A, Wang X, O'Neill FA, Walsh D, Kendler KS (2002) Genetic variation in the 6p22.3 gene DTNBP1, the human ortholog of the mouse dysbindin gene, is associated with schizophrenia. Am J Hum Genet 71:337–348. https://doi.org/10.1086/341750

Stubbs EG, Ash E, Williams CPS (1984) Autism and congenital cytomegalovirus. J Autism Dev Disord 14:183–189. https://doi.org/10.1007/BF02409660

Süssmuth SD, Tumani H, Ecker D, Ludolph AC (2003) Amyotrophic lateral sclerosis: disease stage related changes of tau protein and S100 beta in cerebrospinal fluid and creatine kinase in serum. Neurosci Lett 353(1):57–60

Taganov KD, Boldin MP, Chang K-J, Baltimore D (2006) NF- B-dependent induction of microRNA miR-146, an inhibitor targeted to signaling proteins of innate immune responses. Proc Natl Acad Sci 103:12481–12486. https://doi.org/10.1073/pnas.0605298103

Takai N, Yamaguchi M, Aragaki T, Eto K, Uchihashi K, Nishikawa Y (2004) Effect of psychological stress on the salivary cortisol and amylase levels in healthy young adults. Arch Oral Biol 49:963–968

Talebizadeh Z, Butler MG, Theodoro MF (2008) Feasibility and relevance of examining lymphoblastoid cell lines to study role of microRNAs in autism. Autism Res 1:240–250. https://doi.org/10.1002/aur.33

Taupenot L (2003) The chromogranin–secretogranin family. N Engl J Med 348(12):1134–1149

Thenganatt MA, Jankovic J (2014) Parkinson disease subtypes. JAMA Neurol 71:499–504. https://doi.org/10.1001/jamaneurol.2013.6233

Valenti P, Catizone A, Pantanella F, Frioni A, Natalizi T, Tendini M, Berlutti F (2011) Lactoferrin decreases inflammatory response by cystic fibrosis bronchial cells invaded with *Burkholderia Cenocepacia* iron-modulated biofilm. Int J Immunopathol Pharmacol 24:1057–1068. https://doi.org/10.1177/039463201102400423

van der Strate BW, Beljaars L, Molema G, Harmsen M, Meijer DK (2001) Antiviral activities of lactoferrin. Antivir Res 52:225–239. https://doi.org/10.1016/S0166-3542(01)00195-4

van Horssen J, Witte ME, Schreibelt G, de Vries HE (2011) Radical changes in multiple sclerosis pathogenesis. Biochim Biophys Acta (BBA) - Mol Basis Dis 1812:141–150. https://doi.org/10.1016/j.bbadis.2010.06.011

Vivacqua G, Latorre A, Suppa A, Nardi M, Pietracupa S, Mancinelli R, Fabbrini G, Colosimo C, Gaudio E, Berardelli A (2016) Abnormal salivary total and oligomeric alpha-synuclein in Parkinson's disease. PLoS One 11:e0151156. https://doi.org/10.1371/journal.pone.0151156

Walton E, Liu J, Hass J, White T, Scholz M, Roessner V, Gollub R, Calhoun VD, Ehrlich S (2014) *MB-COMT* promoter DNA methylation is associated with working-memory processing in schizophrenia patients and healthy controls. Epigenetics 9:1101–1107. https://doi.org/10.4161/epi.29223

Wang KKW, Posmantur R, Nath R, McGinnis K, Whitton M, Talanian RV, Glantz SB, Morrow JS (1998) Simultaneous degradation of αII- and βII-Spectrin by Caspase 3 (CPP32) in apoptotic cells. J Biol Chem 273(35):22490–22497

Wang L, Sato H, Zhao S, Tooyama I (2010) Deposition of lactoferrin in fibrillar-type senile plaques in the brains of transgenic mouse models of Alzheimer's disease. Neurosci Lett 481:164–167. https://doi.org/10.1016/j.neulet.2010.06.079

Wang F-Z, Weber F, Croce C, Liu C-G, Liao X, Pellett PE (2008) Human cytomegalovirus infection alters the expression of cellular microRNA species that affect its replication. J Virol 82:9065–9074. https://doi.org/10.1128/JVI.00961-08

Wang P, Xie K, Wang C, Bi J (2014) Oxidative stress induced by lipid peroxidation is related with inflammation of demyelination and neurodegeneration in multiple sclerosis. Eur Neurol 72:249–254. https://doi.org/10.1159/000363515

Warner LE (1999) Functional consequences of mutations in the early growth response 2 gene (EGR2) correlate with severity of human myelinopathies. Hum Mol Genet 8(7):1245–1251

Weiss ES, Wang KKW, Allen JG, Blue ME, Nwakanma LU, Liu MC, Lange MS, Berrong J, Wilson MA, Gott VL, Troncoso JC, Hayes RL, Johnston MV, Baumgartner WA (2009) Alpha II-Spectrin breakdown products serve as novel markers of brain injury severity in a canine model of hypothermic circulatory arrest. Ann Thorac Surg 88(2):543–550

Whitaker JN (2016) Myelin basic protein in cerebrospinal fluid and other body fluids. Mult Scler J 4(1):16–21

Wild EJ, Boggio R, Langbehn D, Robertson N, Haider S, Miller JRC, Zetterberg H, Leavitt BR, Kuhn R, Tabrizi SJ, Macdonald D, Weiss A (2015) Quantification of mutant huntingtin protein in cerebrospinal fluid from Huntington's disease patients. J Clin Invest 125:1979–1986. https://doi.org/10.1172/JCI80743

Wilmot B, Fry R, Smeester L, Musser ED, Mill J, Nigg JT (2016) Methylomic analysis of salivary DNA in childhood ADHD identifies altered DNA methylation in *VIPR2*. J Child Psychol Psychiatry 57:152–160. https://doi.org/10.1111/jcpp.12457

Wolfe MS (2012) The role of tau in neurodegenerative diseases and its potential as a therapeutic target. Australas Sci 2012:1–20

Wright GEB, Niehaus DJH, van der Merwe L, Koen L, Korkie LJ, Kinnear CJ, Drögemöller BI, Warnich L (2012) Association of MB-COMT polymorphisms with schizophrenia-susceptibility and symptom severity in an African cohort. Prog Neuro-Psychopharmacol Biol Psychiatry 39:163–169. https://doi.org/10.1016/j.pnpbp.2012.06.006

Wucherpfennig KW, Zhang J, Witek C, Matsui M, Modabber Y, Ota K, Hafler DA (1994) Clonal expansion and persistence of human T cells specific for an immunodominant myelin basic protein peptide. J Immunol 152:5581–5592

Yardan T, Erenler AK, Baydin A, Aydin K, Cokluk C (2011 Mar) Usefulness of S100B protein in neurological disorders. J Pak Med Assoc 61(3):276–281

Yong VW, Agrawal SM, Stirling DP (2007) Targeting MMPs in acute and chronic neurological conditions. Neurotherapeutics 4(4):580–589

Yrigollen CM, Han SS, Kochetkova A, Babitz T, Chang JT, Volkmar FR, Leckman JF, Grigorenko EL (2008) Genes controlling affiliative behavior as candidate genes for autism. Biol Psychiatry 63:911–916. https://doi.org/10.1016/j.biopsych.2007.11.015

Zamvil SS, Steinman L (1990) The T lymphocyte in experimental allergic encephalomyelitis. Annu Rev Immunol 8(1):579–621

Chapter 8
Salivary Biomarkers in the Diagnosis and Monitoring of Metabolic and Endocrine Diseases

Asta Tvarijonaviciute, Sónia Lucena, Fernando Capela e Silva, and Elsa Lamy

Objectives

The present chapter revises the current knowledge about the use of salivary biomarkers for metabolic and endocrine disease diagnosis and monitoring in humans and different animal species. Currently, most information available comes from human medicine and studies performed with experimental animals such as mouse or rat, and less from other veterinary species. Furthermore, since obesity and its-related pathologies are currently recognised as the biggest worldwide public health crisis and socioeconomic problem in the twenty-first century, this chapter will focus most on this disease.

A. Tvarijonaviciute
Interdisciplinary Laboratory of Clinical Analysis (Interlab-UMU), Regional Campus of International Excellence Campus Mare Nostrum, University of Murcia, Murcia, Spain
e-mail: asta@um.es

S. Lucena
MED - Mediterranean Institute for Agriculture, Environment and Development, University of Évora, Évora, Portugal

Department of Veterinary Medicine, School of Science and Technology, University of Évora, Évora, Portugal
e-mail: slucena@uevora.pt

F. Capela e Silva
MED - Mediterranean Institute for Agriculture, Environment and Development, IIFA - Instituto de Investigação e Formação Avançada, University of Évora, Évora, Portugal

Department of Biology, University of Évora, Évora, Portugal
e-mail: fcs@uevora.pt

E. Lamy (✉)
MED - Mediterranean Institute for Agriculture, Environment and Development, IIFA - Instituto de Investigação e Formação Avançada, University of Évora, Évora, Portugal
e-mail: ecsl@uevora.pt

A. Tvarijonaviciute et al. (eds.), *Saliva in Health and Disease*,
https://doi.org/10.1007/978-3-030-37681-9_8

8.1 Introduction

Endocrine and metabolic diseases are among the most common diseases that can affect humans and animals. In recent decades there has been an increase in the prevalence and incidence of many of them, especially those related to nutrition and metabolism. Metabolic disease is considered a disease or disorder that disrupt normal metabolism on a cellular level. Currently, the most often occurring metabolic disease is obesity and diseases associated with it, among which endocrine disease such as diabetes mellitus (DM). Endocrine diseases comprise disorders that result in altered production of endocrine hormone(s) (also referred as hormone imbalance) or development of lesions (such as nodules or tumors) in the endocrine system, which may or may not affect hormone levels. Close relationship between metabolic and endocrine disease exists since metabolic alterations can finally result in endocrine system disruption and vice versa. For instance, patients suffering from some endocrine diseases such as DM, hyperadrenocortisism or hypothyroidism present obesity.

Obesity is a multicausal disease that due to increased insulin resistance and chronic inflammatory status leads to increased risk of developing DM, cardiovascular diseases, cancer, and renal diseases among others. However, despite human obesity was recognised as a disease in 2008, and a number of studies were undertaken in order to increase knowledge about its pathogenesis, treatment options and prevention, its prevalence is keeping rising and it is considered to be one of the largest contributors to poor health in most countries (NCD Risk Factor Collaboration (NCD-RisC) 2016; Swinburn et al. 2019). Therefore, obesity was stated to be a worldwide public health crisis and socioeconomic problem in the twenty-first century (Hartman et al. 2016; Canfora et al. 2019). In as similar manner, obesity rates are increasing among veterinary species including dogs, cats, horses, resulting not only in worsened health and quality of life of these animals, but also decreased life span (Kealy et al. 2002). In cats and dogs, obesity was recognized to be a disease just in 2018 (https://bit.ly/2vbTTyO), endorsed by a total of 23 veterinary organisations.

In humans, obesity is usually diagnosed by calculating the body mass index (BMI) following the formula: $BMI = BW_{kg}/(height_{m})^2$, where BW refers to body weight. People having BMI equal or above 30 are considered obese (Table 8.1). In animals, the obesity is considered as BW increase in 30% when compared with the ideal BW of the species. However, in many cases, it is very hard or even impossible to know the accurate ideal BW, thus body charts were developed in order to score body condition of an animal. The body condition scores (BCS) are based on the visual evaluation and palpation of the animals (Table 8.1). However, when the possible obesity-related risks have to be evaluated, the serum biochemistry is performed. Traditionally, triglycerides, total cholesterol, high density lipoprotein (HDL) cholesterol, glucose and insulin are determined in serum, since alterations in these analytes were related with the increased risk to develop cardiovascular diseases and diabetes mellitus among others. However, recently new biomarkers such as adipokines, oxidative stress markers, or acute phase proteins have emerged contributing to the pathogenesis of the disease.

Table 8.1 Weight groups according Body Mass Index (BMI) in humans and Body Condition Score (BCS) Charts in animals

	Humans	Animals		
Weight group	BMI	5 point BCS	9 point BCS	Description
Underweight	< 18.5	1–2	1–4	Ribs – easily palpated without or with poor fat coverage; Bone structures are prominent and easily identifiable; Muscle tone and mass – depressed; Subcutaneous fat – little or no; Abdomen – very collected. Hair – poor quality.
Normal-weight	18.5–25	3	5	Ribs – easily palpated with fat coverage; Bone prominences are palpable but not visible; Muscle tone and mass – normal; Subcutaneous fat – little; no accumulations; Abdomen – collected but not pronounced; Hair – good quality.
Overweight	25–30	4	6–7	Ribs – palpated with difficulty due to the accumulation of superimposed fat; Bone prominences cannot be identified; Muscle tone and mass – normal; Subcutaneous fat – in some areas evident accumulation (e.g., tale base); Abdomen – not collected; Hair – good or reduced quality.
Obese	> 30	5	8–9	Ribs – impossible to palpate due to superimposed fat; Bone prominences are hardly palpable but not visible; Muscle tone and mass – normal to depressed; Subcutaneous fat is evident and there are accumulations in the neck, base of the tail and abdominal region; Abdomen – not collected, rounded appearance; Hair – good or reduced quality.

Endocrine disease diagnosis is traditionally performed by determination of the concentrations of the hormone of interest in serum and, frequently, dynamic function test, either *stimulating* or *suppressing* a particular hormonal axis and observing the appropriate hormonal response, are undertaken. However, general biochemistry analysis can permit to suspect ongoing endocrine disease, since alterations in lipids, acute phase proteins, or glucose among others occur in these diseases.

Nevertheless, in both metabolic and endocrine diseases, the use of saliva is getting attention mainly due to its non-invasive nature (see Chap. 2). Furthermore, saliva analysis is leaded not only to evaluate the possible utility of this biofluid as a source of traditional biomarkers, but also for new biomarker identification with the aim to clarify and contribute to increased knowledge about pathophysiology of these diseases.

In the following sections we will review state-of-the-art of data obtained via sialochemistry, proteomics, metabolomics and microbiome analyses in metabolic and endocrine diseases.

8.2 Sialochemistry

8.2.1 Glucose

Humans It was hypothesized that glucose into saliva could enter due to leakage from blood across the basement membrane of salivary glands (Puttaswamy et al. 2017) and due to the capacity to cross the salivary gland epithelium in proportion to its concentration in blood (Abikshyeet et al. 2012). Microvascular alterations in the blood vessels due to DM could also contribute to enter glucose into saliva (Puttaswamy et al. 2017). Different authors have studied salivary glucose concentrations in healthy controls and patients with DM (Smriti et al. 2016; Carramolino-Cuéllar et al. 2017). And, although contradiction exists, saliva was suggested to be an useful biofluid for glucose monitoring, since moderate to positive correlation between salivary and serum glucose concentrations and between salivary glucose and serum glycated hemoglobin (HbA_{1c}) were observed (Abikshyeet et al. 2012; Panchbhai 2012; Gupta et al. 2017).

However, cautions should be taken when analyzing glucose metabolism-related biomarkers due to their stability in saliva samples. It was observed that salivary glucose is rapidly decomposed by oral bacteria and enzymes of saliva (Sandham and Kleinberg 1969a, b), thus samples should be refrigerated at all the moments and analyzed or stored at −80 °C as soon as possible.

Animals One study in dogs indicate that glucose passes to saliva by a simply passive process from plasma (Langley et al. 1963). Intravenous glucose administration revealed delay in peak of around 25 min of salivary glucose with respect to serum, which could be responsible for the lack of correlation between serum and salivary levels (Muñoz-Prieto et al. 2019b). No studies have been reported evaluating salivary glucose concentrations in dogs with diabetes mellitus or other metabolic/endocrine diseases.

8.2.2 Fructosamine

Fructosamine is considered as an index of average blood glucose levels up to 2 weeks that is not affected by the diet. For this reason, this analyte is used as a blood glucose control index.

Humans Nakamoto et al. (2003) described positive correlation between salivary fructosamine glycated protein and serum HbA_{1c} and blood glucose and suggested its possible utility in diabetes diagnosis.

Animals A method for fructosamine determination in canine saliva was satisfactory validated showing adequate precision and accuracy and, as expected, no significant changes in its concentrations after intravenous glucose administration were observed (Muñoz-Prieto et al. 2019b). Further studies are now needed to evaluate the possible utility of salivary fructosamine in cases of diabetes. Nevertheless, it is important to mention, that fructosamine was described to decrease up to 65% even after being stored at −80 °C, thus, samples should be analyzed in fresh (Muñoz-Prieto et al. 2019b). For this reason, some authors suggested determine glycated proteins, such as fructosamine glycated protein, in saliva since it was hypothesized that glycated proteins were more stable (Nakamoto et al. 2003).

8.2.3 Insulin

Humans Insulin-like immunoreactivity in saliva has been reported by different groups of authors (Marchetti et al. 1986; Fekete et al. 1993; Messenger et al. 2003).

There are two main hypotheses regarding the insulin presence in the saliva:

1. Insulin is actively synthesized and secreted by salivary gland. This hypothesis is based mainly on animal studies, in which insulin mRNA have been found in salivary glands and that salivary glands were sensitive to changes in glucose levels (Shubnikova et al. 1984; Kerr et al. 1995; Taouis et al. 1995). Furthermore, synthesis of this hormone was confirmed in humans and animal models through detection of radiolabeled product in salivary gland tissue incubated in vitro with [^{3}H]leucine (Murakami et al. 1982).
2. Insulin enters saliva from blood by ultrafiltration. This hypothesis is supported by studies that detected relation (although delayed) between blood and saliva insulin levels together with the lower salivary insulin concentrations and the observation that salivary insulin was affected by *iv* glucose administration (Messenger et al. 2003).

In subjects with normal glucose tolerance, both normal weight and obese, the concentrations of insulin in saliva increase after oral glucose tolerance test, although with a delay in the peak of approximately 30–45 min (Marchetti et al. 1986; Pasic and Pickup 1988; Fekete et al. 1993). In healthy school-age girls, salivary insulin levels were correlated with serum insulin and BMI, while this correlation was not detected in boys (Tvarijonaviciute et al. 2019). Nevertheless, insulin handling by salivary glands of patients with type I and II diabetes seem to differ (Marchetti et al. 1986; Pasic and Pickup 1988). Interestingly, strong positive correlation between salivary and serum insulin was observed when pooled data of patients with type I diabetes were studied, while the correlation did not exist when comparing salivary and serum insulin concentrations of each individual separately (Pasic and Pickup 1988). For this reason, some authors indicate that the use of salivary insulin levels for diabetes control would not be reasonable (Pasic and Pickup 1988).

Animals Rat salivary glands and saliva were shown to present biologically active insulin-like immunoreactivity that can participate in metabolic regulations via amino acid uptake and lipogenesis (Taouis et al. 1995). Furthermore, as stated above, it was demonstrated the active secretion of insulin by salivary glands in rats and mice (Kerr et al. 1995). Kerr et al. (1995) observed that the behavior of salivary insulin in healthy normal mice differs from the mice with induced diabetes, being much lower in those with diabetes and suggesting that salivary glands can act as extrapancreatic source of insulin in cases of diabetes.

In dogs, intravenous glucose administration resulted in a delayed increase of insulin in saliva with respect to serum of about 10 min (Muñoz-Prieto et al. 2019b). The significant increase was noted in saliva 15 min after injection remaining high 45 min after. This delay, could be the reason of the lack of correlation between serum and saliva insulin concentrations. Unfortunately, no data exist about salivary insulin levels in dogs with metabolic diseases such as obesity or diabetes mellitus.

8.2.4 Lipids

Dysregulation of lipid metabolism is related to development of a number of pathologies including metabolic diseases obesity and diabetes mellitus among others (Gianfrancesco et al. 2018; Hou et al. 2019). In order to evaluate, diagnose or monitor lipid metabolism, in human medicine, triglycerides, cholesterol, low density lipoproteins (LDC) and high density lipoproteins (HDL) are determined in serum. Both HDL and LDL are predominantly involved in cholesterol metabolism. HDL is considered the "good" lipoprotein with protective role on vascular system, while increases LDL were associated with arteriosclerosis. It is important to notice, that the lipid and lipoprotein distribution in serum of humans and different animal species differ, for this reason, caution should be taken when selecting appropriate animal model for human dyslipidemia studies (Table 8.2) (Yin et al. 2012).

Humans Very few information about salivary lipids is available (Matczuk et al. 2017). From eight main groups of lipids, five were identified in saliva (Table 8.3). Salivary lipid profile in different major salivary glands is similar, while considerable

Table 8.2 Species having high density lipoprotein (HDL) or low density lipoprotein (LDL) as a main serum lipoprotein

Main lipoprotein	Species
HDL	Dog, cat, horse, ruminants, rat, mouse
LDL	Humans, Most of primates, rabbit, hamster, Guinea pig, pig, camel, rhinoceros

Table 8.3 Lipid groups present in saliva

Nr.	Group[a]	Described in Saliva
1	Fatty acyls	√
2	Glycerolipids	√
3	Glycerophospholipids	√
4	Sphingolipids	√
5	Sterol lipids	√
6	Prenol lipids	
7	Saccharolipids	
8	Poliketides	

[a]Lipid classification according to International lipid Classification and nomenclature Committee and LIPID MAPS consortium

differences were observed in minor glands (Rabinowitz and Shannon 1975). Although, this is not of big importance when total saliva is used, it should be taken in consideration in patients with salivary gland pathologies. Just for instance, alterations in salivary lipid profile was reported in patients with cystic fibrosis, a disease that involves impaired secretion by the endocrine glands, including salivary glands (Matczuk et al. 2017). Nevertheless, no studies exist evaluating possible salivary lipid profile changes in presence of systemic metabolic diseases.

Animals In animals, only experimental studies in rats were reported describing salivary lipid profiles and their possible alterations in pathologies. For instance, experimental model consisting in diabetes induction in rats with streptozotocin resulted in increased lipids, and in particular stearic and linoleic acids that were normalized after successful treatment with insulin (Anderson and Garrett 1986; Morris et al. 1992; Mahay et al. 2004). However, no studies exist about salivary lipid profiles in naturally occurring pathologies in animals.

8.2.5 Adipokines

Adipokines are proteins mainly synthesized in the adipose tissue, although in the last years their synthesis by other tissues such as salivary gland was confirmed by different researchers (Katsiougiannis et al. 2006). Adiponectin and leptin are two of the most studied adipokines in metabolic diseases due to their close relationship with insulin resistance and inflammation among others (Bastard et al. 2006). The presence of both adiponectin and leptin in saliva has been described by different groups of authors.

Humans Salivary adiponectin concentrations correlate positively with serum and, thus, saliva was indicated as appropriate biofluid for this adipokine measurement in different clinical situations such as insulin resistance and obesity (Desai and Mathews 2014; Nigro et al. 2015; Teke et al. 2019) and could serve as convenient adjunct method in predicting cardio-metabolic risks in the population (Attlee et al. 2019). Salivary leptin concentrations were shown to be higher in overweight individuals and patients with diabetes mellitus as compared with normal weight individuals (Jayachandran et al. 2017; Tvarijonaviciute et al. 2017). However, no association between salivary adiponectin or leptin was detected with metabolic syndrome (Thanakun et al. 2014). Furthermore, cautions must be taken in presence of the gingivitis, since an increase in salivary adiponectin was described (Meriç et al. 2018).

Animals Salivary adiponectin was described in dogs and its correlation with serum was reported (Tvarijonaviciute et al. 2014). Furthermore, salivary adiponectin concentrations were lower, although not statistically significantly, in obese dogs in comparison to normal weight or overweight dogs (Muñoz-Prieto et al. 2019a). However, as occur in humans, salivary adiponectin was shown to be increased in dogs with gingivitis naturally occurring and due to teeth cleaning procedures (Tvarijonaviciute et al. 2014), thus oral health should be evaluated in the dogs when saliva for adiponectin determination is collected.

In pigs, salivary leptin was related with body weight, food ingestion and inflammation (Schmidt et al. 2016).

8.2.6 Inflammatory Biomarkers

Some metabolic diseases, such as obesity or diabetes mellitus, are related with the presence of low grade inflammation. For this reason, acute phase proteins, especially C-reactive protein (CRP) and interleukins are usually used to evaluate pro-inflammatory status of patients with metabolic and endocrine diseases (Bastard et al. 2006).

Humans Utility of CRP and interleukins in saliva were proved to be useful biomarkers in studies of obesity, metabolic syndrome or diabetes mellitus being levels of pro-inflammatory biomarkers elevated in these metabolic pathologies (Naidoo et al. 2012; Dezayee and Al-Nimer 2016; Hartman et al. 2016; Balaji et al. 2017; Janem et al. 2017; Tvarijonaviciute et al. 2019).

Animals The association of salivary S100, a protein participating in the regulation of the immune homeostasis and inflammation (Bao et al. 2012), with metabolic diseases was suggested, since its increase in submandibular glands of rats, after 2 months of induced-hyperglycemia, was observed and linked to the inflammatory response and impaired metabolic and energy production processes (Alves et al. 2013). Furthermore, S100 was higher in saliva of dogs with Obesity Related Metabolic Disease (ORMD) as compared with overweight obese dogs without ORMD (Lucena et al. 2019).

8.2.7 Antioxidants

Humans Presence of pro-oxidant status has been reported in different metabolic diseases. However, studies on salivary oxidative biomarker behavior in metabolic diseases are scarce, although some reported results evidence alterations in levels of these biomarkers including malondialdehyde, uric acid, superoxide dismutase, total oxidant status (TOS) and total antioxidant status (TAS) in saliva of patients with diabetes mellitus, gestational DM, obesity, hepato-metabolic comorbidities (Hartman et al. 2016; Madi et al. 2016; Troisi et al. 2019b; Zygula et al. 2019). In addition, oxidative stress was shown to be a strong inducer of alkaline phosphatase in various tissues (Torino et al. 2016). And its salivary concentrations were higher in patients with chronic periodontitis with type-2 diabetes mellitus than chronic periodontitis without diabetes mellitus and healthy patients (Sridharan et al. 2017; De et al. 2018). The adequate metabolic control and periodontal treatment result in normalisation of levels of oxidative stress biomarkers in saliva (Aral et al. 2017).

Animals Despite the limited number of studies about salivary antioxidants and metabolic diseases in animals, in dogs, the presence of oxidative stress was suggested in dogs with ORMD, since alterations in salivary glutathione S-transferase, superoxide dismutase and Hsp70 were observed in presence of ORMD (Lucena et al. 2019). Furthermore, a 3-month experimental-weight loss resulted in increased salivary levels of copper chaperone ATOX1 and alkaline phosphatase in purebred Beagles (unpublished data), leading to the hypothesis of a reduction in oxidative stress, resulting in elevation of antioxidant defense markers, as it was reported in serum in humans (Bawahab et al. 2017).

Rats with insulin resistance presented higher values of superoxide dismutase, catalase, peroxidase and total antioxidant status in the parotid glands in comparison with the control rats (Zalewska et al. 2014).

8.2.8 Stress Markers

Humans Among stress-related biomarkers, the majority of the studies highlight the disturbance of salivary cortisol, cortisone and salivary alpha amylase levels in metabolic and endocrine diseases. These biomarkers were related with BMI, fasting glucose concentrations and insulin sensitivity among others (Incollingo Rodriguez et al. 2015; Aldossari et al. 2019; Liu et al. 2019).

In endocrine diseases, although varying results have been reported, salivary cortisol and cortisone have been suggested to reflect inappropriate production of cortisol in the organism – both excess (e.g., Cushing's syndrome with a sensitivity of 92–100% and a specificity of 93–100%) and insufficiency (e.g., Addison's disease) and to monitor in a non-invasive manner the response to treatment (Gilbert and Lim

2008; Blair et al. 2017). It is important to notice, that saliva collection for cortisol determination when its inappropriate production is suspected should be performed at midnight to eliminate bias due to circadian rhythm (Viardot et al. 2005; Gilbert and Lim 2008).

For the reasons described above and because of their high stability in saliva (i.e. cortisol is stable in saliva up to 6 weeks at room temperature when preserved with citric acid 10 g/L) salivary stress-related biomarkers were reported as promising biomarkers in studies of metabolic and endocrine diseases (Chen et al. 1992; Hartman et al. 2016; Blair et al. 2017).

Animals In dogs, salivary cortisol usefulness to diagnose hiperadrenocortisims was studied (Wenger-Riggenbach et al. 2010). The authors observed that although salivary cortisol in dogs with diagnosed hiperadrenocortisims was higher, overlap between healthy (n = 21) and diseased dogs (n = 6) were detected (2 healthy dogs showed elevated results and 2 dogs with hiperadrenocortisims showed low results) (Wenger-Riggenbach et al. 2010). Nevertheless, low number of animals was used, thus future studies are required in order to clarify the clinical utility of salivary cortisol determination in dogs. Nevertheless, it is important to highlight that when salivary cortisol is up to be determined, sex and neuter status, age, regular living environment, time in environment before testing, testing environment, owner presence during testing, and collection media should be taken into consideration (Cobb et al. 2016). In addition, methods for sample collection, storage and analysis should be acknowledged when comparing results from different studies (Cobb et al. 2016; Damián et al. 2018).

Furthermore, some of the stress-related biomarkers, such as salivary alpha amylase, were observed while using proteomic approach to be altered in metabolic/endocrine diseases in animals (for more information see section *Proteomics*).

8.2.9 Thyroid Hormones

Humans There is a controversy about the possible use of thyroid hormone measurements in saliva for the diagnosis of hypothyroidism. A report that used a RIA, indicated a good correlation between saliva and serum concentrations of thyroxine (T4) (r = 0.74) and a good agreement between saliva T4 values and the functional state of the thyroid (Putz et al. 2009), being this high correlation later confirmed (Gotovtseva and Korot'ko 2002). Meanwhile, Al-Hindawi et al. (2017) did not observe significant differences in free T4 and TSH in saliva between patients with hypothyroidism and healthy individuals, although the values of saliva were parallels to their values in serum. It is of interest to note that in this later study ELISA kits that usually are less sensitive that RIA were used. Furthermore, samples were stored at −20 °C during a not defined time, fact that could have influenced concentrations of the hormones. The sensitivity of the assays has been indicated as one of the main limitations for the measurement of T4 in saliva since concentrations of T4 in this fluid is much lower than in serum (1/100) (Vining et al. 1983).

Animals Till the date, thyroid hormones were not evaluated in saliva of animals in relation to metabolic and/or endocrine diseases.

8.2.10 Trace Elements

Humans Trace elements, such as magnesium, zinc, and calcium levels in saliva were suggested to be useful biomarkers for differentiating patients with type 2 diabetes mellitus from non-diabetics being higher as in controls (Marín Martínez et al. 2018). Furthermore, the salivary magnesium could serve as a marker of high cardiovascular risk since its levels were associated with abdominal obesity in men (Marín Martínez et al. 2018).

8.3 Proteomics

Proteomics means the analysis of the total protein set expressed by a cell or an organism. Different approaches can be used for proteomic analysis, namely expression proteomics, functional proteomics and structural proteomics. In this part of the chapter, we will focus mainly in expression proteomics studies that have been performed in the context of metabolic diseases. Expression proteomics is mainly used to generate a large qualitative data set with the expression levels of the proteins present in the simple. This provides a global analysis of protein composition, post-translational modifications, and the dynamic nature of protein expression.

Salivary proteomics has been shown particularly valuable, since it allows to have a source of information about proteins from a fluid that is collected under a non-invasive way. This opened new doors for disease biomarker discovery, which can be particularly interesting also in the case of endocrine and metabolic diseases. Among these last, diabetes was the one receiving greater attention in terms of saliva proteome. Although fewer, some studies did also report changes in salivary proteome induced by obesity. A review about what is known in humans and animals will be subsequently presented.

Humans Studies in adults and children showed existing differences in salivary proteome of obese individuals in comparison with normal-weight controls. Just for instance, Rangé et al. (2012) observed higher levels of albumin, α- and β- haemoglobin chains, as well as α-defensins 1,2 and 3 in obese individuals, comparatively to normal weight ones. These authors emphasized the interest of saliva as a biological fluid to monitor inflammatory status in obesity, which is considered a low-grade inflammatory condition. Furthermore, differences between obese and non-obese women were observed at the level of proteins such as α-amylase, zinc-α2 glycoprotein and cystatins, among others (Lamy et al. 2015).

Salivary proteome of obese patients was also studied evaluating the effect that body weight loss could have. Bariatric surgery appeared to have effect at the level of salivary proteins like salivary amylase, cystatins and carbonic anhydrase VI (Lamy et al. 2015), proteins referred as potentially involved in oral food perception (Rodrigues et al. 2017). A different study, where weight loss was obtained after weeks of continuous physical activity and caloric restriction, showed also changes in salivary proteome, namely amylase, carbonic anhydrase VI and cystatins, but in opposite direction as changes detected in the study about the effect of bariatric surgery, suggesting that the weight-loss procedures can differently affect physiological changes, reflected in saliva (Simões and Lamy, *not published*).

Also in children, salivary proteome of overweight individuals present differences from normal-weight ones (Rodrigues et al. 2019). When these differences were accessed by in-gel based methods, namely two-dimensional electrophoresis (2-DE), protein spots identified as zinc-α2 glycoprotein, α-amylase and S-type cystatins were observed increased in obese children (Rodrigues et al. 2019).

Concerning salivary α-amylase, different studies report different associations with obesity: some authors refer decreased copy number of the gene that codifies for this protein (Mejía-Benítez et al. 2015; Pinho et al. 2018) and decreased enzymatic activity (Lasisi et al. 2019); however, some other authors observed increased expression levels of spots identified as containing this salivary enzyme (Lamy et al. 2015). At the same time that these different results may seem contradictory, if we look to the proteoforms of the protein that are increased, is possible to see that are mainly forms with molecular masses lower than the one from the native form of the protein (Lamy et al. 2015). It is possible that these are non-active forms, probably resultant from higher proteolysis in the mouth of obese individuals. But this needs further elucidation. Finally, a recent study points that this relationship between obesity and salivary amylase is not simple and that starch intake can modify it (Rukh et al. 2017). Nevertheless, some authors failed to observe a relationship between salivary proteome and BMI (Mosca et al. 2019).

Several studies about saliva proteome in metabolic diseases were performed in the context of diabetes. By using a gel-free based approach, namely multidimensional liquid chromatography/tandem mass spectrometry (2D-LC-MS/MS), whole saliva from type-2 diabetic individuals was compared with non-diabetic ones, with the identification of 65 proteins with an increased greater than twofold, in diabetic, comparatively to control (Rao et al. 2009). The majority of these proteins are involved in metabolism or immune responses. The usefulness of saliva proteome assessment in diabetes were reinforced by the existence of changes in it according to glycemic control, both in adults (Bencharit et al. 2013) and children (Pappa et al. 2018).

Animals Comparatively to humans, there are less studies about salivary proteomics in animals. In the case of metabolic diseases, it is mainly concerning diabetes, using rodent models and, only more recently, studies in dogs concerned with metabolic dysfunctions and obesity were published.

Among the different salivary proteins observed to be related with obesity and metabolic diseases, in animals, salivary α-amylase secretion has been suggested of interest to predict susceptibility for weight gain (Rodrigues et al. 2015). In the mentioned study, rats with susceptibility to obesity presented higher α-amylase levels prior to weight gain experimentally-induced with high-fat diet. But this relationship between this salivary protein and obesity needs to be further elucidated, since some studies observed decreased levels of this salivary protein after diet-induced obesity (Lasisi et al. 2019). A potential role of leptin in salivary α-amylase secretion has been recently suggested (Lamy et al. 2018), leading to the hypothesis that the changes in this salivary protein, associated with obesity, can be due to leptin action. In fact, decreased levels of expression levels and enzymatic activity of salivary α-amylase were observed when hyperleptinemia was induced in an animal model (Lamy et al. 2018), leading to the hypothesis that obese, by having higher circulating leptin levels could also have lower α-amylase levels in their saliva.

Recently, through a proteomic approach, the presence of salivary amylase in dog saliva samples has been reported (de Sousa-Pereira et al. 2015), being this presence supported by the measurement of its enzymatic activity in saliva from these animals (Contreras-Aguilar et al. 2017). Nevertheless, until the moment, a potential association between its levels and dog obesity was not reported.

Potential changes in salivary proteome of dogs submitted to weight loss were recently studied. An increase of BPIFA1 salivary levels was found in purebred Beagles dogs after a 3-month experimentally-weight loss (unpublished data), which goes in line with studies of obesity and insulin resistance in humans (Guo et al. 2017). This protein has been positively correlated with insulin action and associated with immune system and inflammatory pathways (Gubern et al. 2006). As such, the aforementioned study (unpublished data) goes in line with a possible improvement in insulin sensitivity after weight loss. Besides this, other salivary proteins, whose abundance was changed after weight loss, were proteins related with immune system/inflammation, oxidative stress and glucose metabolism. These were the cases of copper chaperone ATOX1 and alkaline phosphatase, for example, whose increased levels lead to the hypothesis of a reduction in oxidative stress, after weight loss (unpublished data).

Also in the case of experimentally-weight loss in Beagle dogs, the salivary levels of angiopoietin like 5 protein was increased and strongly positively correlated with the percentage of weight loss (data not shown). This protein is involved in glucose metabolism and the observed relationship with body weight loss suggests a positive effect of this process in glucose regulation. These results are in accordance with a reported negative association between the levels of serum angiopoietin like 3 and 4 proteins and body weight, diabetes status, and parameters of glucose control across a wide range of BMI (Cinkajzlová et al. 2018). To the best of the authors' knowledge there were no reports of an association between angiopoietin like 5 and obesity.

The relationship between salivary proteome and metabolic diseases was recently demonstrated in a study performed with Obesity Related Metabolic Disease (ORMD) dogs (Lucena et al. 2019). Among the different proteins observed to be different in ORMD dogs, comparatively to control ones, salivary levels of Glucose-6-phosphate dehydrogenase were found to be increased. The levels of the enzyme glucose-6-phosphate dehydrogenase were already reported to be increased in diabetes (Hamzah et al. 2018). This enzyme is involved in the pentose phosphate pathway and its activity is related with the production of NADPH. It was suggested that, at an initial phase of diabetes, increases levels of this enzyme prevent the oxidative stress known to be associate to the development of diabetes and metabolic syndrome. Furthermore, other authors reported a decreased levels of glucose-6-phosphate dehydrogenase in rats with chronic hyperglycemia, resulting in increased oxidative stress (Xu et al. 2005). Other salivary proteins, referred as antioxidant and/or inflammatory biomarkers were also observed to be changed in dog ORMD, by this proteomic approach, and were referred in previous sections of this chapter.

The salivary proteins kallikreins were referred as associated with diabetes, in rodent model. A significant decrease of Kallikrein proteins after 2 months of induced-hyperglycemia in rats was observed, reinforcing results of the effect of chronic hyperglycemia on the proteome of submandibular glands (Alves et al. 2013). Kallikrein proteins are extracellular matrix protein constituents, which had already been reported having salivary levels decreased in type 2 diabetes human subjects (Rao et al. 2009).

8.4 Metabolomics

Being part of the "omic" sciences, metabolomics is the global assessment and validation of endogenous small-molecule metabolites within a biologic system. Salivary metabolomics gained interest in the last few years, becoming of interest to monitor biological status and for monitoring diseases.

Humans Salivary metabolomics is an emerging area, with a limited number of studies, at the moment. However, recently this approach has been used to study metabolic syndrome and fatty liver in obese children (Troisi et al. 2019a) and diabetes and diabetes-related periodontal disease in adults (Barnes et al. 2014). Distinct salivary metabolic signatures for pediatric obesity and its related fatty liver and metabolic syndrome were observed as defined mainly by energy, amino and organic acid metabolism, as well as in intestinal bacteria metabolism (Troisi et al. 2019a). These metabolic processes were associated with the diet, fatty acid synthase pathways, microbiota and intestinal mucins (Troisi et al. 2019a). In the same line, 69 out of 475 detected metabolites (14%) in saliva were over or under-expressed in saliva of patients with diabetes as compared to healthy controls (Barnes et al. 2014). Oxidative stress and anti-oxidadtive capacity through increased purin degradation signature and decreased redox balance and lipid metabolism through altered ω-3/

ω-6 fatty acid profiles were reported to be the main processes associated with diabetes and periodontal disease (Barnes et al. 2014). These data suggest about potentials that have salivary metabolomics studies to identify early occurring metabolic alterations in a non-invasive way, although future research is necessary to confirm these results and to further study the complex biological pathways, their interactions and their possible changes in metabolic and endocrine diseases.

Animals To date, metabolomics was not performed in saliva of animals to evaluate possible alterations in metabolic diseases. Future studies are needed to fill this knowledge gap.

8.5 Microbiome

Microbiome refers to the entire habitat, including microorganisms (bacteria, archaea, lower and upper eukaryotes, and viruses), their genomes (ie, genes), and environmental conditions (Marchesi and Ravel 2015). However some authors limit the definition of microbiome to the collection of genes and genomes of members of a microbiota (Marchesi and Ravel 2015) and in some literature these terms are often used interchangeably (Ursell et al. 2012).

Interest in the study of microbiome has increased in recent years in humans (Maguire and Maguire 2017; Thomas et al. 2017; Mohajeri et al. 2018) and animals (Deng and Swanson 2015; Trinh et al. 2018), since the evidence is growing that associated microorganisms make essential contributions to health and well-being, and changes in their amounts/proportions are associated with a number of different diseases. In the case of the oral microbiota, it has a significant impact on both the oral and systemic health (Dewhirst et al. 2010; Wade 2013; Verma et al. 2018) and is also a potential diagnostic indicator of several oral and systemic diseases (Gao et al. 2018).

Inter-individual variations were significantly larger than intra-individual variations for most of the dominant genera in the oral microbiota (Monteiro-Da-Silva et al. 2014; Barroso et al. 2015; Sato et al. 2015). Multiple factors and their interactions modulate oral microbiome. These factors are intrinsic or extrinsic to the individual (host) and include:

(i) the genetic composition of the host (ex: ethnicity, gender, circadian rhythm);
(ii) oral environment (ex: saliva composition; (Marsh et al. 2016; Lynge Pedersen and Belstrøm 2019));
(iii) lifestyle, behaviour and diet;
(iv) socioeconomic status (Cornejo Ulloa et al. 2019);
(v) climatic conditions (Li et al. 2014) and geographical location (Shaw et al. 2017) although some authors report that the influence of geographical location on the oral microbiome is not significant (Nasidze et al. 2009).

It should also be mentioned, in order to understand the complexity and the dynamics of oral microbiota, composed of hundreds of taxa interacting across multiple spatial scales, that the different structures and tissues of the oral cavity present distinct microbial populations and there are scientific evidences that most oral microbes are site specialists (Aas et al. 2005; Mark Welch et al. 2016; Welch et al. 2019). Distinct microbial communities appear to be observed between dental, tongue, and salivary samples, with high levels of similarity observed between the tongue and salivary communities (Hall et al. 2017). Furthermore, bacterial communities vary along an ecological gradient from the front to the back of the mouth (Proctor et al. 2018).

Humans Scientific evidences exist that oral/salivary microbiome have a link with the presence of obesity and other metabolic diseases. Si et al. (2017) compared the oral microbiome from subgingival plaque with the gut microbiome and their results support the notion that metabolic disease can influence the non-gut human microbiome. In the same manner, other studies also observed differences in oral microbiota between overweight and normal-weight persons (Goodson et al. 2009; Wu et al. 2018; Mervish et al. 2019; Raju et al. 2019). Furthermore, the overweight was related to the greater diversity of microbial species, although lower total amounts (Mervish et al. 2019). Microbiota diversity and composition were significantly associated with body size and gender in school-age children (Raju et al. 2019). Overall, these studies emphasize the utility of local oral bacteria as potential biomarkers for systemic metabolic disease.

Some studies have confirmed that the oral diseases and diabetes mellitus are closely related (Negrato et al. 2013). According to Ebersole et al. (2008) the increased severity of periodontal disease associated with type 2 diabetes may reflect an alteration in the pathogenic potential of periodontal bacteria and/or a modification of the characteristics of the host's inflammatory response. Furthermore, oral microbiota is an important factor in the development of diabetes, affecting oral bone development and increasing the risk and severity of tooth loss (Xiao et al. 2017), what in periodontitis can compromise the chewing process and the sensory function of periodontal tissues (Borges et al. 2012). Saeb et al. (2019) observed a reduction of the biological and phylogenetic diversity in the diabetes and prediabetes oral microbiota in comparison with that in the normoglycemic oral microbiota and this was associated with an increase in the pathogenic content of the hyperglycemic microbiota.

The results of Tam and colleagues suggested that obesity alters composition and diversity of the oral microbiota in patients with type 2 diabetes mellitus, but the impact of glycemic control on oral microbiota, however, remains to be elucidated (Tam et al. 2018). Detailed information on the different taxa and species can be find in literature (Gao et al. 2018; Lu et al. 2019).

Animals To date there are only a few reports that assessed the characterization of dogs (Dewhirst et al. 2012; Sivakami et al. 2015) and cats (Sturgeon et al. 2014; Dewhirst et al. 2015; Adler et al. 2016) oral microbiomes using modern sequencing technology. Equine subgingival plaque microbiota shares many similarities with the human, canine and feline oral microbiomes (Gao et al. 2016). Nevertheless, contrary to what happens in humans, the evaluation of the relationship between the oral microbiome and metabolic diseases is not yet documented.

There is a growing number of families having pets, mainly dogs and cats. Since many of these animals are treated as family members and are in direct contact with the people, it is expected that the exchange of microorganisms between humans and animals occur, through various routes, including saliva (Nishiyama et al. 2007; Song et al. 2013; Misic et al. 2015). However, dogs and their owners presented appreciably differences in oral microbiome, being their oral microbiotas not correlated with residing in the same household (Oh et al. 2015). Further studies are needed to study the oral microbiome relation with metabolic and endocrine disease in animals and the possible connection with changes in animal-owner microbiota.

8.6 Conclusion

Strong scientific evidences exist suggesting saliva to constitute a promising tool to evaluate metabolic diseases and endocrinopathies-related alterations.

References

Aas JA et al (2005) Defining the normal bacterial flora of the oral cavity. J Clin Microbiol 43:5721–5732. https://doi.org/10.1128/JCM.43.11.5721-5732.2005

Abikshyeet P, Ramesh V, Oza N (2012) Glucose estimation in the salivary secretion of diabetes mellitus patients. Diabetol Metab Syndr Obes Target Ther Dove Press 5:149–154. https://doi.org/10.2147/DMSO.S32112

Adler CJ et al (2016) Diet may influence the oral microbiome composition in cats. Microbiome 4:23. https://doi.org/10.1186/s40168-016-0169-y

Aldossari NM, El Gabry EE, Gawish GEH (2019) Association between salivary amylase enzyme activity and obesity in Saudi Arabia. Medicine 98(23):e15878. https://doi.org/10.1097/MD.0000000000015878

Al-Hindawi S, Luaibi N, Al-Ghurabei B (2017) Possible use of saliva as a diagnostic tool in hypothyroidism. Glob J BioSci Biotechnol 6(3):539–542. Available at: http://scienceandnature.org/GJBB_Vol6(3)2017.php. Accessed 11 Aug 2019

Alves RMP et al (2013) iTRAQ-based quantitative proteomic analysis of submandibular glands from rats with STZ-induced hyperglycemia. J Biochem 153:209–220. https://doi.org/10.1093/jb/mvs142

Anderson LC, Garrett JR (1986) Lipid accumulation in the major salivary glands of streptozotocin-diabetic rats. Arch Oral Biol 31(7):469–475

Aral CA et al (2017) Metabolic control and periodontal treatment decreases elevated oxidative stress in the early phases of type 1 diabetes onset. Arch Oral Biol 82:115–120. https://doi.org/10.1016/j.archoralbio.2017.06.009

Attlee A et al (2019) Relationship of salivary adipocytokines, diet quality, physical activity, and nutrition status in adult Emirati females in United Arab Emirates. Diabetes Metab Syndr Clin Res Rev 13(1):40–46. https://doi.org/10.1016/J.DSX.2018.08.006

Balaji A et al (2017) Salivary Interleukin-6 – a pioneering marker for correlating diabetes and chronic periodontitis: a comparative study. Indian J Dent Res 28(2):133. https://doi.org/10.4103/ijdr.IJDR_167_14

Bao L, Odell AF, Stephen SL, Wheatcroft SB, Walker JH, Ponnambalam S (2012) The S100A6 calcium-binding protein regulates endothelial cell-cycle progression and senescence. FEBS J 279(24):4576–4588

Barnes VM et al (2014) Global metabolomic analysis of human saliva and plasma from healthy and diabetic subjects, with and without periodontal disease. PLoS One. Edited by Ö. Yilmaz. Public Library of Science 9(8):e105181. https://doi.org/10.1371/journal.pone.0105181

Barroso E et al (2015) Stability of saliva microbiota during moderate consumption of red wine. Arch Oral Biol 60:1763–1768. https://doi.org/10.1016/j.archoralbio.2015.09.015

Bastard JP et al (2006) Recent advances in the relationship between obesity, inflammation, and insulin resistance. Eur Cytokine Netw 17(1):4–12. Available at: http://www.ncbi.nlm.nih.gov/pubmed/16613757. Accessed 4 Aug 2019

Bawahab MA, Assiri AS, Maksoud WA, Patel A, Kadoumi O, Zaman GS, Alessih RMK, Haider SS (2017) Effects of weight reduction after sleeve gastrectomy on metabolic variables in Saudi obese subjects in Aseer province of kingdom of Saudi Arabia. Obes Surg 27(8):2005–2014

Bencharit S et al (2013) Salivary proteins associated with hyperglycemia in diabetes: a proteomic analysis. Mol BioSyst NIH Public Access 9(11):2785–2797. https://doi.org/10.1039/c3mb70196d

Blair J et al (2017) Salivary cortisol and cortisone in the clinical setting. Curr Opin Endocrinol Diabetes Obes 24(3):161–168. https://doi.org/10.1097/MED.0000000000000328

Borges Tde F et al (2012) Changes in masticatory performance and quality of life in individuals with chronic periodontitis. J Periodontol 84:325–331. https://doi.org/10.1902/jop.2012.120069

Canfora EE et al (2019) Gut microbial metabolites in obesity, NAFLD and T2DM. Nat Rev Endocrinol Nat Publ Group 15(5):261–273. https://doi.org/10.1038/s41574-019-0156-z

Carramolino-Cuéllar E et al (2017) Salivary glucose as a metabolic control marker in patients with type 2 diabetes. J Biol Regul Homeost Agents 31(2 Suppl 1):181–187. Available at: http://www.ncbi.nlm.nih.gov/pubmed/28691471. Accessed 4 July 2019

Chen YM, Cintrón NM, Whitson PA (1992) Long-term storage of salivary cortisol samples at room temperature. Clin Chem 38(2):304. Available at: http://www.ncbi.nlm.nih.gov/pubmed/1541015. Accessed 11 Aug 2019

Cinkajzlová A et al (2018) Angiopoietin-like protein 3 and 4 in obesity, type 2 diabetes mellitus, and malnutrition: the effect of weight reduction and realimentation. Nutr Diabetes 8(1):21. https://doi.org/10.1038/s41387-018-0032-2

Cobb ML et al (2016) A systematic review and meta-analysis of salivary cortisol measurement in domestic canines. Domest Anim Endocrinol Elsevier 57:31–42. https://doi.org/10.1016/J.DOMANIEND.2016.04.003

Contreras-Aguilar MD et al (2017) Detection and measurement of alpha-amylase in canine saliva and changes after an experimentally induced sympathetic activation. BMC Vet Res 13:266. https://doi.org/10.1186/s12917-017-1191-4

Cornejo Ulloa P, van der Veen MH, Krom BP (2019) Review: modulation of the oral microbiome by the host to promote ecological balance. Odontology 107:437–448. https://doi.org/10.1007/s10266-019-00413-x

Damián JP et al (2018) Serial collection method of dog saliva: effects of different chemical stimulants on behaviour, volume and saliva composition. Open Vet J 8(3):229. https://doi.org/10.4314/ovj.v8i3.1

de Sousa-Pereira P et al (2015) Cross species comparison of mammalian saliva using a LC-MALDI-based proteomic approach. Proteomics 15:1598–1607. https://doi.org/10.1002/pmic.201400083

De A et al (2018) Estimation of salivary and serum alkaline phosphatase level as a diagnostic marker in type-2 diabetes mellitus with periodontal health and disease: a clinico-biochemical study. J Oral maxillofac Pathol JOMFP Wolters Kluwer Medknow Publ 22(3):445. https://doi.org/10.4103/jomfp.JOMFP_212_18

Deng P, Swanson KS (2015) Gut microbiota of humans, dogs and cats: current knowledge and future opportunities and challenges. Br J Nutr 113:S6–S17. https://doi.org/10.1017/S0007114514002943

Desai GS, Mathews ST (2014) Saliva as a non-invasive diagnostic tool for inflammation and insulin-resistance. World J Diabetes Baishideng Publ Group Inc 5(6):730–738. https://doi.org/10.4239/wjd.v5.i6.730

Dewhirst FE et al (2010) The human oral microbiome. J Bacteriol 192:5002–5017. https://doi.org/10.1128/JB.00542-10

Dewhirst FE et al (2012) The canine oral microbiome. PLoS One 7:e36067. https://doi.org/10.1371/journal.pone.0036067

Dewhirst FE et al (2015) The feline oral microbiome: a provisional 16S rRNA gene based taxonomy with full-length reference sequences. Vet Microbiol 175:294–303. https://doi.org/10.1016/j.vetmic.2014.11.019

Dezayee ZI, Al-Nimer MM (2016) Saliva C-reactive protein as a biomarker of metabolic syndrome in diabetic patients. Indian J Dent Res 27(4):388. https://doi.org/10.4103/0970-9290.191887

Ebersole JL et al (2008) Microbiologic and immunologic characteristics of periodontal disease in Hispanic Americans with type 2 diabetes. J Periodontol 79:637–646. https://doi.org/10.1902/jop.2008.070455

Fekete Z et al (1993) Salivary and plasma insulin levels in man. Biochem Mol Biol Int 30(4):623–629. Available at: http://www.ncbi.nlm.nih.gov/pubmed/8401319. Accessed 1 Aug 2019

Gao W et al (2016) In-depth snapshot of the equine subgingival microbiome. Microb Pathog 94:76–89. https://doi.org/10.1016/j.micpath.2015.11.002

Gao L et al (2018) Oral microbiomes: more and more importance in oral cavity and whole body. Protein Cell 9:488–500. https://doi.org/10.1007/s13238-018-0548-1

Gianfrancesco MA et al (2018) Lipid bilayer stress in obesity-linked inflammatory and metabolic disorders. Biochem Pharmacol 153:168–183. https://doi.org/10.1016/j.bcp.2018.02.022

Gilbert R, Lim EM (2008) The diagnosis of Cushing's syndrome: an endocrine society clinical practice guideline. Clin Biochem Rev Aust Assoc Clin Biochem 29(3):103–106. Available at: http://www.ncbi.nlm.nih.gov/pubmed/19107223. Accessed 11 Aug 2019

Goodson JM et al (2009) Is obesity an oral bacterial disease? J Dent Res Int Assoc Dent Res 88(6):519–523. https://doi.org/10.1177/0022034509338353

Gotovtseva LP, Korot'ko GF (2002) Salivary thyroid hormones in evaluation of the functional state of the hypophyseal-thyroid system. Klin Lab Diagn 7:9–11. Available at: http://www.ncbi.nlm.nih.gov/pubmed/12187594. Accessed 11 Aug 2019

Gubern C et al (2006) Natural antibiotics and insulin sensitivity: the role of bactericidal/permeability-increasing protein. Diabetes 55:216–224. https://doi.org/10.2337/diabetes.55.01.06.db05-1108

Guo Y et al (2017) Associations of salivary BPIFA1 protein in chronic periodontitis patients with type 2 diabetes mellitus. Int J Endocrinol 2017:1087017. https://doi.org/10.1155/2017/1087017

Gupta S et al (2017) Correlation of salivary glucose level with blood glucose level in diabetes mellitus. J Oral Maxillofac Pathology JOMFP Wolters Kluwer Medknow Publ 21(3):334–339. https://doi.org/10.4103/jomfp.JOMFP_222_15

Hall MW et al (2017) Inter-personal diversity and temporal dynamics of dental, tongue, and salivary microbiota in the healthy oral cavity. npj Biofilms Microbiomes 3:2. https://doi.org/10.1038/s41522-016-0011-0

Hamzah SA, Hamza LA, Rahman HS (2018) Impact of diabetes mellitus type 2 in the activity of glucose-6-phosphate dehydrogenase in human erythrocyte. Kurdistan Journal of Applied Research 3(1):58–62

Hartman ML et al (2016) Salivary biomarkers in pediatric metabolic disease research. Pediatr Endocrinol Rev PER 13(3):602–611. Available at: http://www.ncbi.nlm.nih.gov/pubmed/27116847. Accessed 27 Feb 2018

Hou C et al (2019) Beneficial effects of pomegranate on lipid metabolism in metabolic disorders. Mol Nutr Food Res 63:1800773. https://doi.org/10.1002/mnfr.201800773

Incollingo Rodriguez AC et al (2015) Hypothalamic-pituitary-adrenal axis dysregulation and cortisol activity in obesity: a systematic review. Psychoneuroendocrinology 62:301–318. https://doi.org/10.1016/j.psyneuen.2015.08.014

Janem WF et al (2017) Salivary inflammatory markers and microbiome in normoglycemic lean and obese children compared to obese children with type 2 diabetes. PLoS One. Edited by Ö. Yilmaz 12(3):e0172647. https://doi.org/10.1371/journal.pone.0172647

Jayachandran T, Srinivasan B, Padmanabhan S (2017) Salivary leptin levels in normal weight and overweight individuals and their correlation with orthodontic tooth movement. Angle Orthod 87(5):739–744. https://doi.org/10.2319/120216-869.1

Katsiougiannis S et al (2006) Salivary gland epithelial cells: a new source of the immunoregulatory hormone adiponectin. Arthritis Rheum Wiley-Blackwell 54(7):2295–2299. https://doi.org/10.1002/art.21944

Kealy RD et al (2002) Effects of diet restriction on life span and age-related changes in dogs. J Am Vet Med Assoc Am Vet Med Assoc 220(9):1315–1320. https://doi.org/10.2460/javma.2002.220.1315

Kerr M et al (1995) Detection of insulin and insulin-like growth factors I and II in saliva and potential synthesis in the salivary glands of mice: effects of type 1 diabetes mellitus. Biochem Pharmacol Elsevier 49(10):1521–1531. https://doi.org/10.1016/0006-2952(95)00017-T

Lamy E et al (2015) Changes in the salivary protein profile of morbidly obese women either previously subjected to bariatric surgery or not. J Physiol Biochem Springer Neth 71(4):691–702. https://doi.org/10.1007/s13105-015-0434-8

Lamy E et al (2018) Effects of hyperleptinemia in rat saliva composition, histology and ultrastructure of the major salivary glands. Arch Oral Biol Pergamon 96:1–12. https://doi.org/10.1016/J.ARCHORALBIO.2018.08.005

Langley LL et al (1963) Secretion of glucose-C14 by dog parotid gland. Arch Oral Biol Pergamon 8(2):127–133. https://doi.org/10.1016/0003-9969(63)90050-5

Lasisi TJ, Shittu STT, Alada AR (2019) Re-establishing normal diet following high fat-diet-induced obesity reverses the altered salivary composition in Wistar rats. J Basic Clin Physiol Pharmacol 30:111–120. https://doi.org/10.1515/jbcpp-2018-0006

Li J et al (2014) Comparative analysis of the human saliva microbiome from different climate zones: Alaska, Germany, and Africa. BMC Microbiol 14:316. https://doi.org/10.1186/s12866-014-0316-1

Liu X et al (2019) Gender-specific independent and combined effects of the cortisol-to-cortisone ratio and 11-deoxycortisol on prediabetes and type 2 diabetes mellitus: from the Henan rural cohort study. J Diabetes Res 2019:1–8. https://doi.org/10.1155/2019/4693817

Lu M, Xuan S, Wang Z (2019) Oral microbiota: a new view of body health. Food Sci Human Wellness 8:8–15. https://doi.org/10.1016/j.fshw.2018.12.001

Lucena S et al (2019) Comparative proteomic analysis of saliva from dogs with and without obesity-related metabolic dysfuntion. J Proteome 201:65–72. https://doi.org/10.1016/j.jprot.2019.04.010

Lynge Pedersen AM, Belstrøm D (2019) The role of natural salivary defences in maintaining a healthy oral microbiota. J Dent 80:S3–S12. https://doi.org/10.1016/j.jdent.2018.08.010

Madi M et al (2016) Status of serum and salivary levels of superoxide dismutase in type 2 diabetes mellitus with oral manifestations: a case control study. Ethiop J Health Sci 26(6):523. https://doi.org/10.4314/ejhs.v26i6.4

Maguire M, Maguire G (2017) The role of microbiota, and probiotics and prebiotics in skin health. Arch Dermatol Res 309:411–421. https://doi.org/10.1007/s00403-017-1750-3

Mahay S, Adeghate E, Lindley MZ, Rolph CE, Singh J (2004) Streptozotocin-induced type 1 diabetes mellitus alters the morphology, secretory function and acyl lipid contents in the isolated rat parotid salivary gland. Mol Cell Biochem 261(1):175–181

Marchesi JR, Ravel J (2015) The vocabulary of microbiome research: a proposal. Microbiome 3:31. https://doi.org/10.1186/s40168-015-0094-5

Marchetti P et al (1986) Salivary insulin concentrations in type 2 (non-insulin-dependent) diabetic patients and obese non-diabetic subjects: relationship to changes in plasma insulin levels after an oral glucose load. Diabetologia 29(10):695–698. Available at: http://www.ncbi.nlm.nih.gov/pubmed/3542670. Accessed 25 June 2019

Marín Martínez L, Molino Pagán D, López Jornet P (2018) Trace elements in saliva as markers of type 2 diabetes mellitus. Biol Trace Elem Res 186(2):354–360. https://doi.org/10.1007/s12011-018-1326-x

Mark Welch JL et al (2016) Biogeography of a human oral microbiome at the micron scale. Proc Natl Acad Sci 113:E791–E800. https://doi.org/10.1073/pnas.1522149113

Marsh PD et al (2016) Influence of saliva on the oral microbiota. Periodontology 2000:80–92. https://doi.org/10.1111/prd.12098

Matczuk J et al (2017) Salivary lipids: a review. Adv Clin Exp Med 26(6):1023–1031. https://doi.org/10.17219/acem/63030

Mejía-Benítez MA et al (2015) Beneficial effect of a high number of copies of salivary amylase AMY1 gene on obesity risk in Mexican children. Diabetologia 58:290–294. https://doi.org/10.1007/s00125-014-3441-3

Meriç P et al (2018) Salivary adiponectin and leptin levels are increased in women with gestational diabetes mellitus and gingival inflammation. Oral Health Prev Dent 16(6):541–547. https://doi.org/10.3290/j.ohpd.a41658

Mervish NA et al (2019) Associations of the oral microbiota with obesity and menarche in inner city girls. J Child Obes iMedPub 4(1):2–10.21767/2572-5394.100068

Messenger B, Clifford MN, Morgan LM (2003) Glucose-dependent insulinotropic polypeptide and insulin-like immunoreactivity in saliva following sham-fed and swallowed meals. J Endocrinol 177(3):407–412. Available at: http://www.ncbi.nlm.nih.gov/pubmed/12773121. Accessed 24 June 2019

Misic AM et al (2015) The shared microbiota of humans and companion animals as evaluated from Staphylococcus carriage sites. Microbiome 3:2. https://doi.org/10.1186/s40168-014-0052-7

Mohajeri MH et al (2018) The role of the microbiome for human health: from basic science to clinical applications. Eur J Nutr 57:1–14. https://doi.org/10.1007/s00394-018-1703-4

Monteiro-Da-Silva F, Araujo R, Sampaio-Maia B (2014) Interindividual variability and intraindividual stability of oral fungal microbiota over time. Med Mycol 52:498–505. https://doi.org/10.1093/mmy/myu027

Morris PA, Prout RES, Proctor GB, Garrett JR, Anderson LC (1992) Lipid analysis of the major salivary glands in streptozotocin-diabetic rats and the effects of insulin treatment. Arch Oral Biol 37(6):489–494

Mosca AC et al (2019) How are macronutrient intake, BMI, ethnicity, age, and gender related to the composition of unstimulated saliva? A case study. J Texture Stud 50:53–61. https://doi.org/10.1111/jtxs.12362

Muñoz-Prieto A, Martínez-Subiela S et al (2019a) A new highly sensitive immunoassay for the detection of adiponectin in serum and saliva of dogs and its application in obesity and canine leishmaniosis. Res Vet Sci WB Saunders 125:374–381. https://doi.org/10.1016/J.RVSC.2019.07.019

Muñoz-Prieto A, Escribano D et al (2019b) Glucose, fructosamine, and insulin measurements in saliva of dogs: variations after an experimental glucose administration. Domest Anim Endocrinol Elsevier 66:64–71. https://doi.org/10.1016/J.DOMANIEND.2018.10.002

Murakami K, Taniguchi H, Baba S (1982) Presence of insulin-like immunoreactivity and its biosynthesis in rat and human parotid gland. Diabetologia Springer-Verlag 22(5):358–361. https://doi.org/10.1007/BF00253582

Naidoo T et al (2012) Elevated salivary C-reactive protein predicted by low cardio-respiratory fitness and being overweight in African children. Cardiovasc J Afr Clinics Cardive Publ (Pty) Ltd 23(9):501–506. https://doi.org/10.5830/CVJA-2012-058

Nakamoto I et al (2003) Correlation between saliva glycated and blood glycated proteins. Environ Health Prev Med Springer Verlag 8(3):95–99. https://doi.org/10.1007/BF02897922
Nasidze I et al (2009) Global diversity in the human salivary microbiome. Genome Res 19:636–643. https://doi.org/10.1101/gr.084616.108
NCD Risk Factor Collaboration (NCD-RisC) (2016) Trends in adult body-mass index in 200 countries from 1975 to 2014: a pooled analysis of 1698 population-based measurement studies with 19·2 million participants. Lancet 387(10026):1377–1396. https://doi.org/10.1016/S0140-6736(16)30054-X
Negrato CA et al (2013) Periodontal disease and diabetes mellitus. J Appl Oral Sci Rev FOB 21(1):1–12. https://doi.org/10.1590/1678-7757201302106
Nigro E et al (2015) Evaluation of salivary adiponectin profile in obese patients. Peptides 63:150–155. https://doi.org/10.1016/j.peptides.2014.11.007
Nishiyama SAB et al (2007) Detection of putative periodontal pathogens in subgingival specimens of dogs. Braz J Microbiol 38:23–28. https://doi.org/10.1590/S1517-83822007000100006
Oh C et al (2015) Comparison of the oral microbiomes of canines and their owners using next-generation sequencing. PLoS One 10:e0131468. https://doi.org/10.1371/journal.pone.0131468
Panchbhai AS (2012) Correlation of salivary glucose level with blood glucose level in diabetes mellitus. J Oral Maxillofac Res 3(3):e3. https://doi.org/10.5037/jomr.2012.3303
Pappa A-M et al (2018) Organic electronics for point-of-care metabolite monitoring. Trends Biotechnol Elsevier Curr Trends 36(1):45–59. https://doi.org/10.1016/J.TIBTECH.2017.10.022
Pasic J, Pickup JC (1988) Salivary insulin in normal and type I diabetic subjects. Diabetes Care 11(6):489–494. https://doi.org/10.2337/diacare.11.6.489
Pinho S, Padez C, Manco L (2018) High AMY1 copy number protects against obesity in Portuguese young adults. Ann Hum Biol 45:435–439. https://doi.org/10.1080/03014460.2018.1490452
Proctor DM et al (2018) A spatial gradient of bacterial diversity in the human oral cavity shaped by salivary flow. Nat Commun 9:681. https://doi.org/10.1038/s41467-018-02900-1
Puttaswamy KA, Puttabudhi JH, Raju S (2017) Correlation between salivary glucose and blood glucose and the implications of salivary factors on the oral health status in type 2 diabetes mellitus patients. J Int Soc Prev Community Dent Wolters Kluwer Medknow Publ 7(1):28–33. https://doi.org/10.4103/2231-0762.200703
Putz Z, Vaňuga A, Velemínský J (2009) Radioimmunoassay of thyroxine in saliva. Exp Clin Endocrinol Diabetes 85(02):199–203. https://doi.org/10.1055/s-0029-1210436
Rabinowitz JL, Shannon IL (1975) Lipid changes in human male parotid saliva by stimulation. Arch Oral Biol Pergamon 20(7):403–406. https://doi.org/10.1016/0003-9969(75)90223-X
Raju SC et al (2019) Gender-specific associations between saliva microbiota and body size. Front Microbiol 10:767. https://doi.org/10.3389/fmicb.2019.00767
Rangé H et al (2012) Salivary proteome modifications associated with periodontitis in obese patients. J Clin Periodontol 39(9):799–806. https://doi.org/10.1111/j.1600-051X.2012.01913.x
Rao PV et al (2009) Proteomic identification of salivary biomarkers of type-2 diabetes. J Proteome Res 8:239–245. https://doi.org/10.1021/pr8003776
Rodrigues L et al (2015) Effects of high fat diet on salivary α-amylase, serum parameters and food consumption in rats. Arch Oral Biol 60:854–862. https://doi.org/10.1016/j.archoralbio.2015.02.015
Rodrigues L et al (2017) Association between salivary leptin levels and taste perception in children. J Nutr Metab Hindawi 2017:1–7. https://doi.org/10.1155/2017/7260169
Rodrigues L et al (2019) Comparison of salivary proteome of children with different sensitivities for bitter and sweet tastes: association with body mass index. Int J Obes Nat Publ Group 43(4):701–712. https://doi.org/10.1038/s41366-018-0289-5
Rukh G et al (2017) Dietary starch intake modifies the relation between copy number variation in the salivary amylase gene and BMI. Am J Clin Nutr 106:256–262. https://doi.org/10.3945/ajcn.116.149831
Saeb ATM et al (2019) Relative reduction of biological and phylogenetic diversity of the oral microbiota of diabetes and pre-diabetes patients. Microb Pathog 128:215–229. https://doi.org/10.1016/j.micpath.2019.01.009

Sandham HJ, Kleinberg I (1969a) The effect of glucose concentration on the interrelation between glucose utilization, pH and carbohydrate storage in a salivary system. Arch Oral Biol 14(6):603–618. https://doi.org/10.1016/0003-9969(69)90184-8

Sandham HJ, Kleinberg I (1969b) Utilization of glucose and lactic acid by salivary sediment. Arch Oral Biol 14(6):597–602. https://doi.org/10.1016/0003-9969(69)90183-6

Sato Y et al (2015) Inter-individual differences in the oral bacteriome are greater than intra-day fluctuations in individuals. PLoS One 10:e0131607. https://doi.org/10.1371/journal.pone.0131607

Schmidt EMS et al (2016) Development and validation of an assay for measurement of leptin in pig saliva. BMC Vet Res BioMed Cent 12(1):242. https://doi.org/10.1186/s12917-016-0871-9

Shaw L et al (2017) The human salivary microbiome is shaped by shared environment rather than genetics: evidence from a large family of closely related individuals. MBio 8(5):pii: e01237-17. https://doi.org/10.1128/mBio.01237-17

Shubnikova EA, Volkova EF, Printseva OYA (1984) Submandibular glands as organs of synthesis and accumulation of insulin-like protein. Acta histochemica 74(2):157–171. Available at: http://www.ncbi.nlm.nih.gov/pubmed/6431732. Accessed 24 June 2019

Si J, Lee C, Ko G (2017) Oral microbiota: microbial biomarkers of metabolic syndrome independent of host genetic factors. Front Cell Infect Microbiol 7. https://doi.org/10.3389/fcimb.2017.00516

Sivakami B, Ambarin Farizah B, Aishwarya T (2015) Vaishnavi, Viveka Devi R, Sakthivel K. Isolation and characterisation of salivary microbiota of street dogs. Journal of Biological and Information Sciences. 4(1):1–5. Available from: http://archive.biolim.org/jbis/read/BOJ001A0029

Smriti K et al (2016) Salivary glucose as a diagnostic marker for diabetes mellitus. J Diabetes Sci Technol Diabetes Technol Soc 10(4):991–992. https://doi.org/10.1177/1932296816637619

Song SJ et al (2013) Cohabiting family members share microbiota with one another and with their dogs. elife 2:e00458. https://doi.org/10.7554/eLife.00458

Sridharan S et al (2017) Salivary alkaline phosphatase as a noninvasive marker for periodontal disease in children with uncontrolled type 1 diabetes mellitus. J Clin Pediatr Dent 41(1):70–74. https://doi.org/10.17796/1053-4628-41.1.70

Sturgeon A et al (2014) Characterization of the oral microbiota of healthy cats using next-generation sequencing. Vet J 201:223–229. https://doi.org/10.1016/j.tvjl.2014.01.024

Swinburn BA et al (2019) The global syndemic of obesity, undernutrition, and climate change: the Lancet Commission report. Lancet (Lond Engl) Elsevier 393(10173):791–846. https://doi.org/10.1016/S0140-6736(18)32822-8

Tam J et al (2018) Obesity alters composition and diversity of the oral microbiota in patients with type 2 diabetes mellitus independently of glycemic control. PLoS One 13:e0204724. https://doi.org/10.1371/journal.pone.0204724

Taouis M et al (1995) Biological activity of immunoreactive insulin-like activity extracted from rat submandibular gland. Am J Physiol Am Physiol Soc 269(2 Pt 1):E277–E282. https://doi.org/10.1152/ajpendo.1995.269.2.E277

Teke E et al (2019) Does metabolic control affect salivary adipokines in type 2 diabetes mellitus? Dent Med Prob 56(1):11–20. https://doi.org/10.17219/dmp/103417

Thanakun S et al (2014) Comparison of salivary and plasma adiponectin and leptin in patients with metabolic syndrome. Diabetol Metab Syndr BioMed Cent 6(1):19. https://doi.org/10.1186/1758-5996-6-19

Thomas S et al (2017) The host microbiome regulates and maintains human health: a primer and perspective for non-microbiologists. Cancer Res 77:1783–1812. https://doi.org/10.1158/0008-5472.CAN-16-2929

Torino C et al (2016) Oxidative stress as estimated by gamma-glutamyl transferase levels amplifies the alkaline phosphatase-dependent risk for mortality in ESKD patients on dialysis. Oxidative Med Cell Longev Hindawi Ltd 2016:8490643. https://doi.org/10.1155/2016/8490643

Trinh P et al (2018) One health relationships between human, animal, and environmental microbiomes: a mini-review. Front Public Health 6:235. https://doi.org/10.3389/fpubh.2018.00235

Troisi J, Belmonte F, Bisogno A, Pierri L et al (2019a) Metabolomic salivary signature of pediatric obesity related liver disease and metabolic syndrome. Nutrients 11:pii: E274. https://doi.org/10.3390/nu11020274

Troisi J, Belmonte F, Bisogno A, Lausi O et al (2019b) Salivary markers of hepato-metabolic comorbidities in pediatric obesity. Dig Liver Dis WB Saunders 51(4):516–523. https://doi.org/10.1016/J.DLD.2018.11.009

Tvarijonaviciute A et al (2014) Measurement of salivary adiponectin concentrations in dogs. Vet Clin Pathol 43(3):416–421. https://doi.org/10.1111/vcp.12169

Tvarijonaviciute A et al (2017) Leptin and NGF in saliva of patients with diabetes mellitus type 2: a pilot study. J Oral Pathol Med 46(9):853–855. https://doi.org/10.1111/jop.12587

Tvarijonaviciute A et al (2019) Saliva as a non-invasive tool for assessment of metabolic and inflammatory biomarkers in children. Clin Nutr. https://doi.org/10.1016/j.clnu.2019.10.034

Ursell LK et al (2012) Defining the human microbiome. Nutr Rev 70:S38–S44. https://doi.org/10.1111/j.1753-4887.2012.00493.x

Verma D, Garg PK, Dubey AK (2018) Insights into the human oral microbiome. Arch Microbiol 200:525–540. https://doi.org/10.1007/s00203-018-1505-3

Viardot A et al (2005) Reproducibility of nighttime salivary cortisol and its use in the diagnosis of hypercortisolism compared with urinary free cortisol and overnight dexamethasone suppression test. J Clin Endocrinol Metab Narnia 90(10):5730–5736. https://doi.org/10.1210/jc.2004-2264

Vining RF, McGinley RA, Symons RG (1983) Hormones in saliva: mode of entry and consequent implications for clinical interpretation. Clin Chem 29(10). Available at: http://clinchem.aaccjnls.org/content/29/10/1752.long. Accessed 11 Aug 2019

Wade WG (2013) The oral microbiome in health and disease. Pharmacol Res Acad 69(1):137–143. https://doi.org/10.1016/J.PHRS.2012.11.006

Welch JLM, Dewhirst FE, Borisy GG (2019) Biogeography of the oral microbiome: the site-specialist hypothesis. Annu Rev Microbiol 73:335–358. https://doi.org/10.1146/annurev-micro-090817-062503

Wenger-Riggenbach B et al (2010) Salivary cortisol concentrations in healthy dogs and dogs with hypercortisolism. J Vet Int Med Wiley 24(3):551–556. https://doi.org/10.1111/j.1939-1676.2010.0494.x

Wu Y et al (2018) Characterization of the salivary microbiome in people with obesity. PeerJ 6:e4458. https://doi.org/10.7717/peerj.4458

Xiao E et al (2017) Diabetes enhances IL-17 expression and alters the oral microbiome to increase its pathogenicity. Cell Host Microbe 22:120–128.e4. https://doi.org/10.1016/j.chom.2017.06.014

Xu Y, Osborne BW, Stanton RC (2005) Diabetes causes inhibition of glucose-6-phosphate dehydrogenase via activation of PKA, which contributes to oxidative stress in rat kidney cortex. Am J Physiol Ren Physiol 289:F1040–F1047. https://doi.org/10.1152/ajprenal.00076.2005

Yin W et al (2012) Plasma lipid profiling across species for the identification of optimal animal models of human dyslipidemia. J Lipid Res Am Soc Biochem Mol Biol 53(1):51–65. https://doi.org/10.1194/jlr.M019927

Zalewska A et al (2014) Antioxidant profile of salivary glands in high fat diet-induced insulin resistance rats. Oral Dis 20(6):560–566. https://doi.org/10.1111/odi.12173

Zygula A et al (2019) Oxidative stress markers in saliva and plasma differ between diet-controlled and insulin-controlled gestational diabetes mellitus. Diabetes Res Clin Pract Elsevier 148:72–80. https://doi.org/10.1016/J.DIABRES.2018.11.021

Chapter 9
Salivary Markers in Inflammatory and Autoimmune Diseases

Alberto Muñoz-Prieto, Eduardo Pons-Fuster, and Pia López-Jornet

9.1 Introduction and Aims

Inflammatory diseases comprise a broad and heterogeneous range of disorders of different origins. Immune-mediated inflammatory diseases characterized by chronic systemic inflammation are caused by immune system disorders that can affect different organs. Autoimmune diseases are of uncertain etiology, though genetic and environmental factors are known to play a role in their etiopathogenesis. They develop when intrinsic immune system damage results in a loss of self-tolerance, giving rise to abnormal reactions against the host tissues that persist over time. According to the latest data of the World Health Organization (WHO), over 80 autoimmune diseases have been identified that affect 3–7% of the world population, and some of them are characterized by similar symptoms. The great majority of those affected (90%) are women of child-bearing age and with an important stress component that tends to act as a symptoms-triggering factor. Autoimmune diseases can be divided into two well differentiated groups: organ-specific autoimmune diseases (i.e., diseases in which the immune system reacts against concrete body organs) and

A. Muñoz-Prieto
Interlab-UMU, Regional Campus of International Excellence Campus Mare Nostrum, University of Murcia, Murcia, Spain

E. Pons-Fuster
Biomedical Research Institute (IMIB-Arrixaca), Faculty of Medicine, Regional Campus of International Excellence Campus Mare Nostrum, University of Murcia, Murcia, Spain

P. López-Jornet (✉)
Department of Oral Medicine, Faculty of Medicine and Dentistry, Regional Campus of International Excellence Campus Mare Nostrum, University of Murcia, Murcia, Spain
e-mail: majornet@um.es

A. Tvarijonaviciute et al. (eds.), *Saliva in Health and Disease*,
https://doi.org/10.1007/978-3-030-37681-9_9

systemic autoimmune diseases (i.e., disorders that can affect any part of the body as well as several systems or multiple organs simultaneously).

Salivary biomarkers are powerful tools that can act as early predictors of diseases and contribute to their prevention and treatment – though they will require confirmation and validation before being applied to clinical practice.

Overall, the present chapter reviews saliva as a potential and effective diagnostic and monitoring tool in patients with inflammatory and immune diseases both in human and veterinary medicine. However, due to the information availability, two different approaches will be used. In human medicine (part 1 of the chapter), the information is organized according to individual oral diseases. Meanwhile, in animals (part 2 of the chapter), due to lower number of available information on the topic about individual diseases, the information is grouped in relation to target salivary biomarker.

9.2 Part 1: Inflammatory and Immunemediated Diseases in Human Medicine

9.2.1 Diseases of the Oral Mucosa

9.2.1.1 Recurrent Aphthous Stomatitis

Recurrent aphthous stomatitis (RAS) is one of the most frequent diseases of the oral mucosa, and is characterized by self-limiting outbreaks of mucosal ulcerations (Hassona and Scully 2016; Escudier et al. 2006). There are three clinical forms:

- Minor RAS: (the most common presentation): between 1–5 ulcers measuring under 1 cm in diameter appear on the non-keratinized mucosa and heal within 10–15 days without leaving scars.
- Major RAS: (the most severe presentation): between 1–2 ulcers measuring over 1 cm in diameter appear on the lips or soft palate and persist for 4–6 weeks, leaving a scar on healing.
- Herpetiform RAS: between 10–100 ulcers measuring 2–3 mm in diameter appear and tend to merge to form larger ulcerations with irregular margins.

The diagnosis of RAS is based on the anamnesis and clinical manifestations, with the exclusion of other related systemic disorders (Tugrul et al. 2016; Freitas et al. 1998). A biopsy in this case is not very useful, since the findings usually correspond to a nonspecific inflammatory process – though histopathological study may be decided if the diagnosis proves uncertain (Eguia-del Valle et al. 2013).

Different studies have found individuals with aphthae to present increased lipid peroxidation as evidenced by a rise in malondialdehyde (MDA) levels. Furthermore, there have been reports of alterations in salivary biomarkers related to oxidative stress (superoxide dismutase [SOD], glutathione peroxidase, catalase), showing that the enzyme and non-enzyme mediated antioxidant defense systems are decreased in patients with aphthae (Saral et al. 2005; Cimen et al. 2003; Karincaoglu et al. 2005).

It has been found that superoxide dismutase, glutathione peroxidase and catalase are increased in saliva but decreased in plasma. In this regard, it has been suggested that the salivary enzyme mediated antioxidant defense systems may be locally stimulated against the aphthae of RAS, thus indicating that the body is able to shift its antioxidant capacity towards those locations where it is needed.

In Behçet's disease, a chronic systemic disorder characterized by oral and genital ulcerations, eye lesions and skin manifestations, increased superoxide production by neutrophils and macrophages in plasma has been observed, with the presence of circulating pro-oxidant substances causing specific chromosomal damage – thereby confirming the intervention of oxidative stress in this disease. Another oxidative stress marker in Behçet's disease is the presence of lipid peroxidation products and cytokines, particularly tumor necrosis factor-alpha (TNFα). Substances reactive to thiobarbituric acid are also significantly increased in these patients (Karincaoglu et al. 2005).

9.2.1.2 Pemphigus and Pemphigoid

Autoimmune blistering diseases are relatively infrequent. The most common presentations are pemphigus vulgaris and pemphigus foliaceus, mucosal and bullous pemphigoid, linear IgA disease and herpetiform dermatitis (Russo et al. 2017).

Pemphigus is characterized by mucocutaneous manifestations in which the elemental lesion is an intraepithelial blister generated by acantholysis. The course of this organ-specific disease is aggressive and chronic, with the production of vesicles and blisters on the skin and mucous membranes as a result of autoantibody action targeted to specific proteins located at the intercellular junctions of the epithelium (Russo et al. 2017; Fuertes De Vega et al. 2014). There are a number of clinical variants of pemphigus. The two principal forms are pemphigus vulgaris (PV) and pemphigus foliaceus (PF). These two disorders in turn are differentiated clinically and histologically, and by the autoantibodies involved. Pemphigus vulgaris is the most frequent presentation, and is characterized by important mucous membrane involvement. Pemphigus foliaceus in turn manifests at skin level. Additional variants are pemphigus vegetans and pemphigus erythematosus. Other recent and much less common forms have also been described, such as paraneoplastic pemphigus (PNP), pemphigus IgA and pemphigus herpetiformis. Oral manifestations are very frequent in pemphigus vulgaris and paraneoplastic pemphigus, but very infrequent in the other types of pemphigus (Esmaili et al. 2015).

The diagnosis of lesions of this kind is based on the combination of the clinical and histological findings, with the use of direct immunofluorescence (DIF) techniques fundamentally applied to affected tissue biopsies.

A BIOCHIP based indirect immunofluorescence technique has recently been introduced for the determination of DSG3 and DSG1 auto-antibodies. Although anti-DSG3 and anti-DSG1 ELISA testing in saliva has already been described for the diagnosis of pemphigus vulgaris, a lack of correlation is observed between the serum and saliva results of the BIOCHIP technique and ELISA testing – thus suggesting that saliva might not be the ideal substrate for the laboratory diagnosis of pemphigus vulgaris (Russo et al. 2017).

Pemphigoid is an autoimmune disease characterized by the presence of auto-antibodies targeted to the basal membrane of the epithelia. Two presentations can be found: bullous pemphigoid and cicatricial pemphigoid. The former affects particularly the skin, with few oral lesions, while in contrast cicatricial pemphigoid mainly affects the mouth, eyes and other mucous membranes, with few skin lesions. Both presentations are characterized by the presence of auto-antibodies targeted to the same antigens, though the location and intensity of these self-damaging reactions within the basal membrane complex probably differ.

The location of the blisters is the decisive clinical element in the diagnosis of autoimmune blistering diseases. Pemphigoid is a chronic disorder characterized by subepithelial blisters that give rise to synechiae and functional sequelae. The diagnosis is based on the clinical and histopathological findings, with the exclusion of other possible blistering disease conditions. A number of antigens (BP180 [the most common], laminin 5 and 6, collagen VII, and 45 kDa, 168 kDa and 120 kDa proteins) have been implicated in the disease. It is characterized by a linear deposition of IgG, IgA and/or C3 along the basement membrane. Furthermore, dual circulating antibodies in the form of IgG and IgA have been shown to be associated with more severe disease. Immunoglobulin G and IgA antibodies may serve as suitable diagnostic markers. The assay of salivary IgA antibodies to NC16a has a diagnostic value similar to that of serum assay (Esmaili et al. 2015; Ali et al. 2016).

9.2.1.3 Oral Lichen Planus

Oral lichen planus (OLP) is a chronic mucocutaneous autoimmune and inflammatory disorder of unknown origin in which the host T lymphocytes attack the cells of the basal layer of the mucosa (Humberto et al. 2018; Nagao et al. 2001; Tvarijonaviciute et al. 2017; Valko et al. 2007; van der Meij et al. 2007). The disease affects about 0.1–4% of the population, and is more common in women than in men. The clinical manifestations are classified into patterns: erosive, ulcerative, atrophic, reticular, papular and plaque form. The diagnosis of OLP is based on clinical and histopathological criteria:

Clinical Diagnostic Criteria (1) Presence of bilateral and more or less symmetrical lesions; (2) Presence of a reticular network of slightly elevated grayish – white lines; (3) Erosive, atrophic, bullous and plaque forms are only accepted as subtypes of the disease in the presence of reticular lesions in other parts of the oral mucosa.

Histopathological Diagnostic Criteria (1) Presence of a well-defined band-like cellular infiltration confined to the superficial portion of the connective tissue and mainly composed of T lymphocytes; (2) Signs of hydropic degeneration of the epithelial basal layer; (3) Absence of epithelial dysplasia.

The possible malignant transformation of lichen planus (OLP) remains the subject of controversy, and further studies are needed in order to identify the factors implicated in the eventual malignization process.

The potential use of saliva for the diagnosis of OLP has been suggested by a number of authors (Agha-Hosseini et al. 2009; Battino et al. 2008; Batu et al. 2016; Iannitti et al. 2012; Lopez-Jornet et al. 2014; Sezer et al. 2007; Darczuk et al. 2016; Souza et al. 2018), since the disease has associated with increased salivary concentrations of certain inflammatory biomarkers (C-reactive protein [CRP], interleukins IL-4, IL-6 and IL-8, interferon gamma [IFN-γ] and TNF-α), stress markers (cortisol) and oxidative stress biomarkers (nitric oxide [NO], etc.) in comparison with healthy controls. All this suggests the presence of an inflammatory state with oxidative stress. In addition to their diagnostic application, measurements of salivary cytokines and NO may have a significant prognostic potential, monitoring the activity of the disease over time, and patient response to treatment.

9.2.2 Salivary Changes in Systemic Diseases

9.2.2.1 Autoimmune Disease Biomarkers: Sjögren's Syndrome

Sjögren's syndrome (SS) is a chronic, autoimmune and inflammatory systemic disorder that affects the exocrine salivary and lacrimal glands, causing dry mouth (xerostomia) and dry eye (xerophthalmia, keratoconjunctivitis sicca) (Fox 2005; Ramos-Casals et al. 2012).

The joint presentation of xerostomia and xerophthalmia is referred to as dry syndrome. Lymphocyte infiltration of the salivary and lacrimal glands is a distinctive feature of SS and may result in destruction of the functional capacity of these glands. Two forms of SS have been established: (a) primary Sjögren's syndrome, characterized by dry syndrome alone; and (b) secondary Sjögren's syndrome, characterized by dry syndrome together with connective tissue disorders such as rheumatoid arthritis (RA) or systemic lupus erythematosus (SLE). Furthermore, approximately one-third of all patients with primary SS also suffer extraglandular alterations, including osteoarthritis and joint and muscle pain, neuropathy, vasculitis and nephritis (Ramos-Casals et al. 2012) (Table 9.1).

As commented above, the disease can exhibit different extraglandular alterations in addition to the symptoms attributable to the exocrine disorders. Sjögren's syndrome can manifest alone (primary SS) or associated to other autoimmune conditions (secondary SS) (Qin et al. 2015). It mainly affects women between the fourth and sixth decades of life, and the multifactorial nature of SS and its tendency to adopt a chronic course over time complicate the diagnosis of the disease. Patients

Table 9.1 Disease conditions associated to Sjögren's syndrome

Associated autoimmune diseases
1. Rheumatoid arthritis
2. Systemic lupus erythematosus
3. Scleroderma
4. Dermatomyositis
5. Hashimoto's thyroiditis
6. Primary biliary cirrhosis
7. Chronic autoimmune hepatitis
8. Mixed connective tissue disease

with SS have a 5% probability of developing lymphoma, i.e., 10–44 times greater than in the normal population (Fox et al. 1984; Koseki et al. 2004).

The most frequent manifestations are dryness of the mouth and eyes. A correct anamnesis and exploration are required, taking into account the fact that extraglandular alterations may be present in up to 65% of all cases. Dry mouth makes speech difficult and complicates the adequate tasting and chewing of food. Xerostomia has a negative impact upon patient quality of life, and predisposes to buccodental disorders such as caries, periodontal disease and infections. The oral mucosa is found to be dry, atrophic and reddened, and the lips and angles of the mouth may appear dry and scaled. The teeth in turn can present multiple caries, with signs of gingival inflammation.

The saliva samples collected from individuals with SS show increased concentrations of Na^+, Cl^-, IgG, lysozyme, and matrix metalloproteinase (MMP) -2 and -9 in the parotid gland saliva. There is also an increase in lactoferrin, IgA, beta-2-microglobulin and albumin in parotid gland saliva and whole saliva, and increased concentrations of kallikrein and cystatin in whole saliva.

Sjögren's syndrome is characterized by an increase in inflammatory proteins and a decrease in acinar (glandular) proteins. Lactoferrin is a product of the ductal and acinar cells of the parotid gland. In this regard, studies of the saliva of individuals with SS have revealed increased lactoferrin levels, with no clear correlation to the magnitude of the lymphocyte infiltrate. Since lactoferrin concentration may also increase in other salivary gland disorders such as parotiditis, it lacks the specificity needed for use a diagnostic marker of SS (Soto-Rojas and Kraus 2002; Shiboski et al. 2012; Jiménez 2010; Pfaffe et al. 2011; Silvestre et al. 2009).

The increase in inflammatory proteins tends to be correlated to chronic inflammation and damage to the salivary glands. The elevation in the expression of beta-2-microglobulin and immunoglobulin light chains λ and κ is believed to reflect the activation and increase in intraglandular immunoglobulin synthesis.

A number of studies on the evolution of the disease and patient response to treatment have found interleukins IL-2 and IL-6 to be increased in these patients, with an apparent decrease in their concentrations following treatment with interferon or pilocarpine. Other studies, such as those of Silvestre et al. (Silvestre et al. 2009), have determined the expression of certain proteins such as fibronectin in patients with SS, relating their concentrations to the different phases of the disease.

Table 9.2 Oxidative stress in saliva of patients with lichen planus

Author	Patients	Measurement method	Results
Battino et al. (2008)	20 lichen planus 20 control	Uric acid, GGT, and albumin levels	Decrease of saliva uric acid and an increase in the TAC
Ergun et al. (2011)	21 LP 20 control	Total antioxidant activity (TAA) and lipid peroxidation	Increased oxidative stress and imbalance in the antioxidant
Batu et al. (2016)	18OLP, 32 patients OLCR 18 healthy controls	Prolidase activity, MDA, sialic acid and advanced oxidation protein products (AOPPs)	Increased prolidase activity and oxidative stress and imbalance in the antioxidant defense system
Agha-Hosseini et al. (2009)	30 lichen planus 30 control	Saliva MDA TAC	Higher MDA, not TAC
Vlková et al. (2012)	16 patients with oral premalignant lesions (leukoplakia, lichen planus, erythroplakia) 16 control	Thiobarbituric acid reacting substances (TBARS), advanced oxidation protein products (AOPP), advanced glycation products (AGEs) and total antioxidant capacity (TAC).	Salivary TBARS and AGEs were significantly higher in patients than in controls. No differences were found in AOPP. TAC and expression of superoxide dismutase were lower in patients than in age-matched controls. Decreased antioxidant status
Agha-Hosseini et al. (2012)	32 OLP, 26 oral carcinoma l (OSCC) 30 control	MDA, the total antioxidant capacity (TAC), and 8-hydroxy-2′-deoxyguanosine (8-OHdG)	OLP and OSCC are more susceptible to an imbalance of antioxidant-oxidative stress status
Abdolsamadi et al. (2014)	36 OLP 36 control	TAC,MDA vitamin A, C and E	MDA were higher and total anti-oxidant capacity was lower decrease vitamin C, E
Lopez Jornet et al. (2014)	70 patients (40 with OLP and 30 control patients)	MDA total anti-oxidant capacity	MDA were higher and total anti-oxidant capacity was lower
Kaur et al. (2016)	40 lichen planus 40 oral leukoplakia, 40 oral submucous fibrosis, 40 oral squamous cell carcinoma (SCC) controls	DNA and lipid damage using salivary 8-hydroxy-2-deoxyguanosine (8-OHdG), malondialdehyde (MDA), and vitamins C and E	Squamous cell carcinoma and pre-cancer patients showed significantly higher levels of salivary 8-OHdG and MDA and lower levels of vitamins C and E when compared to levels in healthy normal subjects.

SOD Superoxide dismutase, *MDA* malondialdehyde, *GSH* reduced glutathione, *TAC* capacidad antiox total, *GPX* peroxidase, *NO* Nitric oxide

Alterations in salivary biomarkers of oxidative stress were also reported in lichen planus (Table 9.2).

9.3 Part 2: Inflammatory and Immunemediated Diseases in Veterinary Medicine

The usefulness of saliva for the measurement of biomarkers of inflammatory or immunomediated diseases in veterinary medicine is not as deeply studied as in human medicine. However, it is interesting to notice that the existence of a major acute phase protein (APP), namely C-reactive protein (CRP) in saliva and its correlation with serum levels was first reported in dogs (Parra et al. 2005a) and few months later in humans (Christodoulides et al. 2005). Nevertheless, an important progress was done in the last years in terms of development and validation of different assays for inflammatory biomarkers quantification in saliva in veterinary species.

The immune system can be evaluated by determination of the biomarkers related to (1) the non-specific innate immune response and (2) the specific acquired immune response (Cerón 2019). The non-specific innate immune response are usually studied by measuring APPs such as CRP, haptoglobin (Hp) and serum A-amyloid (SAA), which reflect general inflammatory status (Cerón 2019). Acquired immune response is usually assessed by the determination of specific antibodies and cytokines (Escribano et al. 2012). In this section, the organization of the reviewer biomarkers of inflammatory and immuno-mediated diseases of different animal species will be done by the specific biomarker and the works related.

9.3.1 C-Reactive Protein

CRP has been widely measured in serum samples of different veterinary species and it is a major positive APP in dogs and pigs. Serum CRP increases are associated with a variety of different diseases in dogs such as acute pancreatitis (Sato et al. 2017), pyometra (Karlsson et al. 2012) and leishmania (Martinez-Subiela et al. 2003). Initial determination of CRP in canine saliva was performed through a time-resolved immunofluorometric assay (TR-IFMA) where higher levels of this biomarker were observed in diseased dogs compared with healthy dogs and positive correlation ($R^2 = 0.698$) between serum and saliva was established (Parra et al. 2005a). Despite these evidences suggesting that salivary canine CRP permits detection of the systemic inflammation, future research is necessary to concrete the behavior of salivary CRP in different inflammatory conditions and evaluate different confounding conditions, such as gingivitis.

The determination of salivary CRP was also proved to be useful to assess the systemic inflammatory status in pigs with naturally occurring diseases (Gutierrez et al. 2009a; Sánchez et al. 2019) and after experimental lipopolysaccharide (LPS) injection (Escribano et al. 2014). The ease of sampling large populations and the reliability of the CRP as a systemic inflammatory marker makes its measurement in saliva as a fundamental tool to monitor the status of these animals which will positively affect public health.

9.3.2 Haptoglobin

Hp is a positive APP produced mainly by hepatocytes although production or expression has also be found in skin, lung, and kidney cells in human beings (Ceron et al. 2005). The main role of Hp is the binding of free hemoglobin (Hb), thus forming Hp-Hb complexes that are non-toxic to living cells (in contrast to free Hb, which has oxidative properties) (Reczyńska et al. 2018). In addition, the binding of Hp to Hb prevents the use of this element by bacteria. In contrast to CRP, Hp is a major APP in cattle and small ruminants but moderate in the canine species (Cray et al. 2009). Increased salivary Hp was reported in dogs with different inflammatory diseases such as leishmaniosis, pyometra and ehrlichiosis (Parra et al. 2005b).

Hp has also been measured successfully in pig saliva even establishing reference limits for salivary levels of Hp in this species of Spanish farms ranging from 0.44–1.38 (percentil 25–75) in conventional pigs and from 0.40–1.03 (percentil 25–75) in Iberian pigs (Sánchez et al. 2019). Increased salivary Hp was observed during chronic course of experimentally induced inflammation (Escribano et al. 2014; Hiss et al. 2003) and in naturally occurring diseases like porcine reproductive and respiratory syndrome virus (PRRS), diarrhea, postweaning multisystemic wasting syndrome, multiple abscesses, or external injuries (Gutierrez et al. 2009b; Gomez-Laguna et al. 2010). Positive significant correlation between serum and salivary concentrations of Hp ($r = 0.79$) were described in pigs (Gutierrez et al. 2009b).

The quantification of salivary Hp together with CRP in dogs and pigs, brings a good opportunity to a wide perspective of study the inflammatory status in this species using non-invasive samples.

9.3.3 Adenosine Deaminase

Adenosine deaminase (ADA) is an enzyme that catalyses the irreversible conversion of adenosine and deoxyadenosine to inosine and deoxyinosine, respectively. ADA are involved in the detoxification of this substrates that are harmful for living cells (Iizuka et al. 1981). This enzyme has been proposed in human beings as a biomarker of chronic inflammation (Mishra et al. 1994). ADA appears in serum and most tissues, especially lymphoid tissues, given that it is necessary for monocyte-to-macrophage differentiation (Macdermott et al. 1980) and for lymphocytes activation (Sullivan et al. 1977). ADA activity in serum can increase as a result of leakage of the enzyme from damaged cells (Bansal et al. 1991).

An automated method for ADA activity measurement was validated for using with canine and porcine saliva samples. Salivary ADA activity was increased in dogs with pyometra and also correlated with serum biomarkers of inflammation, namely CRP (r = 0.408), Hp (r = 0.511) and band neutrophils (r = 0.514, 59). In pigs, salivary ADA activity was elevated in presence of lameness (Tecles et al. 2018b). However, salivary ADA concentrations did not significantly correlated to serum ADA neither in pigs or dogs (Tecles et al. 2018a, b). Salivary ADA was also determined in cows with clinical mastitis and, although no significant differences were observed in comparison with healthy controls, a negative correlation with somatic cell counts in milk was detected (Contreras-Aguilar et al. 2019), suggesting this salivary protein potentially related with this inflammatory disease.

9.3.4 Homocysteine

Homocysteine (Hcy) is an amino acid produced by methionine metabolism, a process with important implication in the methylation of several substances, including DNA (Miller 2003).

One study reported that porcine saliva has higher concentrations of Hcy than human saliva and them were higher in pigs with experimentally induced inflammation (Tecles et al. 2017). However, due to the lack of correlation between salivary and serum Hcy levels and with other inflammatory biomarkers (such as CRP) it was concluded that Hcy in porcine saliva did not reflect inflammatory process and could be related to stress (Tecles et al. 2017; de Oliveira et al. 2004; Chengfeng et al. 2014).

9.3.5 Adipokines

Adipokines are considered a biological active substances produced in adipose tissue and are essential for normal physiological functioning (Pan and Kastin 2007). Adipokines participate in the regulation of many biological processes, including the lipid metabolism, inflammation and immune function. The most studied adipokines are adiponectin (behaving as a negative acute phase protein) and leptin (behaving as a positive acute phase protein).

Although the studies of adipokines in saliva of animals are scarce, some evidences exist that in dogs salivary adiponectin have a potential as biomarker of systemic pathologies, since good correlation between its salivary and serum concentration was detected (r = 0.650) (Tvarijonaviciute et al. 2014). It is important to highlight, that if salivary adiponectin is to be measured, gingiva status should be carefully assessed as increase in its levels due to leakage from blood as a consequence of gingival injury was observed (Tvarijonaviciute et al. 2014).

Salivary concentrations of leptin were increased during experimentally induced inflammation in pigs and was significantly correlated with serum CRP concentrations ($r = 0.398$) (Schmidt et al. 2016).

9.3.6 Other Salivary Inflammatory and Immunomediated Biomarkers Identified Using Proteomics

Proteomics has been applied for detection of salivary biomarkers of local and systemic inflammatory diseases in different veterinary species.

In saliva of dogs with parvovirosis, an infection that curse with a severe inflammation in the digestive system (Schoeman et al. 2013), 86 proteins were significantly differentially expressed in saliva of diseased dogs compared to healthy controls. Among these, the most prominent decrease was observed in cathelicidin antimicrobial peptide (CAMP) and Rho-GDP dissociation inhibitor beta (ARHGDIB), while lipocalin-9 and BPI fold-containing family B member 2 (BPiFB2) were the two most up-regulated proteins (Franco-Martínez et al. 2018). Furthermore, apolipoprotein A-I, a negative biomarker of inflammation in humans (Li et al. 2011), was reduced in animals with parvovirus (Franco-Martínez et al. 2018). In the same line, proteomics-based study in dogs with leishmaniosis, a disease that induce inflammation and alterations in immune-response, showed decreased apolipoprotein A-I in saliva of clinically ill animals (Franco-Martínez et al. 2019).

In healthy pigs, several proteins involved in the immune response or inflammatory activity such as Igs or cystatins that have antimicrobial activities and lipocalin that may act as a scavenger of pro-inflammatory lipids were identified in saliva (Gutiérrez et al. 2011) presenting potential for the follow-up of inflammatory pathologies in this species, although further studies are needed to confirm this hypothesis.

Regarding ruminants, a previous study performed a global survey of the glycoproteins of bovine saliva of healthy animals (Ang et al. 2011). In this work, the authors could identify about 450 proteins with different roles in the organism, but they found a 2.2% of carbohydrate binding proteins such as galactoside binding protein and peptidoglycan recognition protein that are involved in the innate immunity process (Ang et al. 2011). In small ruminants (sheep and goat), the proteomic approach of salivary proteins showed that the second largest group (23 in sheep and 24 in goat) of proteins secreted by parotid gland presented protection functions or were related to immune response (Lamy et al. 2009).

In saliva of horses with inflammation due to different pathologies such as colitis, septic synovitis, septicemia, a proteomic analysis revealed increase in specific APPs such as SAA, Hp, fibrinogen, α1-acid glycoprotein, ceruloplasmin, α2-macroglobulin and α1-antitrypsin (Jacobsen et al. 2014). Moreover, some of the proteins were described for the first time in saliva of horses. The hypothesis is that they could be potential salivary markers of systemic inflammation. The higher expression of APPs and identification of new proteins in saliva of horses with inflammation point-out the potential of saliva as a non-invasive tool for detection and monitoring of health status in this species.

9.4 Conclusions

The use of saliva as a diagnostic tool in inflammatory and immunomediated diseases present high potential in both human and veterinary medicine. However, further extensive research is needed, especially in veterinary medicine, before saliva can be used in daily clinical practice.

References

Agha-Hosseini F, Mirzaii-Dizgah I, Farmanbar N, Abdollahi M (2012) Oxidative stress status and DNA damage in saliva of human subjects with oral lichen planus and oral squamous cell carcinoma. J Oral Pathol Med 41(10):736–740

Agha-Hosseini F, Mirzaii-Dizgah I, Mikaili S, Abdollahi M (2009) Increased salivary lipid peroxidation in human subjects with oral lichen planus. Int J Dent Hyg 7:246–250. https://doi.org/10.1111/j.1601-5037.2009.00365.x

Abdolsamadi H, Rafieian N, Goodarzi MT, Feradmal J, Davoodi P, Jazayeri M, Taghavi Z, Hoseyni S-M, Ahmadi-Motamayel F (2014) Levels of salivary antioxidant vitamins and lipid peroxidation in patients with oral lichen planus and healthy individuals. Chonnam Med J 50(2):58

Ali S, Kelly C, Challacombe SJ, Donaldson ANA, Dart JKG, Gleeson M et al (2016) Salivary IgA and IgG antibodies to bullous pemphigoid 180 noncollagenous domain 16a as diagnostic biomarkers in mucous membrane pemphigoid. Br J Dermatol 174:1022–1029. https://doi.org/10.1111/bjd.14351

Ang CS, Binos S, Knight MI, Moate PJ, Cocks BG, McDonagh MB (2011) Global survey of the bovine salivary proteome: integrating multidimensional prefractionation, targeted, and glycocapture strategies. J Proteome Res 10:5059–5069. https://doi.org/10.1021/pr200516d

Bansal SK, Singh RP, Narang RK, Joshi LD, Bansal A, Agrawal AK (1991) Serum adenosine deaminase in pulmonary tuberculosis, malignancy and non-tubercular respiratory diseases. Indian J Chest Dis Allied Sci 33:189–193

Battino M, Greabu M, Totan A, Bullon P, Bucur A, Tovaru S et al (2008) Oxidative stress markers in oral lichen planus. Biofactors 33:301–310

Batu S, Ofluoglu D, Ergun S, Warnakulasuriya S, Uslu E, Guven Y et al (2016) Evaluation of prolidase activity and oxidative stress in patients with oral lichen planus and oral lichenoid contact reactions. J Oral Pathol Med 45:281–288. https://doi.org/10.1111/jop.12391

Cerón JJ (2019) Acute phase proteins, saliva and education in laboratory science: an update and some reflections. BMC Vet Res 15:197

Ceron JJ, Eckersall PD, Martynez-Subiela S (2005) Acute phase proteins in dogs and cats: current knowledge and future perspectives. Vet Clin Pathol 34:85–99

Chengfeng S, Wei L, Xinxing W, Lei W, Rui Z, Lingjia Q (2014) Hyperhomocysteinemia is a result, rather than a cause, of depression under chronic stress. PLoS One 9:e106625. https://doi.org/10.1371/journal.pone.0106625

Christodoulides N, Mohanty S, Miller CS, Langub MC, Floriano PN, Dharshan P et al (2005) Application of microchip assay system for the measurement of C-reactive protein in human saliva. Lab Chip 5:261–269. https://doi.org/10.1039/b414194f

Cimen MYB, Kaya TI, Eskandari G, Tursen U, Ikizoglu G, Atik U (2003) Oxidant/antioxidant status in patients with recurrent aphthous stomatitis. Clin Exp Dermatol 28:647–650. https://doi.org/10.1046/j.1365-2230.2003.01415.x

Contreras-Aguilar MD, Monkeviciene I, Ceron JJ, Silinskas I, Vallejo-Mateo PJ, Tecles F et al (2019) Biochemical changes in saliva of cows with inflammation: a pilot study. Res Vet Sci 124:383–386. https://doi.org/10.1016/j.rvsc.2019.04.019

Cray C, Zaias J, Altman NH (2009) Acute phase response in animals: a review. Comp Med 59:517–526

Darczuk D, Krzysciak W, Vyhouskaya P, Kesek B, Galecka-Wanatowicz D, Lipska W et al (2016) Salivary oxidative status in patients with oral lichen planus. J Physiol Pharmacol 67:885–894

de Oliveira AC, Suchecki D, Cohen S, D'Almeida V (2004) Acute stressor-selective effect on total plasma homocysteine concentration in rats. Pharmacol Biochem Behav 77:269–273. https://doi.org/10.1016/j.pbb.2003.11.003

Eguia-del Valle A, Martinez-Conde-Llamosas R, Lopez-Vicente J, Uribarri-Etxebarria A, Aguirre-Urizar J-M (2013) Salivary cortisol determination in patients from the Basque Country with recurrent aphthous stomatitis. A pilot study. Med Oral Patol Oral Cir Bucal 18:e207–e211. https://doi.org/10.4317/medoral.18110.

Ergun S, Troşala ŞC, Warnakulasuriya S, Özel S, Önal AE, Ofluoğlu D, Güven Y, Tanyeri H (2011) Evaluation of oxidative stress and antioxidant profile in patients with oral lichen planus. J Oral Pathol Med 40(4):286–293

Escribano D, Gutiérrez AM, Martínez Subiela S, Tecles F, Cerón JJ (2012) Validation of three commercially available immunoassays for quantification of IgA, IgG, and IgM in porcine saliva samples. Res Vet Sci 93:682–687. https://doi.org/10.1016/j.rvsc.2011.09.018

Escribano D, Campos PHRF, Gutierrez AM, Le Floc'h N, Ceron JJ, Merlot E (2014) Effect of repeated administration of lipopolysaccharide on inflammatory and stress markers in saliva of growing pigs. Vet J 200:393–397. https://doi.org/10.1016/j.tvjl.2014.04.007

Escudier M, Bagan J, Scully C (2006) Number VII Behcet's disease (Adamantiades syndrome). Oral Dis 12:78–84. https://doi.org/10.1111/j.1601-0825.2005.01144.x

Esmaili N, Mortazavi H, Kamyab-Hesari K, Aghazadeh N, Daneshpazhooh M, Khani S et al (2015) Diagnostic accuracy of BP180 NC16a and BP230-C3 ELISA in serum and saliva of patients with bullous pemphigoid. Clin Exp Dermatol 40:324–330. https://doi.org/10.1111/ced.12510

Fox RI (2005) Sjogren's syndrome. Lancet (London, England) 366:321–331. https://doi.org/10.1016/S0140-6736(05)66990-5.

Fox RI, Howell FV, Bone RC, Michelson P (1984) Primary Sjogren syndrome: clinical and immunopathologic features. Semin Arthritis Rheum 14:77–105

Franco-Martínez L, Tvarijonaviciute A, Horvatić A, Guillemin N, Cerón JJ, Escribano D et al (2018) Changes in salivary analytes in canine parvovirus: a high-resolution quantitative proteomic study. Comp Immunol Microbiol Infect Dis 60:1–10. https://doi.org/10.1016/j.cimid.2018.09.011

Franco-Martínez L, Tvarijonaviciute A, Horvatić A, Guillemin N, Bernal LJ, Barić Rafaj R et al (2019) Changes in saliva of dogs with canine leishmaniosis: a proteomic approach. Vet Parasitol 272:44–52. https://doi.org/10.1016/j.vetpar.2019.06.014

Freitas JP, Filipe P, Yousefi A, Emerit I, Guerra Rodrigo F (1998) Oxidative stress in Adamantiades-Behcet's disease. Dermatology 197:343–348. https://doi.org/10.1159/000018029

Fuertes De Vega I, Iranzo-Fernández P, Mascaró-Galy JM (2014) Bullous pemphigoid: clinical practice guidelines. Actas Dermosifiliogr 105:328–346. https://doi.org/10.1016/j.adengl.2012.10.022

Gomez-Laguna J, Gutierrez A, Pallares FJ, Salguero FJ, Ceron JJ, Carrasco L (2010) Haptoglobin and C-reactive protein as biomarkers in the serum, saliva and meat juice of pigs experimentally infected with porcine reproductive and respiratory syndrome virus. Vet J 185:83–87. https://doi.org/10.1016/j.tvjl.2010.04.018

Gutierrez AM, Martinez-Subiela S, Eckersall PD, Ceron JJ (2009a) C-reactive protein quantification in porcine saliva: a minimally invasive test for pig health monitoring. Vet J 181:261–265. https://doi.org/10.1016/j.tvjl.2008.03.021

Gutierrez AM, Martinez-Subiela S, Ceron JJ (2009b) Evaluation of an immunoassay for determination of haptoglobin concentration in various biological specimens from swine. Am J Vet Res 70:691–696. https://doi.org/10.2460/ajvr.70.6.691

Gutiérrez AM, Miller I, Hummel K, Nöbauer K, Martínez-Subiela S, Razzazi-Fazeli E et al (2011) Proteomic analysis of porcine saliva. Vet J 187:356–362. https://doi.org/10.1016/j.tvjl.2009.12.020

Hassona Y, Scully C (2016) Salivary changes in oral mucosal diseases. Periodontol 2000 70:111–127. https://doi.org/10.1111/prd.12102

Hiss S, Knura-Deszczka S, Regula G, Hennies M, Gymnich S, Petersen B et al (2003) Development of an enzyme immuno assay for the determination of porcine haptoglobin in various body fluids: testing the significance of meat juice measurements for quality monitoring programs. Vet Immunol Immunopathol 96:73–82

Humberto JSM, Pavanin JV, da Rocha MJA, Motta ACF (2018) Cytokines, cortisol, and nitric oxide as salivary biomarkers in oral lichen planus: a systematic review. Braz Oral Res 32:e82. https://doi.org/10.1590/1807-3107bor-2018.vol32.0082.

Iannitti T, Rottigni V, Palmieri B (2012) Role of free radicals and antioxidant defences in oral cavity-related pathologies. J Oral Pathol Med Off Publ Int Assoc Oral Pathol Am Acad Oral Pathol 41:649–661. https://doi.org/10.1111/j.1600-0714.2012.01143.x

Iizuka H, Koizumi H, Kamigaki K, Aoyagi T, Miura Y (1981) Two forms of adenosine deaminase in pig epidermis. J Dermatol 8:91–95

Jacobsen S, Top Adler DM, Bundgaard L, Sorensen MA, Andersen PH, Bendixen E (2014) The use of liquid chromatography tandem mass spectrometry to detect proteins in saliva from horses with and without systemic inflammation. Vet J 202:483–488. https://doi.org/10.1016/j.tvjl.2014.08.032

Jiménez Y (2010) La saliva como fluido diagnostico. Fisiopatol. las glándulas salivales, Valencia. Medicina Oral SL 2010:347–355

Karincaoglu Y, Batcioglu K, Erdem T, Esrefoglu M, Genc M (2005) The levels of plasma and salivary antioxidants in the patient with recurrent aphthous stomatitis. J Oral Pathol Med Off Publ Int Assoc Oral Pathol Am Acad Oral Pathol 34:7–12. https://doi.org/10.1111/j.1600-0714.2004.00253.x.

Karlsson I, Hagman R, Johannisson A, Wang L, Karlstam E, Wernersson S (2012) Cytokines as immunological markers for systemic inflammation in dogs with pyometra. Reprod Domest Anim 47(Suppl 6):337–341. https://doi.org/10.1111/rda.12034

Kaur J, Politis C, Jacobs R (2016) Salivary 8-hydroxy-2-deoxyguanosine, malondialdehyde, vitamin C, and vitamin E in oral pre-cancer and cancer: diagnostic value and free radical mechanism of action. Clin Oral Investig 20(2):315–319

Koseki M, Maki Y, Matsukubo T, Ohashi Y, Tsubota K (2004) Salivary flow and its relationship to oral signs and symptoms in patients with dry eyes. Oral Dis 10:75–80

Lamy E, da Costa G, Santos R, Capela e Silva F, Potes J, Pereira A et al (2009) Sheep and goat saliva proteome analysis: a useful tool for ingestive behavior research? Physiol Behav 98:393–401. https://doi.org/10.1016/j.physbeh.2009.07.002

Li H, Li C, Wu H, Zhang T, Wang J, Wang S et al (2011) Identification of Apo-A1 as a biomarker for early diagnosis of bladder transitional cell carcinoma. Proteome Sci 9:21. https://doi.org/10.1186/1477-5956-9-21

Lopez-Jornet P, Martinez-Canovas A, Pons-Fuster A (2014) Salivary biomarkers of oxidative stress and quality of life in patients with oral lichen planus. Geriatr Gerontol Int 14:654–659. https://doi.org/10.1111/ggi.12153

Macdermott RP, Tritsch GL, Howard JFF, Medical H (1980) Adenosine deaminase and nucleoside phosphorylase activities in normal human blood mononuclear cell subpopulations. Clin Exp Immunol 42:303–307

Martinez-Subiela S, Bernal LJ, Ceron JJ (2003) Serum concentrations of acute-phase proteins in dogs with leishmaniosis during short-term treatment. Am J Vet Res 64:1021–1026

Miller AL (2003) The methionine-homocysteine cycle and its effects on cognitive diseases. Altern Med Rev 8:7–19

Mishra OP, Gupta BL, Ali Z, Nath G, Chandra L (1994) Adenosine deaminase activity in typhoid fever. Indian Pediatr 31:1379–1384

Nagao T, Warnakulasuriya S, Ikeda N, Fukano H, Yamamoto S, Yano M et al (2001) Serum antioxidant micronutrient levels in oral lichen planus. J Oral Pathol Med Off Publ Int Assoc Oral Pathol Am Acad Oral Pathol 30:264–267

Pan W, Kastin AJ (2007) Adipokines and the blood-brain barrier. Peptides 28:1317–1330. https://doi.org/10.1016/j.peptides.2007.04.023

Parra MD, Tecles F, Subiela SM, Cerón JJ (2005a) C-reactive protein measurement in canine saliva. J Vet Diagn Investig 17:139–144. https://doi.org/10.1177/104063870501700207

Parra MD, Vaisanen V, Ceron JJ (2005b) Development of a time-resolved fluorometry based immunoassay for the determination of canine haptoglobin in various body fluids. Vet Res 36:117–129. https://doi.org/10.1051/vetres:2004054

Pfaffe T, Cooper-White J, Beyerlein P, Kostner K, Punyadeera C (2011) Diagnostic potential of saliva: current state and future applications. Clin Chem 57:675–687. https://doi.org/10.1373/clinchem.2010.153767

Qin B, Wang J, Yang Z, Yang M, Ma N, Huang F et al (2015) Epidemiology of primary Sjogren's syndrome: a systematic review and meta-analysis. Ann Rheum Dis 74:1983–1989. https://doi.org/10.1136/annrheumdis-2014-205375

Ramos-Casals M, Brito-Zeron P, Siso-Almirall A, Bosch X (2012) Primary Sjogren syndrome. BMJ 344:e3821. https://doi.org/10.1136/bmj.e3821

Reczyńska D, Zalewska M, Czopowicz M, Kaba J, Zwierzchowski L, Bagnicka E (2018) Acute phase protein levels as an auxiliary tool in diagnosing viral diseases in ruminants—a review. Viruses 10:1–14. https://doi.org/10.3390/v10090502

Russo I, Saponeri A, Michelotto A, Alaibac M (2017) Salivary samples for the diagnosis of pemphigus vulgaris using the BIOCHIP approach: a pilot study. In Vivo 31:97–99. https://doi.org/10.21873/invivo.11030

Sánchez J, García A, Ruiz JM, Montes AM, Cabezas-Herrera J, Ros-Lara S et al (2019) Porcine breed, sex, and production stage influence the levels of health status biomarkers in saliva samples. Front Vet Sci 6:1–10. https://doi.org/10.3389/fvets.2019.00032

Saral Y, Coskun BK, Ozturk P, Karatas F, Ayar A (2005) Assessment of salivary and serum antioxidant vitamins and lipid peroxidation in patients with recurrent aphthous ulceration. Tohoku J Exp Med 206:305–312. https://doi.org/10.1620/tjem.206.305

Sato T, Ohno K, Tamamoto T, Oishi M, Kanemoto H, Fukushima K et al (2017) Assesment of severity and changes in C-reactive protein concentration and various biomarkers in dogs with pancreatitis. J Vet Med Sci 79:35–40. https://doi.org/10.1292/jvms.16-0009

Schmidt EMS, Escribano D, Martinez-Subiela S, Martinez-Miró S, Hernández F, Tvarijonaviciute A et al (2016) Development and validation of an assay for measurement of leptin in pig saliva. BMC Vet Res 12:1–7. https://doi.org/10.1186/s12917-016-0871-9.

Schoeman JP, Goddard A, Leisewitz AL (2013) Biomarkers in canine parvovirus enteritis. N Z Vet J 61:217–222. https://doi.org/10.1080/00480169.2013.776451

Sezer E, Ozugurlu F, Ozyurt H, Sahin S, Etikan I (2007) Lipid peroxidation and antioxidant status in lichen planus. Clin Exp Dermatol 32:430–434. https://doi.org/10.1111/j.1365-2230.2007.02436.x

Shiboski SC, Shiboski CH, Criswell LA, Baer AN, Challacombe S, Lanfranchi H et al (2012) American College of Rheumatology classification criteria for Sjogren's syndrome: a data-driven, expert consensus approach in the Sjogren's International Collaborative Clinical Alliance cohort. Arthritis Care Res (Hoboken) 64:475–487

Silvestre FJ, Puente A, Bagan JV, Castell JV (2009) Presence of fibronectin peptides in saliva of patients with Sjogren's syndrome: a potential indicator of salivary gland destruction. Med Oral Patol Oral Cir Bucal 14:e365–e370

Soto-Rojas AE, Kraus A (2002) The oral side of Sjogren syndrome. Diagnosis and treatment. A review. Arch Med Res 33:95–106

Souza MM, Florezi GP, Nico M, de Paula F, Paula FM, Lourenco SV (2018) Salivary proteomics in lichen planus: a relationship with pathogenesis? Oral Dis 24:784–792. https://doi.org/10.1111/odi.12837

Sullivan JL, Osborne WR, Wedgewood RJ (1977) Adenosine deaminase activity in lymphocytes. Br J Haematol 37:157–158

Tecles F, Escribano D, Martinez-Miro S, Ceron JJ (2017) Homocysteine measurement in pig saliva, assay validation and changes after acute stress and experimental inflammation models: a pilot study. Res Vet Sci 112:75–80. https://doi.org/10.1016/j.rvsc.2017.01.009

Tecles F, Escribano D, Contreras-Aguilar MD, Rubio CP, Szczubial M, Ceron JJ et al (2018a) Evaluation of adenosine deaminase in saliva and serum, and salivary alpha-amylase, in canine pyometra at diagnosis and after ovariohysterectomy. Vet J 236:102–110. https://doi.org/10.1016/j.tvjl.2018.04.018

Tecles F, Rubio CP, Contreras-Aguilar MD, Lopez-Arjona M, Martinez-Miro S, Martinez-Subiela S et al (2018b) Adenosine deaminase activity in pig saliva: analytical validation of two spectrophotometric assays. J Vet Diagn Investig 30:175–179. https://doi.org/10.1177/1040638717742947

Tugrul S, Kocyigit A, Dogan R, Eren SB, Senturk E, Ozturan O et al (2016) Total antioxidant status and oxidative stress in recurrent aphthous stomatitis. Int J Dermatol 55:e130–e135. https://doi.org/10.1111/ijd.13101

Tvarijonaviciute A, Carrillo-Sanchez JD, Garcia-Martinez JD, Tecles F, Martinez-Subiela S, German AJ et al (2014) Measurement of salivary adiponectin concentrations in dogs. Vet Clin Pathol 43:416–421. https://doi.org/10.1111/vcp.12169

Tvarijonaviciute A, Aznar-Cayuela C, Rubio CP, Ceron JJ, Lopez-Jornet P (2017) Evaluation of salivary oxidate stress biomarkers, nitric oxide and C-reactive protein in patients with oral lichen planus and burning mouth syndrome. J Oral Pathol Med Off Publ Int Assoc Oral Pathol Am Acad Oral Pathol 46:387–392. https://doi.org/10.1111/jop.12522.

Valko M, Leibfritz D, Moncol J, Cronin MTD, Mazur M, Telser J (2007) Free radicals and antioxidants in normal physiological functions and human disease. Int J Biochem Cell Biol 39:44–84. https://doi.org/10.1016/j.biocel.2006.07.001

van der Meij EH, Mast H, van der Waal I (2007) The possible premalignant character of oral lichen planus and oral lichenoid lesions: a prospective five-year follow-up study of 192 patients. Oral Oncol 43:742–748. https://doi.org/10.1016/j.oraloncology.2006.09.006

Vlková B, Stanko P, Minárik G, Tóthová Ľ, Szemes T, Baňasová L, Novotňáková D, Hodosy J, Celec P (2012) Salivary markers of oxidative stress in patients with oral premalignant lesions. Arch Oral Biol 57(12):1651–1656

Chapter 10
Salivary Biomarkers in Kidney Diseases

Mateusz Maciejczyk, Piotr Żukowski, and Anna Zalewska

According to epidemiological studies, kidney diseases belong with atherosclerosis, asthma, cancer, chronic liver disease, type 2 diabetes, heart disease, metabolic syndrome, to the most common contemporary civilization diseases (Webster et al. 2017; Koye et al. 2018). It is estimated that 600 million people suffer from chronic kidney disease (CKD) and that this number is constantly increasing. CKD concerns therefore 10% of global population, and 30–50% in the high-risk group of patients (Hill et al. 2016; Webster et al. 2017). CKD is a particular pathological process as it may be an effect of other lifestyle diseases and also lead to non-renal complications, especially cardiovascular events. Furthermore, CKD may not give any clinical symptoms for a long time, and therefore it is diagnosed often in late stages necessitating renal replacement treatment. The constant increase in the number of sufferers and a high cost of dialysis, make CKD one of the key challenges for twenty-first century medicine (Hill et al. 2016; Webster et al. 2017; Koye et al. 2018).

The pathogenic factors responsible for CKD and its dangerous complications are still not well understood. It is suspected that factors such as activation of the renin angiotensin aldosterone system (RAAS), disturbances in phosphate and vitamin D metabolism, inflammation, oxidative stress, disrupted remodelling of extracellular matrix (ECM) and genetic factors all play a role in the CKD progression (Fogo 2007; López-Novoa et al. 2010; Yang et al. 2010; Fassett et al. 2011; Tomino 2014; Webster et al. 2017; Maciejczyk et al. 2018c). Therefore, new markers of early kidney damage are constantly sought, which would enable early CKD diagnosis and

M. Maciejczyk (✉)
Department of Hygiene, Epidemiology and Ergonomics, Medical University of Bialystok, Bialystok, Poland

P. Żukowski
Department of Restorative Dentistry, Croydon University Hospital, Croydon, UK

A. Zalewska
Experimental Dentistry Laboratory, Medical University of Bialystok, Bialystok, Poland

A. Tvarijonaviciute et al. (eds.), *Saliva in Health and Disease*,
https://doi.org/10.1007/978-3-030-37681-9_10

slow the disease progress (Fassett et al. 2011; Lopez-Giacoman 2015). New more sensitive and specific diagnostic methods are now accompanied with non-invasive material collection.

Latest epidemiological and clinical research proved that saliva can be an alternative laboratory material used in chronic kidney disease diagnostics (Bots et al. 2007; Bibi et al. 2008; Michelis et al. 2008; Peng et al. 2013; Venkatapathy et al. 2014; Anuradha et al. 2015; Pallos et al. 2015; Yajamanam et al. 2016; Celec et al. 2016; Renda 2017; Alsamarai et al. 2018; Tamimi et al. 2018; Kovalčíkova et al. 2018; Maciejczyk et al. 2018c; 2019b). Saliva can be of a high diagnostic use in measuring the classical kidney function biomarkers (e.g., urea, uric acid and creatinine), as well as the salivary oxidative stress parameters or inflammatory indicators. Taking into account the rather small amount of publications in this area, the aim of present chapter is to review the latest literature in the field of saliva usage as an alternative laboratory material in chronic kidney disease diagnostics.

10.1 Chronic Kidney Disease-Definition, Classification and Diagnosis

The medical term "chronic kidney disease" (CKD) was introduced for the first time in 2002 by an American team of nephrologists from the National Kidney Foundation based on analyses in 45 cohorts that included 1,555,332 participants (National Kidney Foundation 2002). According to the current definition given by the Kidney Disease Improving Global Outcome in 2012, CKD is defined as abnormalities of kidney structure or function, present for 3 months, with implications for health (Table 10.1) (Levey et al. 2011; KDIGO 2012; Webster et al. 2017). The current definition draws attention to the fact that not all structural and functional kidney alterations may have pathological consequences.

Glomerular filtration rate (GFR) is regarded as the best kidney function indicator both in health and disease. According to GFR values, different stages of CKD are differentiated (Table 10.2) (Levey et al. 2011; KDIGO 2012; Wouters et al. 2015;

Table 10.1 Diagnostic criteria for chronic kidney disease (either of the following present ≥ 3 months) (Levey et al. 2011; KDIGO 2012)

Reduced glomerular filtration rate (GFR)	GFR < 60 mL/min/1.73 m^2 (GFR stages G3a–G5)
Markers of kidney injury (one or more)	Albuminuria (albumin excretion rate (AER) ≥ 30 mg/24 h; urine albumin-to-creatinine ratio (ACR) ≥ 30 mg/g [≥3 mg/mM]) Electrolyte and other alterations caused by tubular disorders Urine sediment alterations Structural alterations detected by imaging Histological alterations detected by biopsy Kidney transplantation

Abbreviations: *ACR* urine albumin-to-creatinine ratio, *AER* albumin excretion rate, *GFR* glomerular filtration rate

Table 10.2 Stages of chronic kidney disease (G1, G2, G3a, G3b, G4, G5) (Levey et al. 2011; KDIGO 2012)

GFR stage	GFR (mL/min/1.73 m^2)	Terms
G1	≥ 90	Normal or high
G2	60–89	Mildly decreased[a]
G3a	45–59	Mildly to moderately decreased
G3b	30–44	Moderately to severely decreased
G4	15–29	Severely decreased
G5	<15	Kidney failure

GFR glomerular filtration rate
[a]Relative to young adult level

Gaitonde et al. 2017; Webster et al. 2017). The normal GFR value for a young adult is approximately 90 ml/min/1,73 m^2, decreased GFR < 60. GFR of 15 ml/min/1.73 m^2 (GFR category G5) is defined as kidney failure (Levey et al. 2011; KDIGO 2012; Webster et al. 2017). Decreased GFR can be detected by current estimating equations for GFR based on serum creatinine (sCr) or cystatin C but not by sCr or cystatin C alone. This is essential as in the elderly and in the thin, the normal creatinine levels can coincide a distorted kidney function. Indeed, the creatinine concentration is dependent on many factors such as muscular creatine production, its tubular secretion or non-renal creatinine excretion (Fassett et al. 2011; Levey et al. 2011; Webster et al. 2017). Therefore, the estimated GFR is calculated using Cockroft-Gault (C-G) equation, shortened MDRD (Modification of Diet in Renal Disease) or CKD-EPI (Chronic Kidney Disease Epidemiology Collaboration) (Levey et al. 2011; KDIGO 2012; Gaitonde et al. 2017; Webster et al. 2017).

Proteinuria is often one of the first CKD symptoms (Cravedi et al. 2012; Thorp et al. 2012; Webster et al. 2017). Proteinuria is a general term for the presence of increased amounts of protein in the urine. The upper normal value for protein loss is 150 mg of protein in urine for adults and 250 mg for children and adolescents (Levey et al. 2011; Cravedi et al. 2012; Thorp et al. 2012; Webster et al. 2017). Proteinuria may reflect abnormal loss of plasma proteins due to (1) increased glomerular permeability to large molecular weight proteins (albuminuria or glomerular proteinuria, loss of 0.2 to >20 g protein/24 h), (2) incomplete tubular reabsorption of normally filtered low-molecular- weight proteins (tubular proteinuria, loss of 0.2 to 2 g protein/24 h), or (3) increased plasma concentration of low-molecular-weight proteins (overproduction proteinuria, such as immunoglobulin light chains). Proteinuria may also reflect abnormal loss of proteins derived from the kidney (renal tubular cell constituents due to tubular damage, loss of <0.5 g protein/24 h) and lower urinary tract (Fassett et al. 2011; Cravedi et al. 2012; KDIGO 2012; Thorp et al. 2012; Webster et al. 2017).

Albuminuria refers to abnormal loss of albumin in the urine (Johnson et al. 2012; Gaitonde et al. 2017). Albumin is one type of plasma protein found in the urine in normal subjects and in larger quantity in patients with kidney disease. Albumin is a small molecule protein (~66 kDa), which serves as a transporting protein and its main role is to regulate the oncotic pressure of blood (Johnson et al. 2012).

Table 10.3 Albuminuria category (A1, A2, A3) in chronic kidney disease (Levey et al. 2011; KDIGO 2012)

Category	Albumin excretion rate (AER) (mg/24 h)	Albumin-to-creatinine ratio (ACR) (mg/mmoL)	(mg/g)	Terms
A1	< 30	< 3	< 30	Normal to mildly increased
A2	30–300	3–30	30–300	Moderately increased[a]
A3	> 300	> 30	> 300	Severely increased[b]

Abbreviations: *AER* albumin excretion rate, *ACR* albumin-to-creatinine ratio
[a]Relative to young adult level
[b]Including nephrotic syndrome (albumin excretion usually >2200 mg/24 h [ACR > 2220 mg/g; >220 mg/mmol])

Albuminuria is considered not only an indicator of the gravity of kidney damage but is also strongly correlated with CKD progression (Johnson et al. 2012; Gaitonde et al. 2017). It is also regarded as a marker of glomerular damage (Table 10.3) (Levey et al. 2011; KDIGO 2012). The main reason for increased excretion of albumin is imbalance between its filtration and resorption as well as inflammation and bleeding in the urinary tract. The laboratory measurements of albuminuria include the measurement of the albumin concentration in urine and albumin excretion rate (AER) or urinary albumin excretion (UAE) (Johnson et al. 2012; Webster et al. 2017). Another index used to assess albuminuria is the albumin to creatinine ratio (ACR), which is calculated by dividing albumin concentration in milligrams by creatinine concentration in grams in a single sample of urine. In many tests, the ACR value is compatible with the daily albuminuria (Levey et al. 2011; Johnson et al. 2012; Gaitonde et al. 2017; Webster et al. 2017).

According to KDIGO recommendations, CKD is defined as abnormalities of kidney structure or function, present for 3 months, with implications for health. CKD classification is based on cause, GFR category, and albuminuria category (CGA) (Levey et al. 2011; KDIGO 2012; Webster et al. 2017). The CKD cause is determined on the basis of the presence/absence of a systemic disease, the type of histological or anatomical abnormalities suspected or observed in the kidneys (Levey et al. 2011; Gaitonde et al. 2017; Webster et al. 2017). Unlike CKD, acute kidney injury (AKI) is caused by a sudden impairment of kidney function, which is accompanied by a sharp increase of sCr, and also a decrease in urine volume. In general, in AKI patients, the kidney size and blood hemoglobin concentration do not change, as well as the kidney osteodystrophia, extraosseous calcifications or eye changes are not observed (Levey et al. 2011; KDIGO 2012; Gaitonde et al. 2017; Webster et al. 2017).

10.2 Pathogenesis and Epidemiology of Chronic Kidney Disease

Although the exact etiology of chronic kidney disease is not well understood, it is suspected that that activation of the renin angiotensin aldosterone system (RAAS), phosphate and vitamin D metabolism disturbances, inflammation, oxidative stress and genetic factors are significant pathogenic agents of CKD (Fogo 2007; López-Novoa et al. 2010; Yang et al. 2010; Fassett et al. 2011; Tomino 2014; Webster et al. 2017; Maciejczyk et al. 2018c). However, the renin angiotensin aldosterone system plays a key role in blood pressure and kidney perfusion maintenance (Siragy and Carey 2010). Multiple studies proved that overactivation of this system can lead to multiple organ damage including heart, blood vessels, brain and kidneys (Siragy and Carey 2010). Indeed, it is believed that the angiotensin-2 is the main factor responsible for CKD progression (Rüster and Wolf 2006; Siragy and Carey 2010; Webster et al. 2017). This hormone regulates expression of many cell adhesion proteins, cytokines and chemokines and in overactivation of the RAAS, leads to increased adhesion of monocytes, lymphocytes and macrophages to endothelial and smooth muscle cells of blood vessels (Santos et al. 2012; Pacurari et al. 2014). The inflammation mediators of special importance include monocyte chemoattractant protein-1 (MCP-1/CCL2), tumor necrosis factor-α (TNF-α) and transforming growth factor beta-1 (TGF-β-1), which intensify expression of adhesive cell molecules such as vascular adhesion molecule-1 (VCAM-1) and intercellular adhesion molecule-1 (ICAM-1). This leads to the formation of inflammatory infiltration, disturbances in extracellular matrix remodeling (caused by increased production of collagen), and fibrosis of the kidney parenchyma. Furthermore, angiotensin-2 activates proinflammatory NF-κB signalling pathway, which not only increases production of inflammatory cytokines (mainly TNF-α, IL-1 and IL-6), but also intensifies formation of reactive oxygen (ROS) and nitrogen species (RNS) (Pacurari et al. 2014; Putri and Thaha 2014; Tomino 2014; da Silva et al. 2017). The increase of ROS production is also caused by activation of NADPH oxidase, the main intracellular source of free radicals (Fogo 2007; Maciejczyk et al. 2019a). The oxidative stress is deemed responsible for the loss of elasticity of the blood vessels, the drop-in blood flow and decrease in GFR, which intensifies the CKD progression (Zhou et al. 2009, 2012; Webster et al. 2017; Maciejczyk et al. 2018c; Nakanishi et al. 2018).

Many studies have shown that CKD occurs mainly in people over 65 years of age (Harambat et al. 2012; Lopez-Giacoman 2015; Hill et al. 2016; Koye et al. 2018). Increased incidence of CKD was also noted in patients with diabetes mellitus, hypertension, obesity and cardiovascular diseases (mainly atherosclerosis, ischemia, vasculitis, and thrombotic microangiopathy). Other diseases leading to the development of CKD are renal tubular disorders (e.g., renal tubular acidosis, nephrogenic diabetes insipidus, non-albumin proteinuria, renal potassium and magnesium wasting, Fanconi syndrome, cystinuria), tubulointerstitial diseases (urinary tract infections, stones, obstruction, drug toxicity), as well as cystic and congenital diseases (KDIGO 2012; Jha et al. 2013; Hill et al. 2016; Koye et al. 2018). In most cases, however, the exact CKD etiology is not determinable. The histopathological examination of the kidneys is often omitted due to many contraindications to their implementation.

10.3 Saliva-Secretion, Content and Diagnostic Value

Saliva is secreted by large (parotid, submandibular and sublingual) and small salivary glands (Choromańska et al. 2017). Three major salivary glands secrete 90% of whole saliva, minor salivary glands that are scattered throughout the oral mucosa secrete remaining 10% of the saliva (Zalewska et al. 2015; Żukowski et al. 2018a). Around 60% of the unstimulated whole saliva (UWS) is produced by the submandibular gland, while the parotid, sublingual and minor salivary glands produce about 25%, 7–8% and 7–8%, respectively, of the whole saliva (Zalewska et al. 2015; Knaś et al. 2016; Kołodziej et al. 2017b). Stimulation increases parotid saliva secretion by 10–15%, the remaining major and minor salivary glands do not significantly increase their saliva production (Kołodziej et al. 2017a).

Saliva production is a two-stage process. In the first stage acini produce primary isotonic saliva, which contains the same amount of sodium, potassium, chlorine ions as in plasma. The primary saliva contains most of the organic components of the final saliva. During the passage through intercalated ducts sodium, and calcium cations and phosphate and chloride anions are reabsorbed and potassium cations are usually secreted (Varga 2015; Proctor 2016).

It is worth emphasizing that organic components of the saliva are produced in the salivary glands as well as plasma leakage, immune cells and bacteria, viruses and fungi, desquamated epithelial cells or expectorated bronchial and nasal secretions (Carpenter 2013; Maciejczyk et al. 2016). Saliva contains a whole range of antibacterial proteins (peroxidase, myeloperoxidase, lactoferrin, transferrin, lysozyme, mucins, agglutinins, immunoglobulins -mainly IgA) as well as antifungal proteins (staterins) (Humphrey and Williamson 2001; Carpenter 2013; Falkowski et al. 2018). Other proteins such as enzymes (α-amylase, lipase, carbonic anhydrase), hormones (estrogens, testosteron, progesteron, cortisol), lipids (neutral lipids, phospholipids and glycolipids, glycosphingolipids, cholesterol), urea, uric acid (UA), glucose are also present in the saliva (Humphrey and Williamson 2001; Varga 2015; Maciejczyk et al. 2017; Matczuk et al. 2017). Interestingly, creatinine, cystatin C and other kidney function markers such as β_2-microglobulin, kallikrein and neutrophil gelatinase-associated lipocalin (NGAL) have been identified in the saliva amongst other nitrogenous metabolic waste products (Celec et al. 2016; Yajamanam et al. 2016).

Recently, the importance of the antioxidant salivary defence mechanism has been underlined (Knaś et al. 2013; Avezov et al. 2015; Żukowski et al. 2018b). The only enzymatic antioxidant solely produced by the salivary glands is the salivary peroxidase (Px) or the salivary peroxidase system which consists of peroxidase and myeloperoxidase as well as hydrogen peroxide (H_2O_2) and thiocyanate ion (SCN^-) (Battino et al. 2002; Fejfer et al. 2017). Saliva also contains non-enzymatic antioxidants such as uric acid and the glutathione system, but also albumins, ascorbic acid, transferrin, ceruloplasmin and lactoferrin (Battino et al. 2002; Falkowski et al. 2018). It has been proven that the most important salivary antioxidant is uric acid (UA) which determines 70% of its total antioxidant capacity (Nagler et al. 2002; Żukowski et al. 2018b). The main role of uric acid is to capture hydroxyl radicals, lipid peroxides, and singlet oxygen but not superoxide radical. UA reduces free radical formation through creation of stable complexes with iron, copper and manganese, which stabilises the reduced form of vitamin C (Giordano et al. 2015; Żebrowska et al. 2019). However, in high concentrations, UA works as a prooxidant. Reaction with nitric oxide (NO) leads to creation of 6-aminouracil, which results in increased production of peroxynitrite. Uric acid can also lead to generation of other free radicals such as aminocarbonyl radical or alkylated derivatives (Glantzounis et al. 2005; Sautin and Johnson 2008; Maciejczyk et al. 2018a; Borys et al. 2019).

The substances present in the saliva can be generally divided into two groups: - the substances produced uniquely in the salivary glands and the substances transported from plasma to saliva (most substances) (Malathi et al. 2014; Javaid et al. 2016). It has been proven that biomolecules can move to saliva through passive transport (diffusion and ultrafiltration), facilitated diffusion, pinocytosis and also through active transport. Not surprisingly, the concentration of most of salivary substances is correlated to their content in the blood (Varga 2015; Proctor 2016; Fejfer et al. 2017). Therefore, saliva is utilised increasingly in laboratory diagnostics. It has been demonstrated that saliva can be an alternative to blood in the diagnosis of many metabolic diseases (obesity, insulin resistance, diabetes), neurodegenerative diseases (Alzheimer's disease, dementia), autoimmunological diseases (rheumatoid arthritis, Sjogren's syndrome) as well as chronic kidney disease (Yoshizawa et al. 2013; Malathi et al. 2014; Zhang et al. 2016; Kaczor-Urbanowicz et al. 2017; Klimiuk et al. 2019; Kułak-Bejda et al. 2019).

10.4 Salivary Biomarkers of Chronic Kidney Disease

Early diagnosis of CKD allows for effective nephroprotective treatment delaying the initiation of dialysis and reducing the risk of death due to cardiovascular diseases. However, despite the widespread availability of diagnostic methods, CKD is diagnosed very late (mainly in children and the elderly). In the CKD laboratory diagnostics, there are also no markers detecting the early stages of kidney damage (Lopez-Giacoman 2015; Celec et al. 2016).

Laboratory biomarker is defined as a compound that correlates with the physiological or pathological process, measured by validated and commonly available methods and is endowed with high sensitivity and diagnostic specificity. The ideal

CKD marker should therefore correlate with the histopathological picture and be detected already at an early stage of kidney damage. The hope for an early and non-invasive diagnostics of chronic kidney disease may be the use of saliva as an alternative to blood and other body fluids (Fassett et al. 2011; Giordano et al. 2015; Celec et al. 2016). The results of recent studies indicate the high diagnostic usefulness of salivary urea, creatinine, uric acid and the parameters of redox homeostasis and inflammation in diagnosing of CKD in adults and children.

An overview of salivary biomarkers used in CKD diagnostics is summarized in Table 10.4.

10.4.1 Salivary Urea

Urea (~60 Da) is the main product of protein catabolism in the human body. It is formed in hepatocytes in the urea cycle. This compound undergoes filtration in the kidney glomeruli and in 50% reversible resorption in the proximal tubule and partly in the distal tubule. Urea in the kidneys plays a major role in the process of urinary compaction. Its increased concentrations are found in the plasma of patients with chronic kidney disease. In many studies, it has been shown that urinary excretion of urea is directly proportional to the size of GFR. Therefore, the assessment of urea in plasma and urine is used as a classic CKD biomarker (Lopez-Giacoman 2015; Giardino et al. 2017; Vanholder et al. 2017).

In about 90% urea is excreted in the urine, although its presence has also been demonstrated in saliva (Sein and Arumainayagam 1987; Vuletic et al. 2013; Seethalakshmi et al. 2014; Renda 2017). Interestingly, the salivary concentration of urea significantly correlates with its content in blood plasma (Shannon et al. 1977; Akai et al. 1983; Zúñiga et al. 2012). Very high values of the saliva-blood correlation coefficient were noticed not only in patients with CKD ($r = 0.99$), but also in healthy subjects ($r = 0.74$) (Sein and Arumainayagam 1987). In animal studies, salivary gland and subcutaneous salivary glands have been found to contain significantly less urea compared to salivary parotid saliva. The administration of exogenous urea, however, disturbs not only the secretory activity of salivary glands, but also the concentration of urea in saliva (Watanabe et al. 1984).

In many clinical studies, a significant increase in the concentration of urea in the saliva of patients with CKD has been observed compared to controls (Akai et al. 1983; Tomás et al. 2008; Cardoso et al. 2009; Peng et al. 2013; Vuletic et al. 2013; Seethalakshmi et al. 2014; Celec et al. 2016; Renda 2017). However, Kovalčíkova et al. (2018) showed that the level of salivary urea is higher only in an animal model of acute kidney disease (AKI). Although in healthy subjects the plasma urea concentration was much higher than the concentration of urea in saliva, in patients with CKD similar concentrations of urea in the UWS and SWS as well as in the blood plasma were found (Cardoso et al. 2009). At the same time, it was demonstrated that the analytical method used to assess the plasma urea and serum concentration was used in salivary assays (Cardoso et al. 2009). Diagnostic usefulness of salivary urea in CKD diagnostics was also confirmed by ROC (receiver operating

Table 10.4 Salivary biomarkers in the diagnosis of kidney diseases

Biomarker	Subjects	Results	References
Urea	CKD	High correlation between salivary and serum urea	Shannon et al. (1977), Akai et al. (1983), Cardoso et al. (2009) and Zúñiga et al. (2012)
		High correlation between salivary urea and kidney function markers	Tomás et al. (2008), Cardoso et al. (2009), Peng et al. (2013) and Seethalakshmi et al. (2014)
		Higher level of salivary urea compared to healthy controls	Akai et al. (1983), Tomás et al. (2008), Cardoso et al. (2009), Peng et al. (2013), Vuletic et al. (2013) and Renda (2017)
		Lower level of salivary urea after dialysis	Shannon et al. (1977), Klassen and Krasko (2002), Bots et al. (2007) and Seethalakshmi et al. (2014)
Creatinine	CKD	High correlation between salivary creatinine and sCr	Goll and Mookerjee (1978), Lloyd et al. (1996), Chiou et al. (1997) and Seethalakshmi et al. (2014)
		High correlation between salivary creatinine and kidney function markers	Chiou et al. (1997) and Seethalakshmi et al. (2014)
		Higher level of salivary creatinine compared to healthy controls	Goll and Mookerjee (1978), Chiou et al. (1997), Tomás et al. (2008), Seethalakshmi et al. (2014) and Renda (2017)
Uric acid (UA)	CKD	High correlation between salivary and serum UA	Goll and Mookerjee (1978), Bibi et al. (2008), Peluso and Raguzzini (2016) and Maciejczyk et al. (2018c)
		High correlation between salivary UA and kidney function markers	Goll and Mookerjee (1978), Peluso and Raguzzini (2016) and Maciejczyk et al. (2018c)
		Higher level of salivary UA compared to healthy controls	Goll and Mockerjee (1978), Ben-Zvi et al. (2007), Bibi et al. (2008), Peluso and Raguzzini 2016) and Maciejczyk et al. (2018c)
		Lower level of salivary UA after dialysis	Ben-Zvi et al. (2007), Bibi et al. (2008), Hadi and Al-jubouri (2011) and Peluso and Raguzzini (2016)

(continued)

Table 10.4 (continued)

Biomarker	Subjects	Results	References
Electrolytes	CKD	Higher level of salivary Na^+, K^+ and Cl^- and lower level of salivary Ca^{2+} compared to healthy controls	Bots et al. (2007), Tomás et al. (2008), Anuradha et al. (2015) and Bagalad et al. (2017)
		Lower level of salivary Na^+ after dialysis	Shasha et al. (1983) and Bots et al. (2007)
		High correlation between salivary phosphorus and kidney function markers; higher level of salivary phosphate compared to healthy controls	Savica et al. (2007, 2008, 2009)
Albumin	CKD	High correlation between salivary and plasma albumin; high correlation between salivary albumin and kidney function markers; higher level of salivary albumin compared to healthy controls	Maciejczyk et al. (2018c)
	CKD	Correlation between salivary albumin and albumin in peritoneal dialytic fluid; lower level of salivary albumin after dialysis	Bibi et al. (2008)
β_2-microglobulin	CKD	No relationship between salivary and blood β_2-microglobulin; higher level of salivary β_2-microglobulin compared to healthy controls	Michelis et al. (2008), Vahedi et al. (2013) and Assareh et al. (2014)
Cystatin C	CKD	High correlation between salivary and plasma cystatin C; high correlation between salivary cystatin C and kidney function markers; higher level of salivary cystatin C compared to healthy controls	Alsamarai et al. (2018)
Neutrophil gelatinase-associated lipocalin (NGAL)	Acute renal colic	High correlation between salivary and plasma NGAL; high correlation between salivary NGAL and kidney function markers; higher level of salivary NGAL compared to healthy controls	Tamimi et al. (2018)
Kallikrein	Essential hypertension	High correlation between salivary kallikrein and kidney function markers; higher level of salivary kallikrein compared to healthy controls	Röckel et al. (1980) and Schmid and Heidland (1984)

Redox biomarkers	CKD	Correlation between salivary and plasma redox biomarkers; high correlation between salivary redox biomarkers and kidney function markers	Bibi et al. (2008), Peluso and Raguzzini (2016) and Maciejczyk et al. (2018c)
		Disturbances in salivary antioxidant barrier (Px, CAT, SOD, GR, GSH, UA, albumin, TAS) and higher oxidative damage (AGE, AOPP, PC, MDA) compared to healthy controls	Bibi et al. (2008), Hadi and Al-jubouri (2011), Peluso and Raguzzini (2016) and Maciejczyk et al. (2018c)
C-reactive protein (CRP)	CKD	Higher level of salivary CRP in haemodialysed patients compared to healthy controls and non-haemodialysed patients	Pallos et al. (2015)
Cortisol	Adrenal insufficiency	Higher level of salivary cortisol in hypotensive end stage renal disease (ESRD)	Arregger et al. (2008)
N-methyl-2-pyridone-5-carboxamide (Met2PY)	CKD	High correlation between salivary Met2PY and kidney function markers; higher level of salivary Met2PY compared to healthy controls	Rutkowski et al. (2010)

Abbreviations: *AGE* advanced glycation end products, *AOPP* advanced oxidation protein products, *CKD* chronic kidney disease, *CRP* C-reactive protein, *GR* glutathione reductase, *GSH* reduced glutathione, *MDA* malondialdehyde, *NGAL* neutrophil gelatinase-associated lipocalin, *PC* protein carbonyls, *Px* salivary peroxidase, *sCr* serum creatinine, *SOD* superoxide dismutase, *TAS* total antioxidant status, *UA* uric acid characteristic) analysis. Indeed, Cardoso et al. (2009) have demonstrated that optimal cut-off values of salivary urea >7.5 mM are characterized by 100% sensitivity and 100% specificity in ruling out azotemic states (AUC = 1.0, 95% confidence interval = 0.984–1.00). The high diagnostic value of CKD is also confirmed by the fact that the increase in salivary urea concentration reflects the degree of kidney damage (Tomás et al. 2008; Cardoso et al. 2009) and its level decreases after haemodialysis (Klassen and Krasko 2002; Bots et al. 2007). Zuniga et al. (2012) proved that salivary urea correlates negatively with creatinine clearance (r = −0.70). At the same time, Tomas et al. demonstrated that salivary urea is a more sensitive CKD marker than salivary creatinine, especially in the early stages of the disease (Tomás et al. 2008). In addition, Peng et al. showed that the salivary urea concentration does not change depending on the time of day and does not depend on the volume of saliva (Peng et al. 2013), which is consistent with the studies of Cardoso et al. (2009). Thus, the evaluation of salivary urea can be used to monitor the condition of patients around the clock, without the need of repeated blood collection.

Salivary urea is a promising salivary biomarker of renal function. The advantages of its use are demonstrated by the high compatibility between saliva and blood, independence from salivary flow rate, as well as speed, simplicity and low cost of colorimetric determination. However, it should be remembered that the concentration of urea depends not only on the degree of kidney perfusion, the size of diuresis and GFR, but also on age and sex, the rate of its synthesis related to the daily supply of protein in the diet and the degree of protein catabolism (Lopez-Giacoman 2015; Giardino et al. 2017; Vanholder et al. 2017). Also, diseases associated with liver damage may interfere with urea synthesis processes, which may also affect its concentration in saliva. In addition, salivary urea may be metabolised by bacterial urease present in the oral cavity, although the clinical significance of this process is still not well-known (Goll and Mookerjee 1978; Celec et al. 2016).

10.4.2 Salivary Creatinine

Creatinine (~113 Da) is a product of creatine metabolism derived from skeletal muscle and food intake. This compound undergoes filtration in the glomerulus. It is estimated, however, that about 10% of creatinine undergoes active secretion in the proximal tubule. In clinical practice, renal function is most often assessed on the basis of serum creatinine evaluation (sCr). Creatinine is not an ideal CKD biomarker. Its value is influenced by a number of factors: (1) age (higher concentrations in young people), (2) sex (higher concentrations in men), (3) body build (higher concentrations in people with high muscle mass) and (4) diet (eating foods rich in protein causes sCr to grow). Therefore, sCr is used to calculate the estimated GFR (eGFR). Interestingly, in elderly people the filtration can be lowered with normal creatinine concentration, and in the early stages of CKD filtration can be normal despite elevated sCr (Fassett et al. 2011; Ferguson and Waikar 2012; KDIGO 2012; Webster et al. 2017).

In patients with CKD, sCr concentrations increase. In the case of severe CKD, the percentage of creatinine undergoing active secretion in the proximal tubule also increases, which leads to an overestimation of the creatinine clearance value (Lopez-Giacoman 2015; Webster et al. 2017).

Creatinine is also present in saliva. It has been shown that this compound reaches the oral cavity via ultrafiltration and, in healthy people, the concentration of salivary creatinine is about 10–15% of its concentration in blood serum (Chiou et al. 1997; Tomás et al. 2008; Celec et al. 2016). Increased levels of salivary creatinine were reported in patients with CKD compared to controls (Goll and Mookerjee 1978; Chiou et al. 1997; Tomás et al. 2008; Seethalakshmi et al. 2014; Renda 2017). In this group of patients, a high compatibility of creatinine concentration in saliva and blood was also observed ($r = 0.784$, $p < 0.001$) (Lloyd et al. 1996). However, a similar relationship was not reported in healthy subjects with normal sCr level. It is believed that this is due to the relatively high molecular weight of creatinine and its low solubility in fats, which causes this compound to pass poorly from blood serum to saliva (Celec et al. 2016). Indeed, Lloyd et al. have shown that the salivary creatinine method is characterized by low precision and low sensitivity at low sCr concentrations (Lloyd et al. 1996). The optimal concentration of sCr, at which this method shows high accuracy and specificity are concentrations above 120 μmol/L. Considering that in CKD patients high levels of sCr are present (significantly exceeding 120 μmol/L), a high concentration gradient promotes the diffusion of serum creatinine into the saliva (Renda 2017). In fact, the creatinine concentration in the saliva of CKD subjects is significantly higher than in control group, and above 16.8 μmol/L, this parameter significantly differentiates CKD sufferers and healthy people. In the study of Lloyd et al. (1996), all patients were detected with one false positive result (sensitivity = 100%, specificity = 95.7%, efficacy = 97.7%). This is also confirmed by the studies of Venkatapathy et al. (2014) conducted on a group of 105 patients with chronic kidney disease. Linear regression equation and ROC analysis confirmed the high diagnostic value of salivary creatinine determination in CKD patients (AUC = 0.967, sensitivity = 97.14%, specificity = 86.5%). Celec et al. (2016) indicated, however, that the determination of salivary creatinine may be more useful in screening test than in the assessment of kidney damage in CKD patients.

Chiou et al. (1997) have shown that the ratio of sCr to salivary creatinine ranges from 4.5 to 30.0 and shows high interindividual variability. Thus, the salivary creatinine monitoring has limited diagnostic value in accurately predicting creatinine levels in the blood. The disadvantages of salivary creatinine in laboratory diagnostics are also evidenced by the fact that creatinine concentration in saliva varies depending on the salivary flow rate. However, the greatest changes in serum creatinine occur mainly at very low or very high salivary flow rate, which is achieved mainly in experimental conditions (e.g., by electrical nerve stimulation).

Lloyd et al. (1996) suggested that stimulation of salivation by chewing paraffin gum minimizes differences in salivary creatinine level caused by changes in salivary flow rate.

10.4.3 Salivary Uric Acid

Uric acid (~168 Da) is a product of purine metabolism from the diet and catabolism of endogenous nucleic acids. It is formed with the participation of oxidase and xanthine dehydrogenase. In 70% it is eliminated by the kidneys, whereas in 30% it is passively released into the gastrointestinal lumen. The correct UA clearance is 8–10 mL/min. Most UA is reabsorbed in the proximal tubule, of which 50% are re-secreted in the distal tubule. In patients with CKD, the concentration of UA in the blood plasma is significantly increased and the excretion of UA through the gastrointestinal tract also increases (Feig 2009; Giordano et al. 2015; Lytvyn et al. 2015; Tsai et al. 2017).

As stated previously, UA is largely responsible for the antioxidant properties of saliva (Nagler et al. 2002; Choromańska et al. 2017). Therefore, it is not surprising that uric acid is one of the most frequently tested salivary biomarkers (Wang et al. 2015; Javaid et al. 2016; Zhang et al. 2016). This compound is synthesized by the salivary glands, although the largest percentage of UA passes into the oral cavity from blood plasma. In CKD patients, higher concentrations of salivary UA are observed compared to healthy controls (Goll and Mookerjee 1978; Ben-Zvi et al. 2007; Bibi et al. 2008; Yajamanam et al. 2016; Maciejczyk et al. 2018c; Bilancio et al. 2019). Maciejczyk et al. (2018c) showed a compatibility between the salivary and blood levels of uric acid ($r = 0.668$, $p = 0.005$) in the group of children and adolescents with chronic kidney disease. In healthy children, however, the saliva-blood correlation coefficient was significantly lower ($r = 0.436$, $p = 0.016$). In contrast, Goll and Mookerjee (1978) demonstrated a relationship between salivary and plasma UA only in azotemic patients. A positive correlation of salivary and blood UA was observed before and after dialysis. As demonstrated by Ben-Zvi et al. (2007), haemodialysis reduces uric acid levels in both plasma and saliva. Therefore, it is proposed to use salivary UA to assess the effectiveness of dialysis. This is also confirmed by the results of other studies (Bibi et al. 2008; Hadi and Al-jubouri 2011; Peluso and Raguzzini 2016). What's more, Maciejczyk et al. (2018c) indicated that the increase in salivary UA correlates with classical markers of kidney damage (eGFR, albuminuria, protein in urine). These observations indicate the involvement of uric acid in CKD progression. The usefulness of salivary UA determination was also confirmed by ROC analysis. It was shown that this marker, determined in non-stimulated saliva, differentiates healthy people from CKD children with sensitivity = 64.71% and specificity = 66.67% (AUC = 0.82, $p = 0.004$), whereas in stimulated saliva with sensitivity = 88.24% and specificity = 92.31% (AUC = 0.93, $p < 0.0001$) (Maciejczyk et al. 2018c).

Until now, salivary uric acid was mainly used to assess the antioxidant potential of CKD patients (Ben-Zvi et al. 2007; Bibi et al. 2008; Peluso and Raguzzini 2016; Maciejczyk et al. 2018c). Therefore, the increase in salivary UA in CKD patients suggests the strengthening of the antioxidant barrier compared to a healthy group. However, it should be remembered that in the conditions of ROS overproduction, UA may be rapidly degraded and its decomposition products may be cytotoxic. At high concentrations (hyperuricemia) and/or in the presence of Cu^{+}/Cu^{2+} ions, this compound intensifies lipid peroxidation, i.e. behaves as a pro-oxidant (Glantzounis et al. 2005; Sautin and Johnson 2008). Therefore, it is not surprising that many studies have shown that the assessment of uric acid provides information about the increased risk of cardiovascular disease, type 2 diabetes, cancer, as well as chronic kidney disease (Giordano et al. 2015; Lytvyn et al. 2015; Noma et al. 2017). Bearing in mind that the concentration of UA in plasma/blood serum significantly correlates with its content in saliva, salivary uric acid is a potential, very attractive CKD biomarker. Therefore, the need for further studies evaluating the diagnostic usefulness of salivary UA in both the population of children and adults is indicated.

10.4.4 Salivary Electrolytes

The measurements of electrolyte concentrations, such as sodium (Na^{+}), potassium (K^{+}), calcium (Ca^{2+}), magnesium (Mg^{2+}), chlorides (Cl^{-}) and phosphates, are also used to assess renal function. CKD patients showed significantly higher salivary Na^{+}, K^{+} and Cl^{-} concentrations and a decrease in Ca^{2+} level compared to healthy controls (Bots et al. 2007; Tomás et al. 2008; Anuradha et al. 2015; Bagalad et al. 2017). However, the concentration of salivary electrolytes does not correlate with their content in serum / plasma. This fact is not surprising, because the hypotonic final saliva differs in composition from blood plasma due to the reabsorption / secretion processes taking place in the salivary gland ducts.

Shasha et al. (1983) and Bots et al. (2007) demonstrated that haemodialysis decreases salivary sodium to values close to the control group, while salivary potassium levels do not change significantly. This confirms various mechanisms of Na^{+} and K^{+} secretion / reabsorption into the saliva. Interestingly, Savica et al. (2007, 2008, 2009) showed that the level of salivary phosphate was significantly higher in CKD patients and correlated well with the phosphorus serum ($r = 0.42$, $p < 0.0001$) and sCr ($r = 0.72$, $p < 0.0001$). A negative correlation was also observed between salivary phosphorus and eGFR ($r = 0.72$, $p < 0.0001$), which postulates the use of this parameter in the non-invasive diagnosis of hyperphosphatemia. It is assumed that the increased secretion of phosphates into the saliva is the compensatory effect of the reduced ability of the kidneys to excrete phosphorus. However, an increase in the concentration of phosphorus in the saliva increases its absorption on an empty stomach, which in the way of a vicious circle intensifies hyperphosphatemia in patients with CKD.

The low diagnostic usefulness of salivary electrolytes (apart from the lack of correlation between saliva and blood) is also demonstrated by the fact that their content in saliva depends very much on the flow of saliva. It is well-known that in patients with CKD the secretion of the salivary glands is impaired, which may change the composition of salivary secretion.

10.4.5 Salivary Proteins

Albumin is the main protein in blood plasma. The renal catabolism of albumin consists of glomerular filtration and tubular resorption. As previously described, the assessment of plasma albumin and serum creatinine concentration is one of the basic laboratory tests used in CKD diagnostics (Johnson et al. 2012; Gaitonde et al. 2017). This protein is also present in saliva. Increased levels of salivary albumin were reported in non-stimulated and stimulated saliva in children with chronic kidney disease compared to controls (Maciejczyk et al. 2018c). Interestingly, the concentration of albumin in saliva correlates with its plasma content not only in patients with CKD (UWS-plasma: $r = 0.566$, $p = 0.018$, SWS-plasma: $r = 0.483$, $p = 0.015$), but also in healthy people (UWS-plasma: $r = 0.682$, $p = 0.010$). Furthermore, the diagnostic usefulness of salivary albumin was confirmed by ROC analysis. In stimulated saliva, this parameter differentiates healthy children from children with CKD with sensitivity = 93.33% and specificity = 90.91% (AUC = 0.94, $p = 0.0002$) (Maciejczyk et al. 2018c). However, salivary albumin concentration may also change in dialysis patients with chronic kidney disease. Bibi et al. (2008) showed that after dialysis, the peritoneal dialytic fluid (PDF) contains a much higher level of total protein and albumin, which leak from the serum during dialysis. Thus, it is not surprising that salivary albumin concentration is significantly reduced in patients after dialysis. However, the clinical value of salivary albumin should be confirmed in a larger number of patients, including adults.

β_2-microglobulin is one of the uremic toxins constituting the main histocompatibility complex (MHC) class I. This protein occurs on the surface of many body cells (e.g., T and B lymphocytes, macrophages), and its concentration increases with decreasing glomerular filtration (Lopez-Giacoman 2015). Particularly high concentrations of plasma β_2-microglobulin are observed in G5 patients treated with haemodialysis. Excess β_2-microglobulin in these patients is responsible for the amyloidosis leading to progressive damage to many organs (Fassett et al. 2011; Lopez-Giacoman 2015). Increased concentrations of β_2-microglobulins in unstimulated saliva were demonstrated in patients with CKD compared to healthy controls (Michelis et al. 2008; Vahedi et al. 2013; Assareh et al. 2014). However, the salivary-blood correlation coefficient was low in the control group and the study group (Assareh et al. 2014). The concentration of β_2-microglobulins in saliva did not correlate with the level of creatinine and urea in saliva and serum, which indicates the unquestionable defects of salivary β_2-microglobulin as a biomarker of kidney diseases (Michelis et al. 2008; Vahedi et al. 2013; Assareh et al. 2014).

Cystatin C is a non-glycosylated protein belonging to the cystatin family that inhibits endogenous and exogenous cysteine protein peptidases. Complete filtration in the glomerulus and subsequent catabolism in the kidney tubules makes cystatin C a substance which blood concentration depends only on the size of the filtration (Fassett et al. 2011; Lopez-Giacoman 2015). Thus, it is believed that the assessment of plasma cystatin C is much more sensitive in the evaluation of renal function compared to sCr. The results of recent studies indicate a high diagnostic usefulness of salivary cystatin C in the diagnosis of chronic kidney disease (Alsamarai et al. 2018). More than twice higher concentrations of cystatin C were observed in patients with CKD compared to healthy controls. In addition, it was demonstrated that salivary cystatin positively correlates with serum cystatin, sCr and blood urea in both control and patient group (Alsamarai et al. 2018). The advantages of using cystatin C in laboratory diagnostics are also evidenced by the fact that the concentration of this protein in saliva does not depend on the diet and body's composition, and therefore is a better indicator than the assessment of sCr filtration (Fassett et al. 2011; Lopez-Giacoman 2015).

NGAL (neutrophil gelatinase-associated lipocalin), like cystatin C, is a small protein with a molecular weight of approximately 25 kDa. This compound belongs to the lipocalin family and undergoes free glomerular filtration and reabsorption in proximal tubules (Fassett et al. 2011; Lopez-Giacoman 2015). NGAL seems to be a good biomarker of renal function due to the fact that an increase in the concentration of this compound (both in plasma and urine) is observed already after 2 h from the action of the nephrotoxic factor. The increase in the concentration of NGAL significantly overtakes hypercreatininemia and this protein can be an indicator not only of CKD, but also of AKI (Fassett et al. 2011; Lopez-Giacoman 2015). Increased salivary NGAL levels were observed in acute renal colic patients as well as in plasma and urine of patients with AKI and CKD (Tamimi et al. 2018). A significant correlation was found between the concentration of NGAL in saliva and serum, as well as the concentration of NGAL in saliva and sCr of patients with acute phase of renal colic (Tamimi et al. 2018). Thus, there is a need for further studies assessing the diagnostic usefulness of salivary NGAL in CKD diagnosis.

A promising biomarker of kidney function damage seems to be also kallikrein. This protein belongs to the proteolytic enzymes from the group of serine proteases (Fassett et al. 2011; Lopez-Giacoman 2015). It was shown that in a healthy population, the activity of kallikrein in saliva derived from the parotid gland is inversely proportional to salivary flow rate and sodium concentration (Röckel et al. 1980). The increased release of kallikrein into the saliva was demonstrated in rats with genetic hypertension, DOCA-salt hypertension and renovascular hypertension (Röckel et al. 1980; Schmid and Heidland 1984). In addition, increased salivary concentration of kallikrein is found in patients with essential hypertension and renoparenchymal hypertension associated with impaired renal function (Schmid and Heidland 1984). Röckel et al. (1980) showed significant correlation between the concentration of kallikrein in the saliva derived from the parotid gland and the degree of renal failure measured by sCr level. However, the concentration of salivary kallikrein in advanced renal failure did not depend on blood pressure.

10.4.6 Salivary Redox Biomarkers

The role of oxidative stress in the pathogenesis of CKD is often emphasized (Beetham et al. 2015; Modaresi et al. 2015; Sureshbabu et al. 2015). Therefore, it is not surprising that salivary redox biomarkers are more and more often used to assess the degree of CKD.

In a recent study, Maciejczyk et al. (2018c) proved that salivary redox parameters may be potential diagnostic biomarkers of CKD in children. An enhancement of the salivary antioxidant barrier (↑Px, ↑SOD, ↑UA, ↑ albumin) was noted, which indicates the adaptive response of the body in response to increased production of free radicals. In addition, oxidative damage to proteins (↑AGE, ↑AOPP) and lipids (↑ malondialdehyde) in both unstimulated and stimulated saliva as well as in the blood plasma of children with CKD has been demonstrated compared to controls (Maciejczyk et al. 2018c). Importantly, most salivary redox biomarkers correlated with their content in plasma / erythrocytes, indicating the use of saliva as an alternative diagnostic material compared to blood. In addition, with the help of ROC analysis it has been proven that with high sensitivity and specificity, salivary AOPP differentiate children with chronic kidney disease very accurately from healthy children (AUC in UWS 0.92, and in SWS even 0.98). The positive correlation between eGFR, urine albumin and protein, as well as plasma UA is also evidence of the salivary suitability of AOPP. Nevertheless, none of the redox biomarkers has differentiated between various CKD stages. In the 2019 year, Maciejczyk et al. (2019b) have evaluated total antioxidant activity by measuring FRAP (ferric ion reducing antioxidant parameter) and uric acid (UA)-independent FRAP (FRAP-UA) in the saliva, plasma, and urine of CKD children. They showed that the level of total FRAP, FRAP-UA, and UA was statistically elevated in SWS and urine of CKD patients compared to the control group. These biomarkers increase with CKD severity, and their level in SWS reflects their content in urine. Interestingly, salivary total FRAP clearly distinguishes children with mildly to moderately decreased kidney function (stage 1–3 of CKD) from those with severe renal impairment (stage 4–5 of CKD). The diagnostic utility of salivary total FRAP was also confirmed by ROC analysis (AUC = 1, sensitivity = 100%, specificity = 100%). However, Maciejczyk et al. (2018c) suggest that changes in saliva redox profile may result not only from the severity of the disease process, but also from hypofunction of salivary glands in the course of CKD. Indeed, a decrease in unstimulated and stimulated salivary flow rate was observed, as well as a decrease in α-amylase activity in the saliva of children with CKD.

Bibi et al. (2008) also showed the usefulness of salivary redox biomarkers in the monitoring of dialysis patients with chronic kidney disease. UA and total antioxidant status (TAS) were decreased in saliva and increased in PDF of patients after peritoneal dialysis. It is well known that TAS defines the resultant antioxidant capacity of saliva and is the sum of both enzymatic and non-enzymatic antioxidants (Maciejczyk et al. 2018b). However, salivary antioxidant enzymes such as peroxidase and superoxide dismutase were significantly enhanced in CKD patients.

Interestingly, the correlation coefficient of UA in serum and saliva was 0.60, while in saliva and PDF was 0.45 (Bibi et al. 2008). Therefore, the authors suggest monitoring of salivary uric acid to assess the redox status of CKD dialysis patients. This is also confirmed by the results of other studies (Hadi and Al-jubouri 2011; Peluso and Raguzzini 2016). However, not only haemodialysis, but also CKD and CKD with diabetes mellitus significantly increase the oxidative stress biomarkers in both serum and saliva (Ben-Zvi et al. 2007). Therefore, it is advisable to conduct further studies assessing the clinical usefulness of salivary redox biomarkers in a larger number of CKD patients.

10.4.7 Other Salivary Biomarkers

C-reactive protein (CRP) is one of the most sensitive and at the same time routinely determined indicators of inflammation. The CRP assay in haemodialysis patients is useful for assessing the biocompatibility level between the patient's blood and the materials used for the dialysis treatment (Heidari 2012). However, the studies by Pallos et al. (2015) demonstrated that the salivary CRP levels were significantly higher in haemodialysis patients compared to healthy subjects and CKD patients not receiving renal replacement therapy. Similarly, the concentration of salivary immunoglobulins (↑IgG, ↑IgA) was significantly higher after haemodialysis. The results of these studies indicate that inflammation and immune disorders are more intense in the saliva of patients undergoing dialysis than with patients with less severe CKD (Pallos et al. 2015).

In some CKD patients, adrenal insufficiency may be diagnosed, through the assessment of serum cortisol. However, cortisol can also be found and determined in saliva. The assessment of salivary cortisol avoids stress-related errors during blood sampling and allows measurements multiple times a day (to determine the kinetics of changes in cortisol levels). Arregger et al. (2008) demonstrated the usefulness of salivary cortisol for the detection of adrenal insufficiency in hypotensive end stage renal disease (ESRD) patients. ROC analysis indicates that basal salivary cortisol $\leq$4.4 nM and serum cortisol $\leq$232.0 nM allows adrenal hypofunction to be recognized in ESRD patients, however, neither salivary cortisol nor serum cortisol showed 100% sensitivity to the detection of adrenal insufficiency in this group of patients. The advantages of salivary cortisol determination in laboratory diagnostics are also demonstrated by the fact that its concentration in saliva does not depend on the secretory function of salivary glands, as diffusion of cortisol occurs very quickly (Arregger et al. 2008; Celec et al. 2016).

A promising group of salivary biomarkers are uremic toxins. Rutkowski et al. (2010) showed a significantly higher concentration of salivary N-methyl-2-pyridone-5-carboxamide (Met2PY) in patients with chronic kidney disease compared to the control group. This compound belongs to the class of purine nucleotides, which accumulate in the kidneys and lead to the gradual failure of this organ. Salivary Met2PY correlated negatively with eGFR ($r = -0.55$, $p = 0.0001$) and positively with sCr ($r = 0.68$, $p = 0.0001$), which means that the release of Met2PY into the saliva occurs proportionally to the degree of kidney damage. The usefulness of salivary Met2PY is also demonstrated by the fact that in more than 1/3 of the subjects, its concentration in saliva was significantly higher than in serum (Rutkowski et al. 2010).

In patients with CKD, salivary α-amylase activity was also evaluated (Shimazaki et al. 2008; Maciejczyk et al. 2018c). This enzyme hydrolyses the α-1,4-glycosidic α-glycans (mainly in starch and glycogen, as well as in other complex sugars). Although the assessment of salivary α-amylase activity does not seem to be a specific biomarker of renal function, this parameter can very well inform about the degree of salivary gland dysfunction in CKD patients. Indeed, a positive correlation between the flow of unstimulated and stimulated saliva, total protein concentration and α-amylase activity was demonstrated in children with chronic kidney disease (Maciejczyk et al. 2018c).

10.5 The Advantages and Disadvantages of Using Saliva in the Diagnosis of Kidney Diseases

As shown above, saliva is an alternative to other biological fluids used in the diagnosis of CKD and other kidney diseases. Saliva is an easily available bioliquid, collection of which is painless and requires no special equipment or the presence of medical personnel. Importantly, the concentration of many salivary biomarkers correlates with their content in the blood, as well as with the degree of kidney damage, which proves the high salivary attractiveness in CKD laboratory diagnostics (Malathi et al. 2014; Celec et al. 2016; Kaczor-Urbanowicz et al. 2017; Maciejczyk et al. 2018c). In our work, we demonstrated the high clinical usefulness of salivary urea, creatinine, uric acid, oxidative stress biomarkers or salivary proteins (Fig. 10.1). Therefore, it is advisable to develop salivary diagnostic tests which would allow for quick diagnostics of renal function, also at home (e.g., by means of strip tests).

Despite the undoubted advantages, the use of saliva as a diagnostic material also has some limitations. Ionized compounds and/or high molecular weight substances do not pass from the plasma to the oral cavity or reach very low concentrations in the saliva (below the detection threshold of the analytical methods used). In addition, for many salivary biomarkers, no correlation between saliva and blood was detected or their concentration in the saliva depends to a large extent on the rate of salivation (e.g., salivary electrolytes). Also saliva may get contaminated with food, drinks and blood, at the stage of saliva collection, which may result in distortion of the results obtained. It is worth remembering that patients with CKD not only have a reduction in salivary secretion/salivary pH changes and its buffering properties, but also an increased incidence of periodontal disease and caries. Hypofunction of the salivary glands and diseases of the oral cavity can change the qualitative composition of the saliva and also influence the passage of many molecules from the

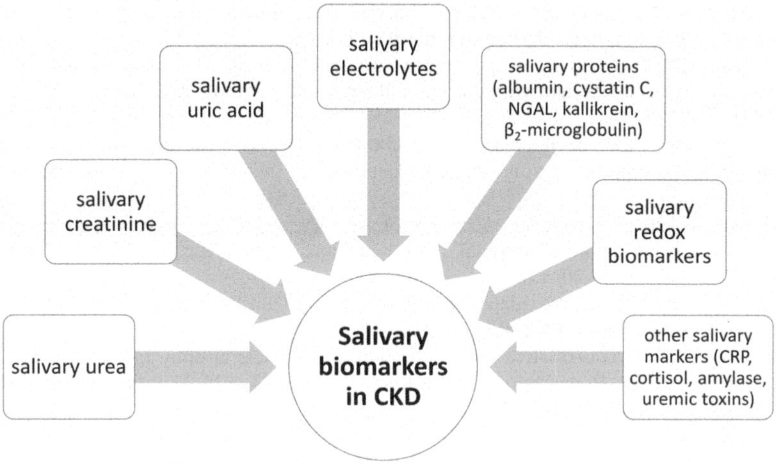

Fig. 10.1 Salivary biomarkers in the diagnosis of chronic kidney disease (CKD). Abbreviations: *CRP* C-reactive protein

plasma to the oral cavity. Therefore, further studies are needed to assess the clinical usefulness of salivary biomarkers on a larger population of patients with kidney diseases.

10.6 Conclusions

1. Saliva is an alternative laboratory material compared to blood and urine in the diagnosis of CKD and other kidney diseases.
2. The concentration of many biomarkers in saliva correlates with their concentration in the blood and the degree of severity of chronic kidney disease.
3. The salivary urea and salivary creatinine appear to be particularly interesting as the CKD diagnostic biomarkers.
4. Clinical usefulness of salivary CKD biomarkers requires further verification in clinical trials on a large population of adults and children.

References

Akai T et al (1983) Salivary urea nitrogen as an index to renal function: a test-strip method. Clin Chem 29(10):1825–1827

Alsamarai ATS et al (2018) Salivary cystatin C as a biochemical marker for chronic renal failure. Eurasian J Anal Chem 13(5). https://doi.org/10.29333/ejac/95168

Anuradha B et al (2015) Oral and salivary changes in patients with chronic kidney disease: a clinical and biochemical study. J Indian Soc Periodontol 19:297–301. https://doi.org/10.4103/0972-124x.154178

Arregger AL et al (2008) Diagnostic value of salivary cortisol in end stage renal disease. Steroids 73:77–82. https://doi.org/10.1016/j.steroids.2007.09.001

Assareh A et al (2014) No relationship between serum and salivary β2 – microglobulin levels in a sample of adult diabetic men with chronic kidney disease without renal replacement therapy. Cell J 16:179–186

Avezov K, Reznick AZ, Aizenbud D (2015) Oxidative stress in the oral cavity: sources and pathological outcomes. Respir Physiol Neurobiol 209:91–94. https://doi.org/10.1016/j.resp.2014.10.007

Bagalad B et al (2017) Diagnostic accuracy of salivary creatinine, urea, and potassium levels to assess dialysis need in renal failure patients. Dental Res J 14:13–18. https://doi.org/10.4103/1735-3327.201138

Battino M et al (2002) The antioxidant capacity of saliva. J Clin Periodontol 29:189–194. https://doi.org/10.1034/j.1600-051X.2002.290301x.x

Beetham KS et al (2015) Oxidative stress contributes to muscle atrophy in chronic kidney disease patients. Redox Rep: Commun Free Radical Res 20(3):126–132. https://doi.org/10.1179/1351000214Y.0000000114.

Ben-Zvi I et al (2007) Effects of diabetes mellitus, chronic renal failure and hemodialysis on serum and salivary antioxidant status. Nephron – Clin Pract. https://doi.org/10.1159/000098320

Bibi G, Green Y, Nagler RM (2008) Compositional and oxidative analysis in the saliva and serum of predialysis chronic kidney disease patients and end-stage renal failure patients on peritoneal dialysis. Ther Apher Dial 12:164–170. https://doi.org/10.1111/j.1744-9987.2008.00564.x

Bilancio G et al (2019) Saliva for assessing creatinine, uric acid, and potassium in nephropathic patients. BMC Nephrol 20(1):242. https://doi.org/10.1186/s12882-019-1437-4

Borys J et al (2019) Glutathione metabolism, mitochondria activity, and nitrosative stress in patients treated for mandible fractures. J Clin Med 8(1). https://doi.org/10.3390/jcm8010127

Bots CP et al (2007) Acute effects of hemodialysis on salivary flow rate and composition. Clin Nephrol 67:25–31. https://doi.org/10.5414/cnp67025

Cardoso EML et al (2009) Assessment of salivary urea as a less invasive alternative to serum determinations. Scand J Clin Lab Invest 69:330–334. https://doi.org/10.1080/00365510802588076

Carpenter GH (2013) The secretion, components, and properties of saliva. Annu Rev Food Sci Technol 4:267–276. https://doi.org/10.1146/annurev-food-030212-182700

Celec P et al (2016) Salivary markers of kidney function – potentials and limitations. Clin Chim Acta 453:28–37. https://doi.org/10.1016/j.cca.2015.11.028

Chiou W et al (1997) Correlation of creatinine levels in saliva and plasma in normal subjects and renal patients. Res Commun Chem Pathol Pharmacol 16:549–556

Choromańska M et al (2017) Antioxidant defence, oxidative stress and oxidative damage in saliva, plasma and erythrocytes of dementia patients. Can salivary AGE be a marker of dementia? Int J Mol Sci 18(10). https://doi.org/10.3390/ijms18102205

Cravedi P, Ruggenenti P, Remuzzi G (2012) Proteinuria should be used as a surrogate in CKD. Nat Rev Nephrol 8:301–306. https://doi.org/10.1038/nrneph.2012.42

da Silva AAS et al (2017) Renin angiotensin system and cytokines in chronic kidney disease: clinical and experimental evidence. Protein Pept Lett 24:799–808. https://doi.org/10.2174/0929866524666170818160809

Falkowski M et al (2018) Whey protein concentrate WPC-80 improves antioxidant defense systems in the salivary glands of 14-month Wistar Rats. Nutrients 10(6):782. https://doi.org/10.3390/nu10060782

Fassett RG et al (2011) Biomarkers in chronic kidney disease: a review. Kidney Int. https://doi.org/10.1038/ki.2011.198

Feig DI (2009) Uric acid: a novel mediator and marker of risk in chronic kidney disease? Curr Opin Nephrol Hypertens 18:526–530. https://doi.org/10.1097/MNH.0b013e328330d9d0

Fejfer K et al (2017) Oxidative modification of biomolecules in the nonstimulated and stimulated saliva of patients with morbid obesity treated with bariatric surgery. Biomed Res Int 2017:1–8. https://doi.org/10.1155/2017/4923769

Ferguson MA, Waikar SS (2012) Established and emerging markers of kidney function. Clin Chem 58:680–689. https://doi.org/10.1373/clinchem.2011.167494

Fogo AB (2007) Mechanisms of progression of chronic kidney disease. Pediatr Nephrol 22:2011–2022. https://doi.org/10.1007/s00467-007-0524-0

Gaitonde DY, Cook DL, Rivera IM (2017) Chronic kidney disease: detection and evaluation. Am Fam Physician

Giardino I et al (2017) Vascular toxicity of urea, a new "old player" in the pathogenesis of chronic renal failure induced cardiovascular diseases. Turk Pediatri Arsivi. https://doi.org/10.5152/TurkPediatriArs.2017.6314

Giordano C et al (2015) Uric acid as a marker of kidney disease: review of the current literature. Dis Markers 2015:1–6. https://doi.org/10.1155/2015/382918

Glantzounis G et al (2005) Uric acid and oxidative stress. Curr Pharm Des 11:4145–4151. https://doi.org/10.2174/138161205774913255

Goll RD, Mookerjee BK (1978) Correlation of biochemical parameters in serum and saliva in chronic azotemic patients and patients on chronic hemodialysis. Ren Fail 2:399–414. https://doi.org/10.3109/08860227809079326

Hadi BAA-R, Al-jubouri RH (2011) Salivary and plasma analysis of oxidative stress biomarkers in end stage renal failure patients. J Baghdad Coll Dent 23:46–50

Harambat J et al (2012) Epidemiology of chronic kidney disease in children. Pediatr Nephrol 27:363–373. https://doi.org/10.1007/s00467-011-1939-1

Heidari B (2012) C-reactive protein and other markers of inflammation in hemodialysis patients. Caspian J Int Med 4(1):611–616

Hill N et al (2016) Global prevalence of chronic kidney disease – a systematic review and meta-analysis. PLoS One 11(7):e0158765

Humphrey SP, Williamson RT (2001) A review of saliva: normal composition, flow, and function. J Prosthet Dent 85:162–169. https://doi.org/10.1067/mpr.2001.113778

Javaid MA et al (2016) Saliva as a diagnostic tool for oral and systemic diseases. J Oral Biol Craniofac Res 6:67–76. https://doi.org/10.1016/j.jobcr.2015.08.006

Jha V et al (2013) Chronic kidney disease: global dimension and perspectives. Lancet. https://doi.org/10.1016/S0140-6736(13)60687-X

Johnson DW et al (2012) Chronic kidney disease and measurement of albuminuria or proteinuria: A position statement. Med J Aust. https://doi.org/10.5694/mja11.11468

Kaczor-Urbanowicz KE et al (2017) Saliva diagnostics – current views and directions. Exp Biol Med 242:459–472. https://doi.org/10.1177/1535370216681550

KDIGO (2012) Clinical practice guideline for the evaluation and management of chronic kidney disease. Kidney Int Suppl 3:4. https://doi.org/10.1038/kisup.2012.76

Klassen JT, Krasko BM (2002) The dental health status of dialysis patients. In: Journal (Canadian Dental Association)

Klimiuk A et al (2019) Salivary redox biomarkers in different stages of dementia severity. J Clin Med 8(6). https://doi.org/10.3390/jcm8060840

Knaś M et al (2013) Oxidative stress and salivary antioxidants. Dental Med Problems 50(4):461–466

Knaś M et al (2016) Oxidative damage to the salivary glands of rats with streptozotocin-induced diabetes-temporal study: oxidative stress and diabetic salivary glands. J Diabetes Res 2016:1–13. https://doi.org/10.1155/2016/4583742

Kołodziej U, Maciejczyk M, Niklińska W et al (2017a) Chronic high-protein diet induces oxidative stress and alters the salivary gland function in rats. Arch Oral Biol 84:6–12. https://doi.org/10.1016/j.archoralbio.2017.09.006

Kołodziej U, Maciejczyk M, Miasko A et al (2017b) Oxidative modification in the salivary glands of high fat-diet induced insulin resistant rats. Front Physiol 8(JAN):20. https://doi.org/10.3389/fphys.2017.00020

Kovalčíkova A et al (2018) Salivary creatinine and urea are higher in an experimental model of acute but not chronic renal disease. PLoS One 13:e0200391. https://doi.org/10.1371/journal.pone.0200391

Koye DN et al (2018) The global epidemiology of diabetes and kidney disease. Adv Chronic Kidney Dis 25:121–132. https://doi.org/10.1053/j.ackd.2017.10.011
Kułak-Bejda A et al (2019) Diagnostic value of salivary markers in neuropsychiatric disorders. Dis Markers 2019:1–6. https://doi.org/10.1155/2019/4360612
Levey AS et al (2011) The definition, classification, and prognosis of chronic kidney disease: a KDIGO controversies conference report. Kidney Int. https://doi.org/10.1038/ki.2010.483
Lloyd JE, Broughton A, Selby C (1996) Salivary creatinine assays as a potential screen for renal disease. Ann Clin Biochem 33:428–431. https://doi.org/10.1177/000456329603300505
Lopez-Giacoman S (2015) Biomarkers in chronic kidney disease, from kidney function to kidney damage. World J Nephrol 4:57–73. https://doi.org/10.5527/wjn.v4.i1.57
López-Novoa JM et al (2010) Common pathophysiological mechanisms of chronic kidney disease: therapeutic perspectives. Pharmacol Ther 128:61–81. https://doi.org/10.1016/j.pharmthera.2010.05.006
Lytvyn Y, Perkins BA, Cherney DZI (2015) Uric acid as a biomarker and a therapeutic target in diabetes. Can J Diabetes 39:239–246. https://doi.org/10.1016/j.jcjd.2014.10.013
Maciejczyk M et al (2016) The significance of matrix metalloproteinases in oral diseases. Adv Clin Exp Med 25(2):383–390. https://doi.org/10.17219/acem/30428
Maciejczyk M et al (2017) Lysosomal exoglycosidase profile and secretory function in the salivary glands of rats with streptozotocin-induced diabetes. J Diabetes Res 2017:1–13. https://doi.org/10.1155/2017/9850398
Maciejczyk M, Matczuk J et al (2018a) Eight-week consumption of high-sucrose diet has a pro-oxidant effect and alters the function of the salivary glands of rats. Nutrients 10(10). https://doi.org/10.3390/nu10101530
Maciejczyk M, Żebrowska E et al (2018b) Redox balance, antioxidant defense, and oxidative damage in the hypothalamus and cerebral cortex of rats with high fat diet-induced insulin resistance. Oxidative Med Cell Longev 2018:1–11. https://doi.org/10.1155/2018/6940515
Maciejczyk M, Szulimowska J et al (2018c) Salivary biomarkers of oxidative stress in children with chronic kidney disease. J Clin Med 7(8). https://doi.org/10.3390/jcm7080209
Maciejczyk M, Żebrowska E, Chabowski A (2019a) Insulin resistance and oxidative stress in the brain: What's new? Int J Mol Sci 20(4):874. https://doi.org/10.3390/ijms20040874
Maciejczyk M, Szulimowska J, Taranta-Janusz K, Werbel K, Wasilewska A, Zalewska A (2019b) Salivary FRaP as a marker of chronic kidney disease progression in children. Antioxidants 8(9):409. https://doi.org/10.3390/antiox8090409
Malathi N, Mythili S, Vasanthi HR (2014) Salivary diagnostics: a brief review. ISRN Dent 2014:1–8. https://doi.org/10.1155/2014/158786
Matczuk J et al (2017) Salivary lipids: a review. Adv Clin Exp Med 26(6):1023–1031. https://doi.org/10.17219/acem/63030
Michelis R et al (2008) Salivary β2-microglobulin analysis in chronic kidney disease and hemodialyzed patients. Blood Purif 25:505–509. https://doi.org/10.1159/000113010
Modaresi A, Nafar M, Sahraei Z (2015) Oxidative stress in chronic kidney disease. Iran J Kidney Dis 9(3):165–179. 2044/766 [pii]
Nagler RM et al (2002) Characterization of the differentiated antioxidant profile of human saliva. Free Radic Biol Med 32:268–277. https://doi.org/10.1016/S0891-5849(01)00806-1
Nakanishi T et al (2018) Misdistribution of iron and oxidative stress in chronic kidney disease. Free Radic Biol Med 133:248–253. https://doi.org/10.1016/j.freeradbiomed.2018.06.025
National Kidney Foundation (2002) K/DOQI clinical practice guidelines for chronic kidney disease: evaluation, classification, and stratification. Am J Kidney Dis 39(2 Suppl 1):S1–S266
Noma K, Kihara Y, Higashi Y (2017) Is serum uric acid a biomarker, but not a mediator in patients with lifestyle and cardiovascular diseases? Int Heart J 58:467–469. https://doi.org/10.1536/ihj.17-301
Pacurari M et al (2014) The renin-angiotensin-aldosterone system in vascular inflammation and remodeling. Int J Inflamm 2014:1–13. https://doi.org/10.1155/2014/689360

Pallos D et al (2015) Salivary markers in patients with chronic renal failure. Arch Oral Biol 60:1784–1788. https://doi.org/10.1016/j.archoralbio.2015.09.008
Peluso I, Raguzzini A (2016) Salivary and urinary total antioxidant capacity as biomarkers of oxidative stress in humans. Pathol Res Int 2016:1–14. https://doi.org/10.1155/2016/5480267
Peng CH et al (2013) Influencing factors for saliva urea and its application in chronic kidney disease. Clin Biochem 46:275–277. https://doi.org/10.1016/j.clinbiochem.2012.10.029
Proctor GB (2016) The physiology of salivary secretion. Periodontology 2000:11–25. https://doi.org/10.1111/prd.12116
Putri AY, Thaha M (2014) Role of oxidative stress on chronic kidney disease progression. Acta Med Indones 46(3):244–252
Renda R (2017) Can salivary creatinine and urea levels be used to diagnose chronic kidney disease in children as accurately as serum creatinine and urea levels? A case-control study. Ren Fail 39:452–457. https://doi.org/10.1080/0886022X.2017.1308256
Röckel A et al (1980) Enhanced parotid kallikrein secretion in essential, adrenal, and renoparenchymal hypertension and in advanced renal failure. Nephron 26:137–144. https://doi.org/10.1159/000181969
Rüster C, Wolf G (2006) Renin-angiotensin-aldosterone system and progression of renal disease. J Am Soc Nephrol 17:2985–2991. https://doi.org/10.1681/asn.2006040356
Rutkowski B et al (2010) Distribution of purine nucleotides in uremic fluids and tissues. J Ren Nutr 20:S7–S10. https://doi.org/10.1053/j.jrn.2010.05.003
Santos PCJL, Krieger JE, Pereira AC (2012) Renin angiotensin system, hypertension, and chronic kidney disease: pharmacogenetic implications. J Pharmacol Sci 120:77–88. https://doi.org/10.1254/jphs.12r03cr
Sautin YY, Johnson RJ (2008) Uric acid: the oxidant-antioxidant paradox. Nucleosides Nucleotides Nucleic Acids 27:608–619. https://doi.org/10.1080/15257770802138558
Savica V et al (2007) Phosphate salivary secretion in hemodialysis patients: implications for the treatment of hyperphosphatemia. Nephron - Physiol 105:p52–p55. https://doi.org/10.1159/000098544
Savica V et al (2008) Salivary phosphate secretion in chronic kidney disease. J Ren Nutr 18:87–90. https://doi.org/10.1053/j.jrn.2007.10.018
Savica V et al (2009) Salivary phosphorus and phosphate content of beverages: implications for the treatment of uremic hyperphosphatemia. J Ren Nutr 19:69–72. https://doi.org/10.1053/j.jrn.2008.10.024
Schmid G, Heidland A (1984) Cation excretion and kallikrein secretion by the rat parotid gland in various hypertensive models. J Cardiovasc Pharmacol 6:S101. https://doi.org/10.1097/00005344-198400061-00018
Seethalakshmi C, Koteeswaran D, Chiranjeevi V (2014) Correlation of serum and salivary biochemical parameters in end stage renal disease patients undergoing hemodialysis in pre and post-dialysis state. J Clin Diagn Res. https://doi.org/10.7860/JCDR/2014/10404.5306
Sein KT, Arumainayagam G (1987) Correlation between serum urea and salivary urea. Clin Chem 33(12):2303 LP–2302304. Available at: http://clinchem.aaccjnls.org/content/33/12/2303.abstract
Shannon IL et al (1977) Human parotid saliva urea in renal failure and during dialysis. Arch Oral Biol 22:83–86. https://doi.org/10.1016/0003-9969(77)90082-6
Shasha SM et al (1983) Salivary content in hemodialysed patients. J Oral Med 38(2):67–70. Available at: http://europepmc.org/abstract/MED/6599836
Shimazaki M et al (2008) Clinical performance of a salivary amylase activity monitor during hemodialysis treatment. Biomark Insights 3:BMI.S997
Siragy HM, Carey RM (2010) Role of the intrarenal renin-angiotensin-aldosterone system in chronic kidney disease. Am J Nephrol 31:541–550. https://doi.org/10.1159/000313363
Sureshbabu A, Ryter SW, Choi ME (2015) Oxidative stress and autophagy: crucial modulators of kidney injury. Redox Biol 4:208–214. https://doi.org/10.1016/j.redox.2015.01.001

Tamimi A et al (2018) Salivary neutrophil gelatinase-associated Lipocalin sampling feasibility in acute renal colic. J Endourol 32:566–571. https://doi.org/10.1089/end.2017.0864
Thorp ML et al (2012) Proteinuria among patients with chronic kidney disease: a performance measure for improving patient outcomes. Jt Comm J Qual Patient Saf 38:277–AP4
Tomás I et al (2008) Changes in salivary composition in patients with renal failure. Arch Oral Biol 53:528–532. https://doi.org/10.1016/j.archoralbio.2008.01.006
Tomino Y (2014) Pathogenesis and treatment of chronic kidney disease: a review of our recent basic and clinical data. Kidney Blood Press Res 39:450–489. https://doi.org/10.1159/000368458
Tsai C-W et al (2017) Serum uric acid and progression of kidney disease: a longitudinal analysis and mini-review. PLoS One 12(1):e0170393. https://doi.org/10.1371/journal.pone.0170393
Vahedi M et al (2013) The relationship between salivary beta-2 microglobulin and uremia intensity in men with chronic renal failure. Cell J 14:276–281
Vanholder R, Gryp T, Glorieux G (2017) Urea and chronic kidney disease: the comeback of the century? (in uraemia research). Nephrol Dial Transplant 33(1):4–12. https://doi.org/10.1093/ndt/gfx039
Varga G (2015) Physiology of the salivary glands. Surgery (United Kingdom) 33:581–586. https://doi.org/10.1016/j.mpsur.2015.09.003
Venkatapathy R et al (2014) Salivary creatinine estimation as an alternative to serum creatinine in chronic kidney disease patients. Int J Nephrol 2014:1–6. https://doi.org/10.1155/2014/742724
Vuletic L et al (2013) Effect of L-arginine dietary supplementation on salivary urea concentration and pH in physically active individuals. Aust Dent J 58:491–497. https://doi.org/10.1111/adj.12114
Wang J et al (2015) Salivary biomarkers of oxidative stress: a critical review. Free Radic Biol Med 85:95–104. https://doi.org/10.1016/j.freeradbiomed.2015.04.005
Watanabe JUN et al (1984) Salivary excretion of urea in dogs. J Pharmacobiodyn 7(5):294–303. https://doi.org/10.1248/bpb1978.7.294.
Webster AC et al (2017) Chronic kidney disease. Lancet 389:1238–1252. https://doi.org/10.1016/S0140-6736(16)32064-5
Wouters OJ et al (2015) Early chronic kidney disease: Diagnosis, management and models of care. Nat Rev Nephrol. https://doi.org/10.1038/nrneph.2015.85
Yajamanam N et al (2016) Utility of saliva as a sample to assess renal function and estimated glomerular filtration rate. Saudi J Kidney Dis Transpl 27:312–319. https://doi.org/10.4103/1319-2442.178549
Yang HC, Zuo Y, Fogo AB (2010) Models of chronic kidney disease. Drug Discov Today Dis Model 7:13–19. https://doi.org/10.1016/j.ddmod.2010.08.002
Yoshizawa JM et al (2013) Salivary biomarkers: toward future clinical and diagnostic utilities. Clin Microbiol Rev. https://doi.org/10.1128/CMR.00021-13
Zalewska A et al (2015) Antioxidant profile, carbonyl and lipid oxidation markers in the parotid and submandibular glands of rats in different periods of streptozotocin induced diabetes. Arch Oral Biol 60(9):1375–1386. https://doi.org/10.1016/j.archoralbio.2015.06.012
Żebrowska E et al (2019) High protein diet induces oxidative stress in rat cerebral cortex and hypothalamus. Int J Mol Sci 20(7). https://doi.org/10.3390/ijms20071547
Zhang CZ et al (2016) Saliva in the diagnosis of diseases. Int J Oral Sci 8:133–137. https://doi.org/10.1038/ijos.2016.38.
Zhou L et al (2009) Accumulation of advanced oxidation protein products induces podocyte apoptosis and deletion through NADPH-dependent mechanisms. Kidney Int 76(11):1148–1160. https://doi.org/10.1038/ki.2009.322
Zhou LL et al (2012) The receptor of advanced glycation end products plays a central role in advanced oxidation protein products-induced podocyte apoptosis. Kidney Int 82(7):759–770. https://doi.org/10.1038/ki.2012.184

Żukowski P et al (2018a) Effect of N-acetylcysteine on antioxidant defense, oxidative modification, and salivary gland function in a rat model of insulin resistance. Oxidative Med Cell Longev 2018:1–11. https://doi.org/10.1155/2018/6581970

Żukowski P, Maciejczyk M, Waszkiel D (2018b) Sources of free radicals and oxidative stress in the oral cavity. Arch Oral Biol 92:8–17. https://doi.org/10.1016/j.archoralbio.2018.04.018.

Zúñiga ME et al (2012) Validation of the salivary urea test as a method to diagnose chronic kidney disease. J Nephrol 25:431–436. https://doi.org/10.5301/jn.5000022

Chapter 11
Salivary Diagnosis of Infectious Diseases

Silvia Martínez-Subiela and Ana Cantos-Barreda

11.1 Introduction

Systemic infectious diseases can be associated in some cases with high mortality rates. Thus, there is a need for early detection and diagnosis in order to initiate an appropriate treatment regime as soon as possible (Farnaud et al. 2010). Nowadays, the diagnosis of infectious diseases is still dependent on the evaluation of blood and/or tissue samples. Although they are effective, these procedures are invasive and expensive, moreover, depending on different clinical conditions, these types of tests may not be accessible for many patients and health care providers (Yoshizawa et al. 2013). For all these reasons saliva-based diagnostics have been the primary focus of investigation for a variety of infectious pathogens for several years (Farnaud et al. 2010).

The method of detection of an infection could be either direct – detection of the pathogen or its nucleic acids (DNA or RNA), or indirect – detection of host salivary antibodies IgA, IgM or IgG against the pathogen. The use of one or other type of methodology would vary depending on the specific pathogen; for example, direct detection of Mycobacterium tuberculosis in saliva by culture was less effective than the detection of bacterial DNA by PCR (17% vs. 98%) (Farnaud et al. 2010).

The aim of this chapter is to review the main methods that have been developed and evaluated in saliva for the diagnostic and monitoring of systemic infectious diseases of humans and domestic animals.

S. Martínez-Subiela (✉) · A. Cantos-Barreda
Interlab-UMU, Regional Campus of International Excellence Campus Mare Nostrum, University of Murcia, Murcia, Spain
e-mail: silviams@um.es

A. Tvarijonaviciute et al. (eds.), *Saliva in Health and Disease*,
https://doi.org/10.1007/978-3-030-37681-9_11

11.2 Salivary Diagnostics of Human Infectious Diseases

A brief summary of saliva-based methods developed for the detection of selected infectious pathogens affecting humans is presented in Table 11.1.

Table 11.1 Overview of selected publications describing methods for the diagnosis of infectious diseases using saliva as a sample

Pathogen	Biomarker	References
HIV	IgG	Scully and Samaranayake (1992), Cordeiro et al. (1993), Scully (1997) and Martínez et al. (1999)
	IgA	Matsuda et al. (1993)
Hepatitis A virus	Total and IgM	Thieme et al. (1992)
	IgG	Ahmed et al. (2011)
	RNA	Mackiewicz et al. (2004)
Hepatitis B virus	Total and Ig M	Scully and Samaranayake (1992) and Thieme et al. (1992)
Hepatitis C virus	Total and Ig M	Scully and Samaranayake (1992) and Thieme et al. (1992)
	Total	Cha et al. (2013)
Dengue virus	IgG	Cuzzubbo et al. (1998), Balmaseda et al. (2003, 2008) and Vázquez et al. (2007)
	IgM	Cuzzubbo et al. (1998), Balmaseda et al. (2003, 2008), Vázquez et al. (2007) and Chakravarti et al. (2007)
	IgA	Balmaseda et al. (2003, 2008), Vázquez et al. (2007) and Yap et al. (2011)
	RNA	Torres et al. (2000), Balmaseda et al. (2008) and Poloni et al. (2010)
Ebola viruses	RNA	Formenty et al. (2006) and Bausch et al. (2007)
Zika virus	RNA	Musso et al. (2015)
Measles virus	IgG	Perry et al. (1993), Garrido Redondo et al. (1997), Gill et al. (2002), Kremer and Muller (2005) and Vainio et al. (2008)
	IgM	Perry et al. (1993), Brown et al. (1994) and Hutse et al. (2010)
	IgA	Garrido Redondo et al. (1997)
	RNA	Jin et al. (2002) and Hutse et al. (2010)
Mumps virus	IgG	Perry et al. (1993) and Vainio et al. (2008)
	IgM	Perry et al. (1993) and Warrener and Samuel (2006)
	RNA	Jin et al. (2002)
Rubella virus	IgG	Parry et al. (1987), Perry et al. (1993), Nokes et al. (1998), Ramsay et al. (1998), Vyse et al. (1999), Christopher Maple and Jones (2002) and Ben Salah et al. (2003)
	IgM	Centers for Disease Control and Prevention (CDC) (2008) and Lambert et al. (2015)
	RNA	Vyse et al. (1999), Jin et al. (2002), Abernathy et al. (2009) and Vauloup-Fellous et al. (2010)
Helicobacter pylori	IgG	Loeb et al. (1997) and Luzza and Pallone (1997)
	DNA	Jiang et al. (1998) and Anand et al. (2014)

11.2.1 Viral Diseases

11.2.1.1 Human Immunodeficiency Virus (HIV)

Two types of human Immunodeficiency Viruses are described, HIV-type 1 (HIV-1) and HIV-type 2 (HIV-2). The main agent of acquired immune deficiency syndrome is HIV-1 and is a related member of the Lentivirus genus of the *Retroviridae* family (Fanales-Belasio et al. 2010). Traditionally, HIV infection have been performed through antibody detection in serum or plasma and thus, require trained personnel for collection and involves a high risk of transmission (Martínez et al. 1999). Saliva tests for HIV detection have been reported to be a noninvasive alternative to the quantification of antibodies in blood (Lawrence 2002). Different assay formats have been developed in which whole saliva was used for detecting antibodies directed against specific HIV viral protein epitopes with high sensitivity (98–100%) and specificity (97 100%) (Scully and Samaranayake 1992; Scully 1997). Furthermore, in one study that uses an enzyme-linked fluorescence technique combined with Western blot, saliva showed better sensitivity and specificity than serum (Martínez et al. 1999). In addition, it has been suggested that salivary detection of IgA may be useful with prognostic purposes, since salivary IgA levels to HIV decrease when infected patients show symptoms, thus indicating the evolution of the infection (Matsuda et al. 1993).

Although different tests have been developed in saliva and oral fluid, there is only one FDA- approved, commercially available testing system, for public use (OraSure®). It detects antibodies against the p24 antigen of HIV and consists of a cotton pad connected to a nylon stick and a vial that contains a preservative solution (Malamud 1997). The samples collected with OraSure® device could have IgG concentrations three to fourfold higher than those usually found in whole saliva (Cordeiro et al. 1993) and the storage solution maintain sample stability overtime at similar levels than initial concentrations (Malamud 1997). Additionally, the storage solution has been reported that inhibits different strains of HIV (Bestwick and Fitchen 1997) further increasing the safety of this method.

11.2.1.2 Hepatitis

Viral hepatitis constitutes an important public health problem all over the world. Hepatitis A (HAV), is caused by a virus of the *Picornaviridae* family and contamination of water supply and food is considered as the main sources of transmission (Oba et al. 2000). This type of hepatitis is one of the most frecuent causes of infectious hepatitis in the world. In contrast, Hepatitis B virus (HBV) and hepatitis C virus (HCV) are frequently related with chronic disease and can ultimately cause severe liver-related complications such as cirrhosis and hepatocellular carcinoma. Although infection is quite common, most patients do not show symptoms which implies a high potential risk of progression and transmission of the disease (Yoshizawa et al. 2013).

Traditional diagnosis and monitoring of viral hepatitis consist mainly on blood-based serological tests determining viral load as well as viral antibodies and antigens; however it has been described that antigens and/or antibodies for hepatitis A, B, and C viruses can be detected in the salivary samples of infected individuals (Amado et al. 2006) and, thus, saliva has been suggested as a useful alternative to serum for the diagnosis of variants of viral hepatitis. HAV can be diagnosed with 100% sensitivity and 98% specificity based on the presence of IgM antibodies in saliva (Thieme et al. 1992). Comparison of serum and saliva levels of antibody to HAV revealed excellent agreement (Thieme et al. 1992). Furthermore, it has been demonstrated that the determination of salivary IgG concentrations is useful to evaluate the efficacy of HAV immunizations (Ahmed et al. 2011). Not only antibody detection but also RNA detection in saliva has been suggested as a useful marker for tracing and monitoring HAV infection in community settings (Mackiewicz et al. 2004).

Similarly to HAV, analysis of saliva is highly sensitive and specific for the diagnosis of viral hepatitis B as well as hepatitis C with a sensitivity and specificity approaching 100% (Scully and Samaranayake 1992; Thieme et al. 1992). These findings suggest a potential role for saliva as a noninvasive mode of HBV and HCV diagnosis and disease state monitoring (Yoshizawa et al. 2013). Even, a commercially available rapid test (OraQuick® HCV test) has been developed and evaluated with saliva showing a sensitivity and specificity of 97.8% and 100%, respectively, by this supporting the utility of rapid testing using oral fluid in various medical and non-medical settings (Cha et al. 2013).

11.2.1.3 Dengue

Dengue virus (DENV) is an arthropod-borne flavivirus mainly transmitted by mosquito vectors (Guzman and Harris 2015). It has been described five antigenically distinct dengue viruses, DENV1-DENV5, that can cause dengue fever and severe dengue (Mustafa et al. 2015). Different tests have been developed for the clinical diagnosis of the disease that detect the virions, nucleic acids, antibodies, or antigenic components of a DENV infection (Wasik et al. 2018).

Literature shows that various antibodies such as IgA, IgM, IgE are detectable in the diagnosis of Dengue on using saliva with variable sensitivities and specificities (from 39% to 100%) (Cuzzubbo et al. 1998; Balmaseda et al. 2003, 2008; Vázquez et al. 2007). It has been described that salivary IgG levels could be used to distinguish between primary and secondary dengue virus infections (Cuzzubbo et al. 1998) and, although detection of IgG in saliva was less sensitive than in serum or filter-paper blood spots, it is considered an acceptable and useful marker for community-based studies, because of its non-invasive nature (Ravi Banavar and Vidya 2014). Therefore, saliva was considered a promising sample for dengue diagnostics (Chakravarti et al. 2007). In addition, the utility of saliva in an assay that detects DENV-specific IgA in the early phase of a 2nd dengue infection shows 100% sensitivity and 97% specificity from the day-one after fever onset with a good correlation to IgA levels in serum (Yap et al. 2011). Besides the specific antibodies, DENV RNA has also been found in saliva by RT-PCR (Torres et al. 2000). Furthermore, in case reports analyzing acute infections, the detection of DENV in saliva by RT-PCR as well as the application of filter-paper for saliva sampling, prove the usefulness of these non-invasive samples (Balmaseda et al. 2008; Poloni et al. 2010).

11.2.1.4 Ebola

Ebola viruses contain a single-stranded RNA genome that encodes seven viral proteins. Different methods for detecting Ebola infection and/or disease have been developed for use in clinical laboratory settings that can be grouped into three categories: serologic tests, antigen tests that detect viral proteins, and molecular tests that detect viral RNA (Strong et al. 2006; Broadhurst et al. 2016). Antigen detection and molecular tests have demonstrated to be very effective for acute diagnosis, however, serology is minimally useful as a diagnostic tool in the acute setting (Broadhurst et al. 2016).

Ebola virus, as well as other viral hemorrhagic fevers, are mostly detected in blood derived samples such as whole blood, plasma or serum although different studies reported Ebola detection also in body fluids including saliva (Niedrig et al. 2018).

There are only one reported study in which antibodies against Ebola has been investigated in saliva but it failed to detect antibodies in the oral fluid specimens obtained from seropositive patients as based on serum analyses (Formenty et al. 2006). In contrast, RT-PCR was effective to detect RNA of the virus in saliva (Formenty et al. 2006; Bausch et al. 2007). Furthermore, higher mortality was reported among patients with RT-PCR-positive saliva, suggesting that it can be an indicator of a poor prognosis (Bausch et al. 2007). Although detection of Ebola virus by RT-PCR in oral fluid specimens is sufficiently reliable as a diagnostic tool, especially in the investigation of the outbreak of Ebola, blood samples are still necessary to analyse the biological status and immune response of the patient and to optimize treatment plan (Shanbhag 2015).

Further studies are necessary to improve laboratory methods to detect suspected cases early and to design more-sensitive screening of the disease. However, it has been suggested that the use of oral fluid samples could make earlier detection of outbreaks much easier (Shanbhag 2015; Niedrig et al. 2018).

11.2.1.5 Zika

Zika virus (ZIKV) is an arthropod-borne virus (arbovirus) of the family *Flaviviridae* and genus *Flavivirus* (Musso and Gubler 2016). Laboratory Zika fever diagnosis is challenging because there is no “gold standard” diagnosis tool. The cross reactivity of antibodies between Flaviviruses, limits the use of serology, viral culture is not routinely performed and there is no antigenic detection test available (Musso et al. 2015).

ZIKV RNA has been identified in saliva samples increasing the molecular detection rate of ZIKV in acute cases, but ZIKV did not persist for a longer time frame in saliva as in urine or semen (Musso et al. 2015). The inconsistency of saliva makes urine the most reliable and utilized secondary sample type ('WHO | Laboratory testing for Zika virus infection' 2016; Bingham et al. 2016) and a combination of samples (blood/urine and saliva) has been recommended to increase the sensitivity of the virus detection (Musso et al. 2015; Zhang et al. 2016).

11.2.1.6 Measles, Mumps and Rubella

The detection of measles, mumps and rubella (MMR) antibodies in saliva represent the most advanced application to the diagnostic utility of saliva in case of infectious diseases since a MMR salivary surveillance program has already been successfully running in the United Kingdom from 1994 (Madar et al. 2002).

Measles virus (MV) is a negative single-stranded RNA virus, belonging to the genus *Morbillivirus*, of the family *Paramyxoviridae* (Bellini et al. 1994). Laboratory confirmation of measles cases is an essential aspect of surveillance at all stages of control programs because clinical diagnosis is unreliable (Featherstone et al. 2003). The mainstay of laboratory confirmation is the detection of measles-specific IgM antibodies in serum samples (Centers for Disease Control and Prevention (CDC) 2008). Several studies have shown that saliva samples may be adequate substitutes to serum for the detection of measles specific IgG and IgM antibodies (Perry et al. 1993; Brown et al. 1994; Garrido Redondo et al. 1997; Gill et al. 2002; Kremer and Muller 2005; Vainio et al. 2008; Hutse et al. 2010) and viral RNA (Jin et al. 2002; Hutse et al. 2010). Saliva based assays for measles elimination program in Europe have been increasingly introduced as a good alternative to blood being recommended by WHO (Ramsay et al. 1997; Centers for Disease Control and Prevention (CDC) 2008).

The causative agent of mumps, the mumps virus, belongs to the family *Paramyxoviridae*, subfamily *Paramyxovirinae*, genus *Rubulavirus* (Maple 2015). Different radioimmunoassays and enzyme immunoassays have been developed to detect specific IgG or IgM (Perry et al. 1993; Warrener and Samuel 2006; Vainio et al. 2008) in saliva and detection of mumps RNA in oral fluid samples collected during the first 14 days after onset of symptoms is also possible (Jin et al. 2002). In comparison with blood based detection methodologies, most of the mumps oral fluid assays reported had acceptable sensitivity (79–94%) and specificity (94–100%) (Maple 2015).

The rubella virus is a positive-sense, single-stranded RNA virus that belongs to the family *Togaviridae* and is the only member of the genus *Rubivirus* (Maple 2015). Different radioimmunoassays and enzyme immunoassays (GACRIA, GACELISA, EIA) have been developed to detect IgG in saliva with variable reported sensitivity and specificity (Parry et al. 1987; Perry et al. 1993; Nokes et al. 1998; Ramsay et al. 1998; Vyse et al. 1999; Ben Salah et al. 2003) but a reduced assay sensitivity with age was reported for IgG in some cases (Nokes et al. 1998; Ramsay et al. 1998). In addition, a rubella IgG time-resolved fluorescence immunoassay was described with promising results in saliva (Christopher Maple and Jones 2002). Saliva-based assays for rubella IgM detection were more established (Centers for Disease Control and Prevention (CDC) 2008; Lambert et al. 2015) and used in rubella surveillance and control (Maple 2015). In addition to antibody testing, oral fluids, if properly collected and stored can be useful to detect rubella virus genome by RT-PCR and to complement the results of antibody testing (Vyse et al. 1999; Jin et al. 2002; Vauloup-Fellous et al. 2010). Furthermore, one study (Abernathy et al. 2009), showed that rubella RT-PCR performed in saliva can detect more cases than IgM analysis of oral fluid samples or even serum collected in the initial 2 days after the outbreak. However, the combination of rubella RT-PCR and serology allow the confirmation of the highest number of rubella cases.

11.2.1.7 Other Viral Diseases

An antibody capture radioimmunoassay (GACRIA) to detect IgG to Epstein-Barr virus (EBV) viral capsid antigen (VCA) in saliva has been developed and has sufficient sensitivity to be used for epidemiological screening and enable testing for anti-EBV VCA on a wide scale (Vyse et al. 1997). The salivary IgA response against Rotavirus in newborn infants was found to be a better marker of rotavirus infection than the serum antibody response (Aiyar et al. 1990). PCR based identification of Herpes simplex virus type-1 (HSV-1) DNA in saliva is a useful method for the early detection of its reactivation that is involved in the pathogenesis of Bell's palsy (Lazarini et al. 2006).

11.2.2 Bacterial Diseases

Helicobacter pylori (*H. pylori*) is a Gram-negative, microaerophilic bacterial pathogen that usually grows in the stomach mucus. *H. pylori* infection is the strongest risk factor for developing gastric and duodenal ulcers (peptic ulcer disease) in humans and, it may also play a role in gastric cancer (Kountouras and Walt 1998; Gisbert 2015). Attempts have been made to use saliva as a diagnostic aid for peptic ulcer disease. ELISA assays for detection of salivary IgG antibodies against *H. pylori* have been developed (Loeb et al. 1997; Luzza and Pallone 1997). However, despite some interesting results the assays used had limited diagnostic value and should be improved. The use of PCR is more effective for the detection of *H. pylori* in the saliva (Jiang et al. 1998; Anand et al. 2014).

Saliva from patients with a variety of other disorders including Pneumococcal pneumonia, shigellosis, pigeon breeders disease, neurocysticercosis and Lyme disease have been evaluated for the presence of specific antibodies, with variable results (Kaufman and Lamster 2002; Kumar Nagarajappa and Bhasin 2015).

11.3 Salivary Diagnostics of Animal Infectious Diseases

A brief summary of saliva-based methods developed for the detection of selected infectious pathogens affecting domestic species is presented in Table 11.2.

Table 11.2 Saliva-based biomarkers investigated for selected pathogens affecting domestic species

Species	Pathogen	Sample	Biomarker	References
Dog	*Helicobacter* spp.	Saliva	DNA	Ekman et al. (2013)
		Oral swab		Recordati et al. (2007), Chung et al. (2014) and Jankowski et al. (2016, 2017)
	Leishmania infantum	Saliva	IgG2, IgA	Cantos-Barreda et al. (2017)
		Oral swab	DNA	Lombardo et al. (2012), de Almeida Ferreira et al. (2013) and Aschar et al. (2016)
	Rabies virus	Oral swab	RNA	Wacharapluesadee et al. (2012)
		Saliva	Rabies virus antigen RNA	Kasempimolporn et al. (2011) and Zhang et al. (2017)Saengseesom et al. (2007) and Kasempimolporn et al. (2011)
Cat	Feline calicivirus (FCV)	Oropharingeal swab	RNA	Helps et al. (2002), Abd-Eldaim et al. (2009) and Druet and Hennet (2017)
	Feline immunodeficiency virus (FIV)	Oral swab	Salivary antibodies against FIV, RNA, proviral DNA	Chang-Fung-Martel et al. (2013), Westman et al. (2016) and Miller et al. (2017)
		Saliva	RNA Proviral DNA	Matteucci et al. (1993)
	Feline leukemia virus (FeLV)	Oral swab	FeLV p27 antigen	Westman et al. (2017)
			Proviral DNA	Cavalcante et al. (2018)
Cattle	Foot-and-mouth-disease (FMD) virus	Saliva	IgA	Archetti et al. (1995)
	Schmallenberg virus	Saliva	IgG, IgA	Lazutka et al. (2015)
Pig	African swine fever virus (ASFV)	Saliva	Specific antibodies	Mur et al. (2013)
			RNA	Lung et al. (2018)
	Actinobacillus pleuropneumoniae	Saliva	IgM, IgG, IgA	González et al. (2017)
			DNA	Cheong et al. (2017)
	Classical swine fever virus (CSFV)	Saliva	IgG and IgA	Panyasing et al. (2018)
			RNA	Dietze et al. (2017), Huang et al. (2017), Petrini et al. (2017) and Lung et al. (2018)

(continued)

Table 11.2 (continued)

Species	Pathogen	Sample	Biomarker	References
	Erysipelothrix rhusiopathiae	Saliva	IgM, IgG, DNA	Giménez-Lirola et al. (2013)
	FMD virus	Saliva	RNA	Lung et al. (2018)
	Haemophilus parasuis	Saliva	DNA	Cheong et al. (2017)
	Influenza A virus (IAV)	Saliva	IAV nucleoprotein antibodies; RNA DNA/RNA	Gerber et al. (2017)
			RNA	Ramírez et al. (2012), Decorte et al. (2015), Biernacka et al. (2016) and Hernández-García et al. (2017)
	Mycoplasma hyopneumoniae	Saliva	DNA	Cheong et al. (2017), Hernández-García et al. (2017) and Pieters et al. (2017)
	Mycoplasma hyorhinis	Saliva	DNA	Cheong et al. (2017)
	Pasteurella multocida	Saliva	DNA	Cheong et al. (2017)
	Porcine circovirus type 2 (PCV2)	Saliva	RNA	Ramírez et al. (2012)
			DNA	Cheong et al. (2017) and Hernández-García et al. (2017)
	Porcine reproductive and respiratory syndrome (PRRS) virus	Saliva	PRRSV-antibodies	Langenhorst et al. (2012), Olsen et al. (2013), Kuiek et al. (2015) and Biernacka et al. (2016)
			IgG and IgA	Decorte et al. (2014) and Gerber et al. (2014)
			IgG	Kittawornrat et al. (2013) and Ouyang et al. (2013)
			IgG, IgA and IgM	Olsen et al. (2013)
			RNA PRRSV-specific neutralizing antibodies	Ramírez et al. (2012), Biernacka et al. (2016), Cheong et al. (2017) and Hernández-García et al. (2017)
		Oral swab and saliva	PRRSV-specific antibodies	Sattler et al. (2015)
	Streptococcus suis	Saliva	DNA	Cheong et al. (2017)
	Swine vesicular disease virus (SVDV)	Oral fluid	RNA	Lung et al. (2018)

11.3.1 Dogs

11.3.1.1 *Helicobacter* spp. Infection

The presence of gastric *Helicobacter* spp. is relatively frequent in dogs with gastritis and chronic vomiting (61–100%), and also in clinically healthy dogs (67–86%). *H. heilmannii* is the most prevalent species of gastric *Helicobacter* spp. in the saliva of dogs (73.3%) (Jankowski et al. 2016). This domestic animal constitutes a reservoir of *Helicobacter* spp. and a risk factor for human non-*pylori Helicobacter* spp. infection (Meining et al. 1998).

Diagnostic methods for *Helicobacter* spp. infection have been traditionally divided into invasive and non-invasive methods. Regarding invasive methods, *Helicobacter* spp. organisms from biopsied samples are usually visualized in stains, and the culture of the biopsied sample is considered the "gold standard" (Fox et al. 1995). About the non-invasive methods, saliva samples and oral swabs have been used as specimens for detection of *Helicobacter* spp. infection in dogs by PCR (Recordati et al. 2007; Ekman et al. 2013; Chung et al. 2014; Jankowski et al. 2016, 2017). Although *Helicobacter* spp. was found in the saliva of a high percentage of dogs with gastritis (76.6%), this percentage was still lower than that obtained on gastric biopsies (100%) (Jankowski et al. 2017).

11.3.1.2 Leishmaniosis

Canine leishmaniosis (CanL) is a zoonotic disease caused by the protozoan parasite *Leishmania infantum* (syn. *L. chagasi* in the New World) in the Mediterranean basin, China, and Central and South America. Domestic dogs constitute the main reservoir of infection for humans, which can develop visceral leishmaniosis (VL) (Moreno and Alvar 2002; Gramiccia and Gradoni 2005). *Leishmania*-infected dogs can remain asymptomatic or develop visceral disease due to immune-complexes deposition. Skin lesions are the most frequent clinical sign found on a physical examination (Baneth and Aroch 2008; Solano-Gallego et al. 2009).

Diagnosis of CanL had been traditionally performed by detection of specific serum antibodies against *Leishmania* spp. or detection of *Leishmania* spp. DNA in different tissues. The most frequently used quantitative serological techniques for the detection of anti-*Leishmania* antibodies are based on the enzyme-linked immunosorbent assay (ELISA) and the immunofluorescence antibody test (IFAT). Immunochromatographic tests provide rapid qualitative results but show a lack of sensitivity (Solano-Gallego et al. 2011). Detection of *Leishmania* spp. DNA by PCR allows high-sensitivity diagnosis (until 0.001 parasites per PCR reaction) (Francino et al. 2006). Lymph node, bone marrow, spleen and skin biopsies provide high-sensitivity results in detection of *Leishmania* spp. infection in both symptomatic or asymptomatic dogs (Maia and Campino 2008; Miró et al. 2008). However, these samples are obtained through invasive procedures that cause stressful situations for the animal and less acceptance by the owner.

Recently, a novel high-sensitive assay for the serological diagnosis of CanL based on the time-resolved immunofluorescence (TR-IFMA) have been developed and validated. This TR-IFMA quantifies the anti-*Leishmania* IgG2 and IgA antibody levels in saliva from dogs with CanL. This assay shows an adequate precision, analytical sensitivity, and accuracy; and greater differences between *Leishmania*-seropositive and *Leishmania*-seronegative dogs than a commercial ELISA in serum. Determination of anti-*Leishmania* IgA levels has less diagnostic value than IgG2. These findings highlight the potential of measuring anti-*Leishmania* IgG2 in canine saliva to diagnose CanL taking advantage of a high-sensitive method and a non-invasive specimen (Cantos-Barreda et al. 2017). Moreover, the potential use of measuring levels of anti-*Leishmania* IgG2 in saliva for treatment monitoring of CanL have been reported (Cantos-Barreda et al. 2018). In addition, alternatively to the use of invasive samples for detection of *Leishmania* spp. DNA by PCR, some authors have investigated the diagnostic utility of using oral swabs. However, the sensitivity of the qPCR in oral swabs is lower than in invasive samples such as bone marrow, lymph node or skin biopsies due to the low parasite burden in mucosae (de Almeida Ferreira et al. 2013; Lombardo et al. 2012; Aschar et al. 2016).

11.3.1.3 Rabies

Rabies is an infectious disease caused by a virus of the family *Rhabdoviridae* genus *Lyssavirus* mainly transmitted to humans by the bite of infected dogs in which this virus can cause fatal encephalitis (rabies) (Walker et al. 2019). Control through vaccination of dogs is effective in reducing the incidence of rabies (Rattanavipapong et al. 2019).

According to the World Health Organization (WHO 2005), the microscopic examination of the brain tissue stained using the direct fluorescent antibody (DFA) assay is the “gold standard” for post-mortem diagnosis of rabies. This technique involves open the skull of the dead animal to collect the brain. Furthermore, the fact that the brain tissue must be intact constitutes a limitation of the technique (Kamolvarin et al. 1993).

The presence of rabies virus in the saliva of infected dogs have been reported, even prior to the appearance of clinical signs (Vaughn et al. 1965). It has been shown that reverse transcriptase polymerase chain reaction (RT-PCR) from oral swabs or saliva exhibited high sensitivity (84.6–87%), but low than using brain tissue (100%) (Saengseesom et al. 2007; Wacharapluesadee et al. 2012). Detection of rabies virus antigen in the saliva has also been used for serological diagnosis of rabies. Zhang et al. (2017) developed an ELISA assay that detected positive rabies virus antigen in six of eight saliva samples from rabid dogs. In addition, Kasempimolporn et al. (2011) validated a rapid immunochromatographic assay for saliva samples with high sensitivity (93%) and specificity (94.4%) regarding the “gold standard” fluorescent antibody test (FAT) on brain smears. Moreover, this study also reported a nested polymerase chain reaction (nested-PCR) using saliva for the detection of rabies virus RNA with a sensitivity and specificity of 100% compared to the FAT results.

11.3.2 Cats

11.3.2.1 Feline Calicivirus Infection

Feline calicivirus (FCV) is a highly infectious pathogen of cats belonging to the family *Caliciviridae*. FCV mainly produce oral and upper respiratory tract disease, frequently occurring oral ulcerations and ocular and nasal discharges. However, vaccination against FCV has been reduced the incidence of disease (Radford et al. 2007).

Conjunctival and oropharyngeal swabs have been traditionally used to diagnose FCV through virus isolation. Application of RT-PCR for FCV RNA detection is reported to be as much sensitive as virus isolation (Helps et al. 2002) and has been extensively used to determine the FCV load in oropharyngeal swabs (Abd-Eldaim et al. 2009; Druet and Hennet 2017).

11.3.2.2 Feline Immunodeficiency

Feline immunodeficiency is an infectious disease affecting domestic cats all around the world caused by a lentivirus of the *Rhabdoviridae* family (Pedersen et al. 1989). Feline immunodeficiency virus (FIV) is primarily transmitted by biting and produces progressive immunosuppression as a consequence of affection of immune system cells. Consequently, cats become more prone to suffer opportunistic infections or cancer, and the outcome can be fatal (Yamamoto et al. 1988; Miller et al. 2017).

Detection of FIV antibodies using immunochromatographic tests is the most common diagnostic tool for FIV diagnosis in the veterinary clinic. However, these commercial snap tests are validated to be performed in serum, plasma or anticoagulated whole blood (Chang-Fung-Martel et al. 2013).

Saliva of FIV-infected cats contains infectious virus particles as well as salivary IgG antibodies that significantly increases over time (Miller et al. 2017), which could facilitate serological diagnosis using saliva samples. The ability of commercial immunochromatographic tests commercially validated in serum, plasma or whole blood specimens to diagnose FIV-infected cats via detection of salivary FIV antibodies has been evaluated. Despite the good specificity (98–100%), sensitivity is found to be lower in saliva samples (44–96%) than in the validated samples (Chang-Fung-Martel et al. 2013; Westman et al. 2016). Detection of FIV RNA in the saliva of cats by different types of PCR has also been reported (Matteucci et al. 1993; Westman et al. 2016; Miller et al. 2017).

11.3.2.3 Feline Leukemia

Feline leukemia virus (FeLV) belongs to the *Retroviridae* family and induces anemia and immunosuppression in infected cats (Cavalcante et al. 2018).

Antigen detection of the viral capsid protein (p27) has been traditionally used to detect cats with FeLV using serum, plasma or whole blood samples (Westman et al. 2017). Detection of this antigen has been usually performed using immunochromatographic tests (Hartmann et al. 2007; Sand et al. 2010).

Westman et al. (2016) evaluated the accuracy of three point-of-care FeLV p27 antigen tests commonly used for serological diagnosis of FeLV-infected cats. However, results derived from this study revealed that sensitivity for all the tests evaluated using saliva (54%) was lower than using whole blood (57–60%), suggesting that saliva is less suitable for p27 antigen detection in FeLV diagnosis. Proviral FeLV DNA was also detected by qPCR from oral swabs (Cavalcante et al. 2018).

11.3.3 Cattle

11.3.3.1 Foot-and-Mouth-Disease

Foot-and-mouth-disease (FMD) is a highly contagious disease caused by the FMD virus, an *Apthovirus* of the *Picornaviridae* family affecting all cloven-foot animals, including domestic ruminants and pig. Characteristic clinical signs are acute febrile reaction and the formation of vesicles in the mouth, tongue, hooves, and nipples (Alexandersen et al. 2003).

Archetti et al. (1995) reported a study in which oropharyngeal and saliva samples were tested by two types of ELISA tests in order to determine the best specimen and ELISA test for FMD-specific mucosal antibodies assessment. For diagnosis, the tissue of choice is epithelium or vesicular fluid. When epithelial tissue is not available, oesophageal-pharyngeal fluid samples can be collected for virus isolation (OIE 2004).

11.3.3.2 Schmallenberg Virus Infection

Schmallenberg virus (SBV) is a recently emerged orthobunyavirus that causes diarrhea, fever, malformations in offspring if fetal infection, and reduces milk yield in adult ruminants (Muskens et al. 2012; Conraths et al. 2013).

Detection of SBV-specific antibodies in tank milk samples using ELISA was widely used to determine the herd-levels exposure to SBV (Daly et al. 2015). However, testing milk samples not include males and young cattle in the analysis. In order to overcome this fact and taking advantage of saliva samples characteristics, Lazutka et al. (2015) developed an ELISA to detect IgG and IgA specific-SBV antibodies in saliva. Their IgG results from saliva showed close agreement with those obtained in serum and milk samples.

11.3.4 Pig

11.3.4.1 African Swine Fever

African swine fever (ASF) caused by a DNA virus member of the family *Asfarviridae*, the ASF virus (ASFV), is one of the most complex and lethal swine diseases and causes fever, abortus, erythema, pneumonia, pericarditis, or kidney hemorrhages, among other main lesions that can be observed in pigs with ASF (Sánchez-Vizcaíno et al. 2015).

Virological and serological detection is recommended to diagnose ASF. The most commonly used techniques for virological detection are virus isolation and haemadsorption (HAD) tests -which are considered the gold standard-, PCR, and direct immunofluorescence. The ELISA is the most commonly used serological test for ASF serological diagnosis. The samples of choose to diagnose ASF include serum, blood, spleen, lymph nodes, kidney, lung and bone marrow (Malmquist and Hay 1960; Sánchez-Vizcaíno et al. 2015).

ASFV-specific antibodies can also be detected in oral fluid samples of experimentally infected pigs with ASFV by ELISA and immunoperoxidase technique (IPT) (Mur et al. 2013). Results from that study lead to the conclusion that oral fluid samples could be a suitable alternative to blood as specimen to detect the presence of ASFV-specific antibodies.

11.3.4.2 Classical Swine Fever

Classical swine fever (CSF) is one of the most important diseases of swine and the causative agent is a virus (CSFV) from the family *Flaviviridae* genus *Pestivirus* (Blome et al. 2017; Petrini et al. 2017). This disease is characterized by unspecific clinical presentation like fever, anorexia, gastrointestinal symptoms, general weakness, and conjunctivitis (Petrov et al. 2014).

Detection of CSFV-specific antibodies is the best option for CSFV surveillance, however, the costs of collection blood samples are a limitation (Petrini et al. 2017). Alternatively, CSFV-specific IgG and IgA have been detected in oral fluid samples, reporting a cost-effective system for screening of populations for CSFV (Panyasing et al. 2018). Moreover, usage of oral fluid or oral swab samples for detection of CSFV RNA has been reported useful the diagnosis of this disease (Huang et al. 2017; Petrini et al. 2017; Lung et al. 2018). Dietze et al. (2017) reported that a rope-based oral fluid sampling method is equally adequate to detect CSFV RNA than the traditionally used oropharyngeal swabs sampling, which is more laborious and time-consuming. In addition, Petrini et al. (2017) reported that the probability of CSFV detection in oral fluids was identical or even higher than in blood samples.

11.3.4.3 Foot-and-Mouth-Disease and Swine Vesicular Disease

Lung et al. in a research published in 2018 reported an automated and integrated multiplex assay, including a multiplex microfluidic CARD (Chemistry and Reagent Devices) and a RT-PCR assay, able to successfully detect FMDV and swine vesicular disease virus in oral fluid from pigs. No FMD viral RNA was recovered from saliva in vaccinated pigs (Parida et al. 2007).

11.3.4.4 Influenza A

The Influenza A virus (IAV) subtypes H1N1, H1N2 and H3N2 cause explosive outbreaks of respiratory disease in swine. RT-PCR and virus isolation on nasal swab samples are currently the most used methods to detect IAV (Decorte et al. 2015). However, the biomarkers of the presence of IAV (nucleoprotein antibodies, RNA and/or DNA) have been detected in saliva and oral fluid specimens of infected pigs (Ramírez et al. 2012; Decorte et al. 2015, Biernacka et al. 2016; Gerber et al. 2017; Hernández-García et al. 2017).

11.3.4.5 Porcine Respiratory Disease Complex

Porcine respiratory disease complex (PRDC) is one of the most important diseases affecting pig production. Diseases related to PRDC lead to lung damage which could results in reduced feed efficiency, poor growth performance and higher medication and management costs (Fablet et al. 2012). Consequently, pig welfare is affected (Sørensen et al. 2006). Coinfection of several bacterial and viral pathogens are frequent in the PRDC, including *Actinobacillus pleuropneumoniae* (APP), *Pasteurella multocida* (PM), *Streptococcus suis* (SS), *Haemophilus parasuis* (HPS), *Mycoplasma hyopneumoniae* (MHP), *Mycoplasma hyorhinis* (MHR), Porcine reproductive and respiratory syndrome (PRRS) virus, porcine circovirus type 2 (PCV2), and swine influenza virus (SIV), and can be detected in saliva specimens (Cheong et al. 2017; González et al. 2017). PCV2, MHP, MHR and PM are detected most frequently among the PRDC affected swine (Hansen et al. 2010).

Pathogens affecting lung in swine have been traditionally detected in samples obtained through invasive and time-consuming procedures, such as blood, tissues or organs collected from slaughterhouses (Fablet et al. 2012). More recently studies have been proposed the used of oral fluids (OF) collected using cotton ropes as a suitable sample source for the detection, diagnosis, surveillance, and monitoring of various swine respiratory pathogens (Cheong et al. 2017).

PRRS antibodies and total antibodies have been detected in pig saliva (Langenhorst et al. 2012; Kittawornrat et al. 2013; Olsen et al. 2013; Ouyang et al. 2013; Decorte et al. 2014; Gerber et al. 2014; Kuiek et al. 2015; Sattler et al. 2015; Biernacka et al. 2016). DNA and/or specific immunoglobulins such as ApxIV-specific IgM, IgG and IgA can be detected in saliva of pigs with PRDC (Cheong et al. 2017; González et al. 2017). PRRSV, PCV2, PM, HPS, APP, MHP, MHR, and SS were detected in the OF of asymptomatic pigs from Korean farms by real-time PCR, nested-PCR or PCR in a survey of porcine respiratory disease complex (Cheong et al. 2017). In this study, the majority of ropes arrived to the laboratory in good conditions, however, some ropes (5.3%) were not enough saturated with OF. As a consequence, these potential samples were discarded. While there are many advantages of OF analysis, such as reducing the number of samples required, labor intensity, costs, and time, some weakness could be observed. In the study performed by Cheong et al. (2017) it was reported that some depressed pigs or some pig that were not confident with the rope did not show interest in chewing the rope and, consequently, the rope was not saturated with OF and excluded from the study. In these cases, the rope may not represent the total population in the pen. However, results suggest that OF-bases analysis may be a potentially useful technique for individual monitoring of the major pathogens involved in PRDC.

11.3.4.6 Swine Erysipelas

Swine erysipelas is caused by the gram-positive facultative anaerobic bacterium *Erysipelothrix rhusiopathiae*. It is an important disease of pigs associated with mortality and poor growth performance due to skin disease (Opriessnig et al. 2011).

E. rhusiopathiae infection is diagnosed postmortem and the gold standard is the isolation of the bacteria from suspect lesions (Bender et al. 2009). However, detection of anti-*Erysipelothrix* IgM and IgG by ELISA and fluorescent microbead-based immunoassay (FMIA) and detection of *E. rhusiopathiae* DNA by real-time PCR in oral fluid from pigs experimentally infected have been used for diagnosis of swine erysipelas (Giménez-Lirola et al. 2013). Results from this study suggest that the use of oral fluid as a sample for detection of erysipelas could be useful in early outbreak detection.

11.4 Conclusions

Since the salivary antibody testing can be performed in a clinic or even at a home, rural and field settings, in some countries it is recommended and used for rapid screening of several infectious diseases ('Global programme on AIDS. Recommendations for the selection and use of HIV antibody tests.' 1992). If the screening test is positive, the person is referred to a health care provider for counselling and additional diagnostic blood tests. Because the test of saliva can be performed outside a formal hospital setting, and the collection of oral fluid is less invasive, less painful, less expensive (i.e., no trained personnel required), and safer, it becomes the sample of choice in community surveys requiring high community participation allowing serological testing on a broader front. As well as in humans, in veterinary medicine, the application of oral fluid-based tests also facilitates monitoring, surveillance and diagnosis of infectious diseases in animal populations.

References

Abd-Eldaim MM et al (2009) Development and validation of a TaqMan real-time reverse transcription-PCR for rapid detection of feline calicivirus. Arch Virol 154(4):555–560. https://doi.org/10.1007/s00705-009-0337-5

Abernathy E et al (2009) Confirmation of rubella within 4 days of rash onset: comparison of rubella virus RNA detection in oral fluid with immunoglobulin M detection in serum or oral fluid. J Clin Microbiol 47(1):182–188. https://doi.org/10.1128/JCM.01231-08

Ahmed M et al (2011) Testing hepatitis A virus antibody in oral fluid among the prospective vaccinees foster the need of new oral HAV rapid test. Indian J Med Microbiol Medknow Publ Media Pvt Ltd 29(1):72. https://doi.org/10.4103/0255-0857.76532

Aiyar J et al (1990) Rotavirus-specific antibody response in saliva of infants with rotavirus diarrhea. J Infect Dis 162(6):1383–1384. https://doi.org/10.1093/infdis/162.6.1383

Alexandersen S et al (2003) The pathogenesis and diagnosis of foot-and-mouth disease. J Comp Pathol 129(1):1–36. https://doi.org/10.1016/S0021-9975(03)00041-0

Amado LA et al (2006) Detection of hepatitis A, B, and C virus-specific antibodies using oral fluid for epidemiological studies. Mem Inst Oswaldo Cruz Fundação Oswaldo Cruz 101(2):149–155. https://doi.org/10.1590/S0074-02762006000200006

Anand PS, Kamath KP, Anil S (2014) Role of dental plaque, saliva and periodontal disease in Helicobacter pylori infection. World J Gastroenterol Baishideng Publ Group Inc 20(19):5639–5653. https://doi.org/10.3748/wjg.v20.i19.5639

Archetti IL et al (1995) Detection of foot-and-mouth disease virus-infected cattle by assessment of antibody response in oropharyngeal fluids. J Clin Microbiol 33(1):79–84

Aschar M et al (2016) Value of the oral swab for the molecular diagnosis of dogs in different stages of infection with Leishmania infantum. Vet Parasitol 225:108–113. https://doi.org/10.1016/j.vetpar.2016.06.005

Balmaseda A et al (2003) Diagnosis of dengue virus infection by detection of specific immunoglobulin M (IgM) and IgA antibodies in serum and saliva. Clin Diagn Lab Immunol 10(2):317–322. https://doi.org/10.1128/cdli.10.2.317-322.2003

Balmaseda A et al (2008) Evaluation of immunological markers in serum, filter-paper blood spots, and saliva for dengue diagnosis and epidemiological studies. J Clin Virol 43(3):287–291. https://doi.org/10.1016/j.jcv.2008.07.016

Baneth G, Aroch I (2008) Canine leishmaniasis: a diagnostic and clinical challenge. Vet J 175(1):14–15. https://doi.org/10.1016/j.tvjl.2006.11.011

Bausch DG et al (2007) Assessment of the risk of Ebola virus transmission from bodily fluids and fomites. J Infect Dis Narnia 196(s2):S142–S147. https://doi.org/10.1086/520545

Bellini WJ, Rota JS, Rota PA (1994) Virology of measles virus. J Infect Dis Narnia 170(Suppl 1):S15–S23. https://doi.org/10.1093/infdis/170.Supplement_1.S15

Bender JS et al (2009) Comparison of conventional direct and enrichment culture methods for Erysipelothrix spp. from experimentally and naturally infected swine. J Vet Diagn Investig 21(6):863–868. https://doi.org/10.1177/104063870902100617

Ben Salah A et al (2003) Validation of a modified commercial assay for the detection of rubella-specific IgG in oral fluid for use in population studies. J Virol Methods 114(2):151–158. Available at: http://www.ncbi.nlm.nih.gov/pubmed/14625050. Accessed 19 Aug 2019

Bestwick RK, Fitchen JH (1997) Virucidal activity of an oral fluid human immunodeficiency virus-1 antibody preservative. Am J Med Elsevier 102(4):26–29. https://doi.org/10.1016/S0002-9343(97)00035-1

Biernacka K et al (2016) Detection of porcine reproductive and respiratory syndrome virus (PRRSV) and influenza a virus (IAV) in oral fluid of pigs. Res Vet Sci 109:74–80. https://doi.org/10.1016/j.rvsc.2016.09.014

Bingham AM et al (2016) Comparison of test results for Zika virus RNA in urine, serum, and saliva specimens from persons with travel-associated Zika virus disease – Florida, 2016. MMWR Morb Mortal Wkly Rep 65(18):475–478. https://doi.org/10.15585/mmwr.mm6518e2

Blome S et al (2017) Classical swine fever – an updated review. Viruses 9(4):1–24. https://doi.org/10.3390/v9040086

Broadhurst MJ, Brooks TJG, Pollock NR (2016) Diagnosis of Ebola virus disease: past, present, and future. Clin Microbiol Rev Am Soc Microbiol J 29(4):773–793. https://doi.org/10.1128/CMR.00003-16

Brown DWG et al (1994) Salivary diagnosis of measles: a study of notified cases in the United Kingdom, 1991-3. BMJ 308(6935):1015–1017. https://doi.org/10.1136/bmj.308.6935.1015

Cantos-Barreda A et al (2017) Quantification of anti-Leishmania antibodies in saliva of dogs. Vet Parasitol 242:54–58. https://doi.org/10.1016/j.vetpar.2017.05.017

Cantos-Barreda A et al (2018) Changes in the concentration of anti-Leishmania antibodies in saliva of dogs with clinical leishmaniosis after short-term treatment. Vet Parasitol 254:135–141. https://doi.org/10.1016/j.vetpar.2018.03.014

Cavalcante LTF et al (2018) Clinical and molecular features of feline foamy virus and feline leukemia virus co-infection in. Viruses 10:1–22. https://doi.org/10.3390/v10120702

Centers for Disease Control and Prevention (CDC) (2008) Recommendations from an ad hoc Meeting of the WHO Measles and Rubella Laboratory Network (LabNet) on use of alternative diagnostic samples for measles and rubella surveillance. MMWR Morb Mortal Wkly Rep 57(24):657–660. Available at: http://www.ncbi.nlm.nih.gov/pubmed/18566565. Accessed 19 Aug 2019

Cha YJ et al (2013) Performance evaluation of the oraquick hepatitis C virus rapid antibody test. Ann Lab Med 33(3):184. https://doi.org/10.3343/alm.2013.33.3.184

Chakravarti A, Matlani M, Jain M (2007) Immunodiagnosis of dengue virus infection using saliva. Curr Microbiol 55(6):461–464. https://doi.org/10.1007/s00284-007-9040-5

Chang-Fung-Martel J et al (2013) A door-to-door prevalence study of feline immunodeficiency virus in an Australian suburb. J Feline Med Surg 15(12):1070–1078. https://doi.org/10.1177/1098612X13491959

Cheong Y et al (2017) Survey of porcine respiratory disease complex-associated pathogens among commercial pig farms in Korea via oral fluid method. J Vet Sci 18(3):283–289. https://doi.org/10.4142/jvs.2017.18.3.283

Christopher Maple PA, Jones CS (2002) Time-resolved fluorometric immunoassay for rubella antibody – a useful method for serosurveillance studies. Vaccine 20(9–10):1378–1382. https://doi.org/10.1016/s0264-410x(01)00474-1

Chung T-H et al (2014) Determination of the prevalence of Helicobacter heilmannii-like organisms type 2 (HHLO-2) infection in humans and dogs using non-invasive genus/species-specific PCR in Korea. J Vet Med Sci 76(1):73–79. https://doi.org/10.1292/jvms.13-0223

Conraths FJ, Peters M, Beer M (2013) Schmallenberg virus, a novel orthobunyavirus infection in ruminants in Europe: potential global impact and preventive measures. N Z Vet J 61(2):63–67. https://doi.org/10.1080/00480169.2012.738403

Cordeiro ML, Turpin CS, McAdams SA (1993) A comparative study of saliva and orasure® oral fluid. Ann NY Acad Sci Wiley 694(1):330–331. https://doi.org/10.1111/j.1749-6632.1993.tb18380.x

Cuzzubbo AJ et al (1998) Detection of specific antibodies in saliva during dengue infection. J Clin Microbiol 36(12):3737–3739. Available at: http://www.ncbi.nlm.nih.gov/pubmed/9817913. Accessed 8 Aug 2019

Daly JM et al (2015) Comparison of Schmallenberg virus antibody levels detected in milk and serum from individual cows. BMC Vet Res 11(1):1–4. https://doi.org/10.1186/s12917-015-0365-1

de Almeida Ferreira S et al (2012) Canine skin and conjunctival swab samples for the detection and quantification of Leishmania infantum DNA in an endemic urban area in Brazil. PLoS Negl Trop Dis 6(4):e1596. https://doi.org/10.1371/journal.pntd.0001596

de Almeida Ferreira et al (2013) Nasal, oral and ear swabs for canine visceral leishmaniasis diagnosis: new practical approaches for detection of Leishmania infantum DNA. PLoS NTDs 7(4):e2150. https://doi.org/10.1371/journal.pntd.0002150

Decorte I et al (2014) Detection of total and PRRSV-specific antibodies in oral fluids collected with different rope types from PRRSV-vaccinated and experimentally infected pigs. BMC Vet Res 10(1):134. https://doi.org/10.1186/1746-6148-10-134

Decorte I et al (2015) Detection and isolation of swine influenza a virus in spiked oral fluid and samples from individually housed, experimentally infected pigs: potential role of porcine oral fluid in active influenza a virus surveillance in swine. PLoS One 10(10):e0139586. https://doi.org/10.1371/journal.pone.0139586

Dietze K et al (2017) Rope-based oral fluid sampling for early detection of classical swine fever in domestic pigs at group level. BMC Vet Res 13:5. https://doi.org/10.1186/s12917-016-0930-2

Druet I, Hennet P (2017) Relationship between feline calicivirus load, oral lesions, and outcome in feline chronic gingivostomatitis (caudal stomatitis): retrospective study in 104 cats. Front Vet Sci 4:209. https://doi.org/10.3389/fvets.2017.00209

Ekman E, Fredriksson M, Trowald-Wigh G (2013) Helicobacter spp. in the saliva, stomach, duodenum and faeces of colony dogs. Vet J 195(1):127–129. https://doi.org/10.1016/j.tvjl.2012.05.001

Fablet C et al (2012) Infectious agents associated with respiratory diseases in 125 farrow-to-finish pig herds: a cross-sectional study. Vet Microbiol 157(1–2):152–163. doi: https://doi.org/10.1016/j.vetmic.2011.12.015

Fanales-Belasio E et al (2010) HIV virology and pathogenetic mechanisms of infection: a brief overview. Annali dell'Istituto superiore di sanita 46(1):5–14. https://doi.org/10.4415/ANN_10_01_02

Farnaud SJC et al (2010) Saliva: physiology and diagnostic potential in health and disease. Sci World J Hindawi 10:434–456. https://doi.org/10.1100/tsw.2010.38

Featherstone D, Brown D, Sanders R (2003) Development of the global measles laboratory network. J Infect Dis 187(s1):S264–S269. https://doi.org/10.1086/368054

Formenty P et al (2006) Detection of Ebola virus in oral fluid specimens during outbreaks of Ebola virus hemorrhagic fever in the Republic of Congo. Clin Infect Dis Narnia 42(11):1521–1526. https://doi.org/10.1086/503836

Fox JG et al (1995) Helicobacter pylori-induced gastritis in the domestic cat. Infect Immun 63(7):2674–2681

Francino O et al (2006) Advantages of real-time PCR assay for diagnosis and monitoring of canine leishmaniosis. Vet Parasitol 137(3–4):214–221. https://doi.org/10.1016/j.vetpar.2006.01.011

Garrido Redondo M et al (1997) Value of salivary antibodies for determining seropositivity to measles, rubella, and mumps in children and adults. An Esp Pediatr 47(5):499–504. Available at: http://www.ncbi.nlm.nih.gov/pubmed/9586291. Accessed 20 Aug 2019

Gerber PF et al (2014) Comparison of commercial enzyme-linked immunosorbent assays and fluorescent microbead immunoassays for detection of antibodies against porcine reproductive and respiratory syndrome virus in boars. J Virol Methods 197:63–66. https://doi.org/10.1016/j.jviromet.2013.12.001

Gerber PF et al (2017) Using oral fluids samples for indirect influenza a virus surveillance in farmed UK pigs. Vet Med Sci 3(1):3–12. https://doi.org/10.1002/vms3.51

Gill J et al (2002) Susceptibility of young offenders to measles and rubella: an antibody prevalence study using oral fluid samples. Commun Dis Public Health 5(4):314–317. Available at: http://www.ncbi.nlm.nih.gov/pubmed/12564248. Accessed 20 Aug 2019

Giménez-Lirola LG et al (2013) Improving ante mortem diagnosis of Erysipelothrix rhusiopathiae infection by use of oral fluids for bacterial, nucleic acid, and antibody detection. J Microbiol Methods 92(2):113–121. https://doi.org/10.1016/j.mimet.2012.11.014

Gisbert JP (2015) Enfermedades relacionadas con la infección por Helicobacter pylori. Gastroenterol Hepatol 38:39–48. https://doi.org/10.1016/S0210-5705(15)30018-2
'Global programme on AIDS. Recommendations for the selection and use of HIV antibody tests.' (1992) Releve epidemiologique hebdomadaire 67(20):145–149. Available at: http://www.ncbi.nlm.nih.gov/pubmed/1599842. Accessed 20 Aug 2019
González W et al (2017) Detection of Actinobacillus pleuropneumoniae ApxIV toxin antibody in serum and oral fluid specimens from pigs inoculated under experimental conditions. J Vet Res (Poland) 61(2):163–171. https://doi.org/10.1515/jvetres-2017-0021
Gramiccia M, Gradoni L (2005) The current status of zoonotic leishmaniases and approaches to disease control. Int J Parasitol 35:1169–1180. https://doi.org/10.1016/j.ijpara.2005.07.001
Guzman MG, Harris E (2015) Dengue. Lancet 385(9966):453–465. https://doi.org/10.1016/S0140-6736(14)60572-9
Hansen MS et al (2010) An investigation of the pathology and pathogens associated with porcine respiratory disease complex in Denmark. J Comp Pathol 143(2–3):120–131. https://doi.org/10.1016/j.jcpa.2010.01.012
Hartmann K et al (2007) Quality of different in-clinic test systems for feline immunodeficiency virus and feline leukaemia virus infection. J Feline Med Surg 9(6):439–445. https://doi.org/10.1016/j.jfms.2007.04.003
Helps C et al (2002) Melting curve analysis of feline calicivirus isolates detected by real-time reverse transcription PCR. J Virol Methods 106(2):241–244. https://doi.org/10.1016/S0166-0934(02)00167-2
Hernández-García J et al (2017) The use of oral fluids to monitor key pathogens in porcine respiratory disease complex. Porcine Health Manag 3(1). https://doi.org/10.1186/s40813-017-0055-4
Huang YL et al (2017) Competitive replication kinetics and pathogenicity in pigs co-infected with historical and newly invading classical swine fever viruses. Virus Res 228:39–45. https://doi.org/10.1016/j.virusres.2016.11.022
Hutse V et al (2010) Oral fluid for the serological and molecular diagnosis of measles. Int J Infect Dis Elsevier 14(11):e991–e997. https://doi.org/10.1016/J.IJID.2010.06.009
Jankowski M et al (2016) Detection of Helicobacter spp. in the saliva of dogs with gastritis. Pol J Vet Sci 19(1):133–140. https://doi.org/10.1515/pjvs-2016-0017
Jankowski M et al (2017) An evaluation of the usefulness of invasive and non-invasive methods used to diagnose Helicobacter spp. infections in dogs. Pol J Vet Sci 20(3):491–499. https://doi.org/10.1515/pjvs-2017-0059
Jiang C et al (1998) Identification of H. pylori in saliva by a nested PCR assay derived from a newly cloned DNA probe. Dig Dis Sci 43(6):1211–1218. https://doi.org/10.1023/a:1018847522200
Jin L, Vyse A, Brown DWG (2002) The role of RT-PCR assay of oral fluid for diagnosis and surveillance of measles, mumps and rubella. Bull World Health Organ World Health Organ 80(1):76–77. Available at: http://www.ncbi.nlm.nih.gov/pubmed/11884979. Accessed 14 Aug 2019
Kamolvarin N et al (1993) Diagnosis of rabies by polymerase chain reaction with nested primers. J Infect Dis 167(1):207–210. https://doi.org/10.1093/infdis/167.1.207
Kasempimolporn S et al (2011) Evaluation of a rapid immunochromatographic test strip for detection of in dog saliva samples. J Vet Diagn Investig 23(6):1197–1201. https://doi.org/10.1177/1040638711425576
Kaufman E, Lamster IB (2002) The diagnostic applications of saliva – a review. Crit Rev Oral Biol Med 13(2):197–212. Available at: http://www.ncbi.nlm.nih.gov/pubmed/12097361. Accessed 20 Aug 2019
Kittawornrat A et al (2013) Kinetics of the porcine reproductive and respiratory syndrome virus (PRRSV) humoral immune response in swine serum and oral fluids collected from individual boars. BMC Vet Res 9(1):61. https://doi.org/10.1186/1746-6148-9-61
Kountouras J, Walt RP (1998) Diagnostic tests for Helicobacter pylori reply. Gut 42(6):899–901. https://doi.org/10.1136/gut.42.6.899c

Kremer JR, Muller CP (2005) Evaluation of commercial assay detecting specific immunoglobulin G in oral fluid for determining measles immunity in vaccinees. Clin Vaccine Immunol 12(5):668–670. https://doi.org/10.1128/CDLI.12.5.668-670.2005

Kuiek AM et al (2015) Comparison of serum and oral fluid antibody responses after vaccination with a modified live (MLV) porcine reproductive and respiratory syndrome virus (PPRSV) vaccine in PRRS endemic farms.Trop. Anim Health Prod 47(7):1337–1342. https://doi.org/10.1007/s11250-015-0868-6

Kumar Nagarajappa A, Bhasin N (2015) Review article saliva as a detective biofluid. Available at: https://pdfs.semanticscholar.org/4e3e/63ab1002729ced2a564bc19b091f3f253c6d.pdf?_ga=2.166594110.1868805039.1547122656-931044907.1521642616. Accessed 10 Jan 2019

Lambert N et al (2015) Rubella. Lancet 385(9984):2297–2307. https://doi.org/10.1016/S0140-6736(14)60539-0

Langenhorst RJ et al (2012) Development of a fluorescent microsphere immunoassay for detection of antibodies against porcine reproductive and respiratory syndrome virus using oral fluid samples as an alternative to serum-based assays. Clin Vaccine Immunol 19(2):180–189. https://doi.org/10.1128/CVI.05372-11

Lawrence HP (2002) Salivary markers of systemic disease: noninvasive diagnosis of disease and monitoring of general health. J Can Dent Assoc 68(3):170–174. Available at: http://www.ncbi.nlm.nih.gov/pubmed/11911813. Accessed 7 Aug 2019

Lazarini PR et al (2006) Herpes simplex virus in the saliva of peripheral Bell's palsy patients. Braz J Otorhinolaryngol 72(1):7–11. Available at: http://www.ncbi.nlm.nih.gov/pubmed/16917546. Accessed 24 Sept 2019

Lazutka J et al (2015) Saliva as an alternative specimen for detection of Schmallenberg virus-specific antibodies in bovines. BMC Vet Res 11(1). https://doi.org/10.1186/s12917-015-0552-0

Loeb MB et al (1997) Evaluation of salivary antibodies to detect infection with *Helicobacter pylori*. Can J Gastroenterol 11(5):437–440. https://doi.org/10.1155/1997/294081

Lombardo G et al (2012) Veterinary parasitology detection of Leishmania infantum DNA by real-time PCR in canine oral and conjunctival swabs and comparison with other diagnostic techniques. Vet Parasitol 184(1):10–17. https://doi.org/10.1016/j.vetpar.2011.08.010

Lung O et al (2018) Fully automated and integrated multiplex detection of high consequence livestock viral genomes on a microfluidic platform. Transbound Emerg Dis 66(1):144–155. https://doi.org/10.1111/tbed.12994

Luzza F, Pallone F (1997) Helisal saliva assay. Gut 41(5):725. Available at: http://www.ncbi.nlm.nih.gov/pubmed/9414994. Accessed 3 Sept 2019

Mackiewicz V et al (2004) Detection of hepatitis A virus RNA in saliva. J Clin Microbiol Am Soc Microbiol (ASM) 42(9):4329–4331. https://doi.org/10.1128/JCM.42.9.4329-4331.2004

Madar R, Straka S and Baska T (2002) Detection of antibodies in saliva – an effective auxiliary method in surveillance of infectious diseases. Bratisl Lek Listy 103(1):38–41. Available at: http://bmj.fmed.uniba.sk/2002/10301-08.pdf. Accessed 13 Feb 2019

Maia C, Campino L (2008) Methods for diagnosis of canine leishmaniasis and immune response to infection. Vet Parasitol 158(4):274–287. https://doi.org/10.1016/j.vetpar.2008.07.028

Malamud D (1997) Oral diagnostic testing for detecting human immunodeficiency virus-1 antibodies: a technology whose time has come. Am J Med 102(4):9–14. https://doi.org/10.1016/S0002-9343(97)00032-6

Malmquist WA, Hay D (1960) Hemadsorption and cytopathic effect produced by african swine fever virus in swine bone marrow and buffy coat cultures. Am J Vet Res 21:104–108

Maple PAC (2015) Application of oral fluid assays in support of mumps, rubella and varicella control programs. Vaccines Multidiscip Digit Publ Inst (MDPI) 3(4):988–1003. https://doi.org/10.3390/vaccines3040988

Martínez PM et al (1999) Human immunodeficiency virus antibody testing by enzyme-linked fluorescent and western blot assays using serum, gingival-crevicular transudate, and urine samples. J Clin Microbiol 37(4):1100–1106. Available at: http://www.ncbi.nlm.nih.gov/pubmed/10074532. Accessed 7 Aug 2019

Matsuda S et al (1993) Characteristics of IgA antibodies against HIV-1 in sera and saliva from HIV-seropositive individuals in different clinical stages. Scand J Immunol Wiley 38(5):428–434. https://doi.org/10.1111/j.1365-3083.1993.tb02584.x

Matteucci D et al (1993) Detection of feline immunodeficiency virus in saliva and plasma by cultivation and polymerase chain reaction. J Clin Microbiol 31(3):494–501

Meining A, Kroher G, Stolte M (1998) Animal reservoirs in the transmission of helicobacter heilmannii: results of a questionnaire-based study. Scand J Gastroenterol 33(8):795–798. https://doi.org/10.1080/00365529850171422

Miller C et al (2017) Pathogenesis of oral FIV infection. PLoS One 12(9):1–24. https://doi.org/10.1371/journal.pone.0185138

Miró G et al (2008) Canine leishmaniosis – new concepts and insights on an expanding zoonosis: part two. Trends Parasitol 24(8):371–377. https://doi.org/10.1016/j.pt.2008.05.003

Moreno J, Alvar J (2002) Canine leishmaniasis: epidemiological risk and the experimental model. Trends Parasitol 18(9):399–405. https://doi.org/10.1016/S1471-4922(02)02347-4

Mur L et al (2013) Potential use of oral fluid samples for serological diagnosis of African swine fever. Vet Microbiol 165(1–2):135–139. https://doi.org/10.1016/j.vetmic.2012.12.034

Muskens J et al (2012) Diarrhea and loss of production on Dutch dairy farmas caused by the Schmallenberg virus. Tijdschr Diergeneeskd 137(2):112–115

Musso D, Gubler DJ (2016) Zika virus. Clin Microbiol Rev 29(3):487–524. https://doi.org/10.1128/CMR.00072-15

Musso D et al (2015) Detection of Zika virus in saliva. J Clin Virol Elsevier 68:53–55. https://doi.org/10.1016/J.JCV.2015.04.021

Mustafa MS et al (2015) Discovery of fifth serotype of dengue virus (DENV-5): a new public health dilemma in dengue control. Med J Armed Forces India 71(1):67–70. https://doi.org/10.1016/j.mjafi.2014.09.011

Niedrig M et al (2018) Find the right sample: a study on the versatility of saliva and urine samples for the diagnosis of emerging viruses. BMC Infect Dis BioMed Cent 18(1):707. https://doi.org/10.1186/s12879-018-3611-x

Nokes DJ et al (1998) A comparison of oral fluid and serum for the detection of rubella-specific antibodies in a community study in Addis Ababa, Ethiopia. Tropical Med Int Health 3(4):258–267. Available at: http://www.ncbi.nlm.nih.gov/pubmed/9623926. Accessed 19 Aug 2019

Oba IT et al (2000) Detection of hepatitis a antibodies by ELISA using saliva as clinical samples. Rev Inst Med Trop Sao Paulo 42(4):197–200. https://doi.org/10.1590/S0036-46652000000400004

OIE Manual of diagnostic tests and vaccines for terrestrial animals (mammals, birds, and bees) (2004) World Organisation for Animals Health, p 111. Available at: https://www.oie.int/doc/ged/D6460.PDF

Olsen C et al (2013) Effect of collection material and sample processing on pig oral fluid testing results. Vet J 198(1):158–163. https://doi.org/10.1016/j.tvjl.2013.06.014

Opriessnig T et al (2011) Erysipelothrix rhusiopathiae : genetic characterization of Midwest US isolates and live commercial vaccines using pulsed-field gel electrophoresis. J Vet Diagn Investig 16(2):101–107. https://doi.org/10.1177/104063870401600202

Ouyang K et al (2013) Development and validation of an assay to detect porcine reproductive and respiratory syndrome virus-specific neutralizing antibody titers in pig oral fluid samples. Clin Vaccine Immunol 20(8):1305–1313. https://doi.org/10.1128/CVI.00276-13

Panyasing Y et al (2018) Detection of classical swine fever virus (CSFV) E2 and E^{rns} antibody (IgG, IgA) in oral fluid specimens from inoculated (ALD strain) or vaccinated (LOM strain) pigs. Vet Microbiol 224:70–77. https://doi.org/10.1016/j.vetmic.2018.08.024

Parida S et al (2007) Reduction of foot-and-mouth disease (FMD) virus load in nasal excretions, saliva and exhaled air of vaccinated pigs following direct contact challenge. Vaccine 25(45):7806–7817. https://doi.org/10.1016/j.vaccine.2007.08.058

Parry JV, Perry KR, Mortimer PP (1987) Sensitive assays for viral antibodies in saliva: an alternative to tests on serum. Lancet (Lond Engl) 2(8550):72–75. https://doi.org/10.1016/s0140-6736(87)92737-1

Pedersen NC et al (1989) Feline immunodeficiency virus infection. Vet Immunol Immunopathol 21(1):111–129. https://doi.org/10.1016/0165-2427(89)90134-7
Perry KR et al (1993) Detection of measles, mumps, and rubella antibodies in saliva using antibody capture radioimmunoassay. J Med Virol 40(3):235–240. Available at: http://www.ncbi.nlm.nih.gov/pubmed/8355022. Accessed 14 Aug 2019
Petrini S et al (2017) Detection of classical swine fever virus infection by individual oral fluid of pigs following experimental inoculation. J Vet Diagn Investig 29(2):254–257. https://doi.org/10.1177/1040638716686393
Petrov A et al (2014) Comparative analyses of host responses upon infection with moderately virulent classical swine fever virus in domestic pigs and wild boar. Virol J 11(1):1–6. https://doi.org/10.1186/1743-422X-11-134
Pieters M, Daniels J, Rovira A (2017) Comparison of sample types and diagnostic methods for in vivo detection of mycoplasma hyopneumoniae during early stages of infection. Vet Microbiol 203:103–109. https://doi.org/10.1016/j.vetmic.2017.02.014
Poloni TR et al (2010) Detection of dengue virus in saliva and urine by real time RT-PCR. Virol J BioMed Cent 7:22. https://doi.org/10.1186/1743-422X-7-22
Radford AD et al (2007) Review article Feline calicivirus. Vet Res 38:319–335. https://doi.org/10.1051/vetres
Ramírez A et al (2012) Efficient surveillance of pig populations using oral fluids. Prev Vet Med 104(3–4):292–300. https://doi.org/10.1016/j.prevetmed.2011.11.008
Ramsay M, Brugha R, Brown D (1997) Surveillance of measles in England and Wales: implications of a national saliva testing programme Bull World Health Organ World Health Organ 75(6):515–521. Available at: http://www.ncbi.nlm.nih.gov/pubmed/9509623. Accessed 19 Aug 2019
Ramsay ME et al (1998) Salivary diagnosis of rubella: a study of notified cases in the United Kingdom, 1991–4. Epidemiol Infect 120(3):315–319. https://doi.org/10.1017/S0950268898008838
Rattanavipapong W, Thavorncharoensap M, Youngkong S (2019) The impact of transmission dynamics of rabies control: systematic review. Vaccine 37 Suppl 1, A154–A165. https://doi.org/10.1016/j.vaccine.2018.11.035
Ravi Banavar S, Vidya GS (2014) Diagnostic efficacy of saliva for dengue – a reality in near future? A piloting initiative. J Clin Diag Res JCDR Res Publ Pvt Ltd 8(3):229–232. https://doi.org/10.7860/JCDR/2014/7521.4169
Recordati C et al (2007) Detection of Helicobacter spp. DNA in the oral cavity of dogs. Vet Microbiol 119(2–4):346–351. https://doi.org/10.1016/j.vetmic.2006.08.029
Saengseesom W et al (2007) Real-time PCR analysis of dog cerebrospinal fluid and saliva samples for ante-mortem diagnosis of rabies. Southeast Asian J Trop Med Public Health 38(1):53–57
Sánchez-Vizcaíno JM et al (2015) An update on the epidemiology and pathology of African swine fever. J Comp Pathol 152(1):9–21. https://doi.org/10.1016/j.jcpa.2014.09.003
Sand C et al (2010) Evaluation of a new in-clinic test system to detect feline immunodeficiency virus and feline leukemia virus infection. Vet Clin Pathol 39(2):210–214. https://doi.org/10.1111/j.1939-165X.2009.00196.x
Sattler T, Wodak E, Schmoll F (2015) Evaluation of the specificity of a commercial ELISA for detection of antibodies against porcine respiratory and reproductive syndrome virus in individual oral fluid of pigs collected in two different ways. BMC Vet Res 11(1):70. https://doi.org/10.1186/s12917-015-0388-7
Scully C (1997) HIV topic update: salivary testing for antibodies. Oral Dis 3(4):212–215. Available at: http://www.ncbi.nlm.nih.gov/pubmed/9643214. Accessed 5 Aug 2019
Scully C, Samaranayake LP (1992) Clinical virology in oral medicine and dentistry. Cambridge University Press, Cambridge
Shanbhag VL (2015) The need of an hour for saliva diagnostic research in detection of Ebola viral disease. Commun Acquir Infect Medknow Publ Media Pvt Ltd 2(1):27. https://doi.org/10.4103/2225-6482.153867

Solano-Gallego L et al (2009) Directions for the diagnosis, clinical staging, treatment and prevention of canine leishmaniosis. Vet Parasitol 165(1–2):1–18. https://doi.org/10.1016/j.vetpar.2009.05.022

Solano-Gallego L et al (2011) LeishVet guidelines for the practical management of canine leishmaniosis. Parasit Vectors 4(1):86. https://doi.org/10.1186/1756-3305-4-86

Sørensen JT et al (2006) Animal production systems in the industrialised world. Rev Sci Tech 25(2):493–503

Strong JE et al (2006) Filoviruses and arenaviruses. In: Manual of molecular and clinical laboratory immunology, 7th edn. American Society of Microbiology, Washington, pp 774–790. https://doi.org/10.1128/9781555815905.ch87

Thieme T et al (1992) Clinical evaluation of oral fluid samples for diagnosis of viral hepatitis. J Clin Microbiol Am Soc Microbiol J 30(5):1076–1079. Available at: http://www.ncbi.nlm.nih.gov/pubmed/1316364. Accessed 8 Aug 2019

Torres JR, Liprandi F, Goncalvez AP (2000) Acute parotitis due to dengue virus. Clin Infect Dis 31(5):e28–e29. https://doi.org/10.1086/317454

Vainio K et al (2008) Detection of measles- and mumps-specific IgG antibodies in paired serum and oral fluid samples from Norwegian conscripts. Eur J Clin Microbiol Infect Dis 27(6):461–465. https://doi.org/10.1007/s10096-008-0460-3

Vaughn JB, Gerhardt P, Newell KW (1965) Excretion of street rabies virus in the saliva of dogs. JAMA 193(5):363. https://doi.org/10.1001/jama.1965.03090050039010

Vauloup-Fellous C et al (2010) Phylogenetic analysis of rubella viruses involved in congenital rubella infections in France between 1995 and 2009. J Clin Microbiol 48(7):2530–2535. https://doi.org/10.1128/JCM.00181-10

Vázquez S et al (2007) Kinetics of antibodies in sera, saliva, and urine samples from adult patients with primary or secondary dengue 3 virus infections. Int J Infect Dis 11(3):256–262. https://doi.org/10.1016/j.ijid.2006.05.005

Vyse AJ et al (1997) Detection of IgG antibody to Epstein-Barr virus viral capsid antigen in saliva by antibody capture radioimmunoassay. J Virol Methods 63(1–2):93–101. https://doi.org/10.1016/s0166-0934(96)02118-0

Vyse AJ et al (1999) Detection of rubella virus-specific immunoglobulin G in saliva by an amplification-based enzyme-linked immunosorbent assay using monoclonal antibody to fluorescein isothiocyanate. J Clin Microbiol 37(2):391–395. Available at: http://www.ncbi.nlm.nih.gov/pubmed/9889225. Accessed 19 Aug 2019

Wacharapluesadee S et al (2012) Detection of rabies viral RNA by TaqMan real-time RT-PCR using non-neural specimens from dogs infected with rabies virus. J Virol Methods 184(1–2):109–112. https://doi.org/10.1016/j.jviromet.2012.05.013

Walker PJ et al (2019) ICTV virus taxonomy profile: Rhabdoviridae. J Gen Virol 99(4):447–448. https://doi.org/10.1099/jgv.0.001020

Warrener L, Samuel D (2006) Evaluation of a commercial assay for the detection of mumps specific IgM antibodies in oral fluid and serum specimens. J Clin Virol Elsevier 35(2):130–134. https://doi.org/10.1016/J.JCV.2005.10.008

Wasik D, Mulchandani A, Yates MV (2018) Salivary detection of dengue virus NS1 protein with a label-free immunosensor for early dengue diagnosis. Sensors (Basel, Switz) Multidiscip Digit Publ Inst (MDPI) 18(8):E2641. https://doi.org/10.3390/S18082641

Westman ME et al (2016) Diagnosing feline immunodeficiency virus (FIV) infection in FIV-vaccinated and FIV-unvaccinated cats using saliva. Comp Immunol Microbiol Infect Dis 46:66–72. https://doi.org/10.1016/j.cimid.2016.03.006

Westman ME et al (2017) Comparison of three feline leukaemia virus (FeLV) point-of-care antigen test kits using blood and saliva. Comp Immunol Microbiol Infect Dis 50:88–96. https://doi.org/10.1016/j.cimid.2016.11.014

WHO Expert consultation on rabies (2005) Second report. WHO. World Health Organization technical report series No. 982, 27. Available at: https://apps.who.int/iris/bitstream/handle/10665/272364/9789241210218-eng.pdf

'WHO | Laboratory testing for Zika virus infection' (2016) WHO. World Health Organization. Available at: https://www.who.int/csr/resources/publications/zika/laboratory-testing/en/. Accessed 13 Aug 2019

Yamamoto JK et al (1988) Pathogenesis of experimentally induced feline immunodeficiency virus infection in cats. Am J Vet Res:1246–1258

Yap G, Sil BK, Ng L-C (2011) Use of saliva for early dengue diagnosis. PLoS Negl Trop Dis Public Libr Sci 5(5):e1046. https://doi.org/10.1371/JOURNAL.PNTD.0001046

Yoshizawa JM et al (2013) Salivary biomarkers: toward future clinical and diagnostic utilities. Clin Microbiol Rev Am Soc Microbiol 26(4):781–791. https://doi.org/10.1128/CMR.00021-13

Zhang J et al (2016) Early detection of Zika virus infection among travellers from areas of ongoing transmission in China: table 1. J Travel Med Narnia 23(5):taw047. https://doi.org/10.1093/jtm/taw047

Zhang JM et al (2017) Incidence of human rabies and characterization of rabies virus nucleoprotein gene in dogs in Fujian Province, Southeast China, 2002–2012. BMC Infect Dis 17(1). https://doi.org/10.1186/s12879-017-2698-9

Chapter 12
Salivary Markers in Systemic and Oral Cancer

Mario Pérez-Sayáns, Cintia M. Chamorro-Petronacci, Pilar Gándara-Vila, and Alejandro I. Lorenzo-Pouso

12.1 Introduction

Saliva is one of the most valuable body fluids after blood. Although its function may seem to originally be confined to the local environment, the alterations in the salivary composition, quality and quantity, may reflect numerous pathologies and systemic diseases and/or cause them in a bidirectional way (Baum 1993).

Cancer takes the lives of approximately 620,000 people in the world each year. Most of these patients pass away as a result of a late diagnosis. In 2016, it is estimated that there were 15.5 million cancer survivors in the United States and this figure is expected to rise to 20.3 million by 2026. This is possible due to the development of new therapeutic approaches, but above all, due to an early diagnosis (Siegel et al. 2017).

Cancer diagnostic techniques are based on clinical signs, imaging tests, laboratory tests and anatomopathological methods. However, these methods may not be sufficient in some cases to diagnose some of the tumours that affect individuals (Mcpherson and Pincus 2017). In this line, new and less invasive diagnostic

M. Pérez-Sayáns (✉) · P. Gándara-Vila
Oral Medicine Unit, MedOralRes Group. Faculty of Medicine and Dentistry, University of Santiago de Compostela, Santiago de Compostela, Spain

Foundation of Health Research Institute of Santiago de Compostela (FIDIS), University Hospital of Santiago de Compostela, Santiago de Compostela, Spain

Faculty of dentistry, University of Santiago de Compostela, Santiago de Compostela, Spain
e-mail: Mario.perez@usc.es

C. M. Chamorro-Petronacci · A. I. Lorenzo-Pouso
Oral Medicine Unit, MedOralRes Group. Faculty of Medicine and Dentistry, University of Santiago de Compostela, Santiago de Compostela, Spain

Foundation of Health Research Institute of Santiago de Compostela (FIDIS), University Hospital of Santiago de Compostela, Santiago de Compostela, Spain

A. Tvarijonaviciute et al. (eds.), *Saliva in Health and Disease*,
https://doi.org/10.1007/978-3-030-37681-9_12

methods have been developed in recent times to minimise the cost involved and to speed up the tests and results. One such example is the liquid biopsy (Smith et al. 2017). The liquid biopsy presents a snapshot of the disease from the primary tumour location and/or from a distance, and it can be used in repeated sampling of tumour markers in order to adapt the therapy in accordance with the patient's response to the treatment; this is also known as personalised treatment or precise treatment (Crowley et al. 2013).

The American Society of Clinical Oncology has published a list of numerous general biomarkers in blood and other fluids with potential clinical applicability over the past few decades: Alpha-photoprotein, Anaplastic Lymphoma Kinase, BCR-ABL, Beta-2-microglobulin, Beta-HCG, Bladder Tumour Antigen, BRAF, CA 15-3/CA 27-29, CA 19-9, CA 125, Calcitonin, Carcinoembryonic Antigen, Chromogranin A, EGFR Receptor: HER1, HE-4, HER2m, Hormone Receptors, Chorionic Gonadotropin, Immunoglobulins, Free Light Chains, KRAS, Lactate dehydrogenase, Specific Neurons Enolase, NMP22, PSA, Prostatic Acid Phosphatase, S-100, soluble mesothelin-related Peptides and the Thyroglobulin (Ludwig and Weinstein 2005; Hayes et al. 1996; Schnipper et al. 2015). Nevertheless, in reality not all of them meet the ideal requirements of a biomarker.

With regards to oral cancer, the results are even more controversial given that it has not been possible to standardise its routine usage. The tissue biopsy, along with the haematoxylin-eosin technique, are still the routine and standard techniques for these patients, which is why a less invasive method must be designed and used as this will benefit of both patients and specialists for the diagnosis and monitoring of such patients (Brinkman and Wong 2006; Epstein et al. 2002; Joseph 2002; Robinson and Mickelson 2006; Bagan and Scully 2008; Brinkman and Wong 2006; Joseph 2002; Robinson and Mickelson 2006). Without a doubt, saliva is one of the fluids with the highest application potential for this purpose, and the following section will be developed based on this fact.

12.2 Salivary Biomarkers

The salivary examination, a non-invasive alternative to the serum test, is an effective way of diagnosing and establishing the prognosis for several diseases such as cancer, and likewise it can also be used to monitor the patient's therapy. Patients who suffer from oral cancer are closely monitored in order to detect recurrences. The diagnostic tools used in those patients are extremely important, in particular for use in high-risk populations (patients who present premalignant areas), patients "healed" after 5 years of monitoring and patients with a clinical history of cancer. Home tests are becoming a reality that allows many patients to avoid having to travel, by determining genomic and proteomic targets (Nagler 2009).

12.2.1 *Salivary Markers in Potentially Malignant Oral Lesions*

12.2.1.1 Leucoplakia

Oral leucoplakia is one of the most frequent, potentially malignant lesions and according to the literature it has a processing capacity of 0.13–34% (Gandara-Vila et al. 2018). It is defined as "a white plaque of questionable risk, having excluded (other) known diseases or disorders that do not increase the risk of cancer" (Warnakulasuriya et al. 2007). Two main clinical types are recognized, homogeneous (white, flat and thin) and non-homogeneous (speckled, nodular and verrucous) (Warnakulasuriya et al. 2007) (Figs. 12.1 and 12.2). The non-homogeneous lesions carry a much higher risk of malignant transformation (Gandara-Vila et al. 2018). Proliferative verrucous leucoplakia (Fig. 12.3) is a seldom variant characterized by persistent progression to malignancy (Flores et al. 2016). It is not associated with specific symptomatology, the patients may complain of roughness or tightness in the area.

Currently, diagnosis is carried out by means of a biopsy in order to exclude other entities and assess the degree of dysplasia and this is the only tool which can predict changes towards malignant transformation (Warnakulasuriya et al. 2007, 2011).

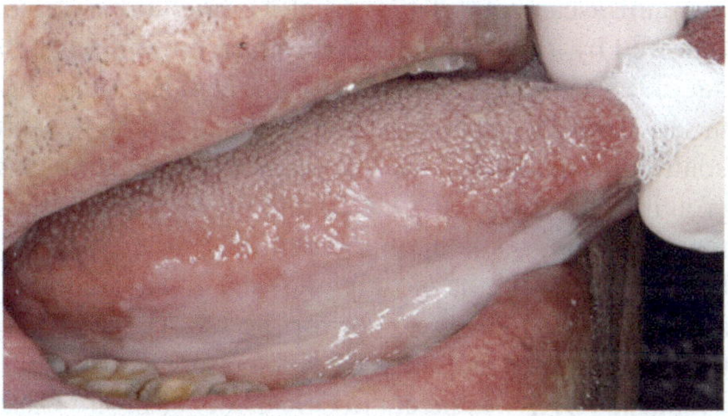

Fig. 12.1 Oral homogeneous leucoplakia in lateral tongue

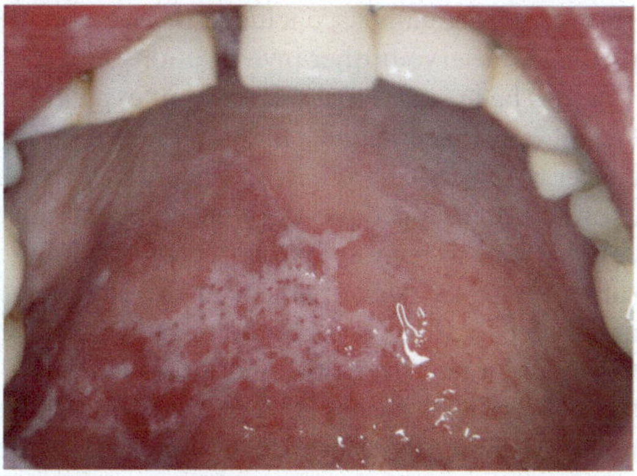

Fig. 12.2 Non-homogeneous type in hard palate

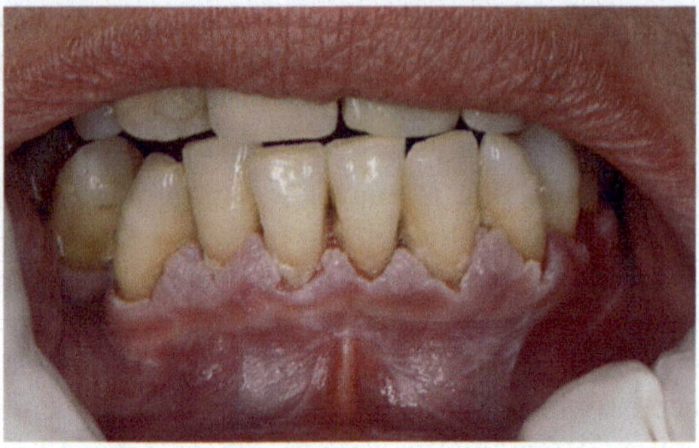

Fig. 12.3 Verrucous proliferative leucoplakia in gingiva

Over the last few years, several authors have highlighted the possibility of using saliva as a diagnostic fluid and they have started to develop research focused on the salivary proteomics and genomics (Cuevas-Cordoba and Santiago-Garcia 2014; Wang et al. 2015a, b).

When reviewing the literature, we find two research fields; one that aims to identify a biomarker that helps us differentiate the presence of leucoplakia in healthy patients and another one which focuses on searching for a molecule which can predict the transformation of leucoplakia into oral cancer. A summary of all studies, salivary biomarkers and results can be observed in the Table 12.1.

Camisasca et al. (2017) recently carried out a phase I biomarker discovery study followed by a pre-validation phase. An immunohistochemistry staining was carried out of the samples obtained through the leucoplakia biopsy with CK10. They found in saliva that 22 spots were highly abundant, 16 appeared in just one group: 7 in the control group and 9 in the leucoplakia group. In this last group, the most interesting proteins found were the CK10 and the lysosome precursor, and in the quantitative intensity analysisthe apolipoprotein A-1 and cystanine SN precursor (decreased in leucoplakia) were the most interesting, given that the rest were proteins that are usually present in saliva, except for the CK-10, which was described in relation to a type of leucoplakia named by Kobayashi (Kobayashi et al. 2012) as orthokeratotic dysplasia (OKD), adjacent to OSCC foci or in situ Carcinomas. But when compared results in saliva with immunohistochemistry CK-10 these associations were not observed. The authors concluded that keratin 10, apolipoprotein A1 and lysozyme precursor deserve further investigation.

It has been stated that the aberrant expression of microRNAs could act as "drivers" of oral cancer, contributing to the progression of premalignant lesions towards cancer. These can directly reach the saliva from the tumour microenvironment or the exosomes that impede its enzymatic degradation by the RNases, by means of extracellular transport. A study carried out by Yang (2013) evaluated the relevance of salivary mRNA as biomarkers in the malignant transformation of leucoplakia by using qRT-PCR to compare the mRNA levels between lesions with mild dysplasia that transformed over time and those which did not transform. Yang found that in the leucoplakia tissues that had developed a carcinoma, miR-10b, miR-660, miR-708 and miR-30e were overexpressed, while miR-145, miR-99b, miR181c and

Table 12.1 Summary of salivary biomarkers studies

Author	Biomarkers studies	Results
Camisasca et al. (2017)	Alpha-amylase, Keratin-10, IgM Kappa IIIb, Lysozyme Precursor, Polymeric immunoglobulin receptor, serum albumin, Ig gamma-1 chain c region, Ig Kappa chain (homosapiens), albumin isoform CRA, Cahin a human serum albumin in A complex with Myristic Acid and Triiodobenzoic acid, immunoglobulin kappa light chain VLJ region. Ig alpha-2 chain C region, apolipoprotein A-1, cystatin-SN precursor(cystatin-1), serum albumin, Mature metal cheltase catalytic antibody with hapten, chain A.	Keratin 10 and lysozyme precursor as well as apolipoprotein A-1 and cystatin SN precursor, were the most interesting one between control and leucoplakia group.
Yang et al. (2013)	miR-10b, miR-660, miR-708 and miR-30e, miR-145, miR-99b, miR181c and miR-197	miR-10b, miR-145, miR-99b, miR-708 and miR-181c were significantly different in saliva of progressive leucoplakia compared with non-progressive leucoplakia patients
Michailidou et al. (2016)	IL-1B, IL-8, OAZ and SAT microRNAs	No differences found between leucoplakia patients and controls
Liu et al. (2012a, b)	miR-31	Augmented in patients with oral cancer but not in oral verrucous leucoplakia
Sharma et al. (2011), Panneer and Sadaksharam (2015) and Brailo et al. (2006)	IL-6	Augmented in leucoplakia and oral cancer but not in healthy control group
Brailo et al. (2012)	IL-6, IL-1β and TNF-α	IL-6 no presented differences in leucoplakia IL-1β was lower in leucoplakia than in controls and higher in oral cancer TNF-α no presented differences
Punyani and Sathawane (2013)	IL-8	No differences between controls and precancerous lesions

(continued)

Table 12.1 (continued)

Author	Biomarkers studies	Results
Ajona et al. (2015)	Cd4	No differences between leucoplakia and oral cancer
Shetty et al. (2015)	Cu, Zn, Fe	Cu higher in leucoplakia than controls
Ayinampudi and Narsimhan (2012)	Cu/Zn	Lower in leucoplakia than control
Shetty et al. (2015)	Cu/Zn	Higher in leucoplakia than control
Kaur et al. (2016)	8-OHdG, vitamin C and vitamin E with MDA	Vitamin C and E levels were lower in patients with leucoplakia
Patel and Metgud (2015)	Enzyme lactate dehydrogenase	Higher in leucoplakia and oral cancer versus controls
Jaeger et al. (2015)	Epidermal growth factor	No differences between leucoplakia and controls
Hoffman et al. (2011)	Endothelin-1	No differences between leucoplakia and oral cancer
Flores et al. (2016)	283 proteins identified 31 showed statistical difference	AGT and DPP1 could be used as potential biomarkers

miR-197 were under expressed. Differences in the saliva between the expression of miR-10b, miR-145, miR-99b, miR-708 and miR-181c were also found in the two groups. While miR-197 and miR-30e were close to the statistical significance (although they did not reach it), miR-660 was also under expressed in the leucoplakias that did not transform, although its relation was not significant, verifying in this way that the affected miRNAs were present both in the saliva and the tissue, although in smaller quantities.

Conversely, Michailidou et al. (2016) studied the presence of IL-1B, IL-8, OAZ and SAT microRNAs in the saliva of a group of 20 patients who suffered from leucoplakia with dysplasia. When comparing the results to those of healthy patients, he did not find any relation between the levels of these independent or combined mRNAs and the presence of leucoplakia.

Liu et al. (2012a, b) studied miR-31, observing that its levels increased significantly in those patients who suffered from cancer at any stage, including the smallest lesions. However, it did not increase in the saliva of those patients, included in the study, who suffered from verrucous leucoplakia. Liu also found that it was more abundant in the saliva than in the plasma and that once the oral carcinoma had been eliminated, a considerable reduction of miR-31 was observed in the saliva which could stem from the cancerous tissue. This led Liu to the conclusion that this microRNA could be a good detector of the presence of a carcinoma on the leucoplakia lesion.

Cytokine IL-6 is another of the possible described salivary markers, which has been studied by several authors (Sharma et al. 2011; Panneer and Sadaksharam 2015; Brailo et al. 2006), and a higher concentration of IL-6 has been detected in patients diagnosed with leucoplakia in comparison with the healthy control group, and likewise it was higher in patients suffering from oral cancer.

Sharma et al. (2011) proposes it as a possible high-risk marker of malignant transformation of leucoplakia, given that he observes that the IL-6 concentration in the saliva increases as the dysplasia increases in the lesions. However, it must be considered that this can also be increased due to tobacco intake or a periodontal disease. Nevertheless, Brailo et al. (2006) does not relate these increased levels with tobacco consumption.

Once again these results contradict those obtained by Brailo et al. (2012) years later, in another study in which he did not find a high IL-6 concentration in patients suffering from leucoplakia. Furthermore, he studied the IL-1β and the TNF-α concentrations in patients who were suffering from leucoplakia and found low levels of IL-1β in LO patients in comparison with the controls, and these levels were higher in patients with oral cancer. With regards to the TNF-α, no significant differences were found among the three groups. In contrast to the aforementioned literature written by other authors who have observed that the salivary IL-6, IL-1β and TNF-α concentration can be affected by periodontal disease, he has not found any differences with regards to the periodontal health of the different groups.

With regards to the cytokines, Punyani and Sathawane (2013) studied the IL-8 levels and found no differences between the concentration of these levels in precancerous lesions and in the control group, however he did detect differences in the oral cancer group when compared with previous groups.

Therefore, we can observe an increase in the concentration of pro-inflammatory cytokines in oral cancer, which could reflect or warn against the development of oral cancer in leucoplakia lesions. However, more studies are needed in order to identify and eliminate the factors which could influence the cytokine values.

The complement activation product (Cd4) has been listed as a lung cancer prognostic marker. For this reason, Ajona et al. (2015) decided to assess it in order to check the concentration level in saliva. However, he did not find a statistically significant relationship between patients suffering from oral cancer and those suffering from leucoplakia.

The level of micronutrients such as copper, zinc and iron have been studied in the saliva of this type of patients (Shetty et al. 2015), and the findings have shown that Cu levels were higher in the saliva of patients suffering from oral leucoplakia than in the controls described several years ago by Ayinampudi and Narsimhan (2012), however, these results contradict themselves when comparing the Cu/Zn relationship, given that Shetty et al. (2015) has observed an increase in said index and Ayianmpudi et al. (2012) has detected a reduction.

The role of the oxidative stress and the leucoplakia action mechanism have also been studied. Kaur et al. (2016) found that the salivary-8-hydroxy-2-doxyguanosine malondialdehyde (8-OHdG), vitamin C and vitamin E with malondialdehyde (MDA) levels were higher and that the vitamin C and E levels were lower in patients with leucoplakia. Furthermore, when used together, these significatively improved the sensitivity and specificity as potential pre-cancer and oral cancer biomarkers Abdolsamadi et al. (2014).

Patel and Metgud (2015) and Shetty et al. (2012) carried out studies with regards to the LDH (Enzyme Lactate Dehydrogenase) in saliva and they both found that the levels increased in the leucoplakia group with respect to the healthy mucous and that the levels increased again in those patients who were suffering from oral cancer.

Jaeger et al. (2015) studied the epidermal growth factor in the saliva of OL and healthy patients using the ELISA technique, reaching the conclusion that it was not possible to consider it an OL biomarker at the moment. Likewise, he did not find a connection when he correlated it to the tissue immunohistochemical levels.

Hoffmann et al. (2011) assessed the role of the endothelin-1 levels in oral leucoplakia, however he observed that the levels in saliva did not allow for it to be a good marker of malignant transformation.

A recent study of proliferative verrucous leucoplakia (Flores et al. 2016) suggests that both angiotensinogen (AGT) and dipeptidyl peptidase 1 (DPP1) may be involved in developmental mechanisms for proliferative verrucous leucoplakia (PVL). 15 PVL patients and 15 controls were included in this study and the samples were analysed using LC-MS/MS. The 6 most abundant proteins in the PVL group were identified. After the logistic regression study had been carried out, small amounts of 2 proteins were found which could be used as potential biomarkers: the angiotensin (AGT) and the dipeptidyl peptidase 1 (DPP1).

However, more studies that follow this same methodology and that include a larger number of patients would be required in order to clarify the role of the biomarkers proposed in the oral leucoplakia study and their evolution towards malignant transformation.

12.2.1.2 Oral Lichen Planus

Oral lichen planus (OLP) is a common chronic inflammatory autoimmune disorder that affects the oral mucosa. With a prevalence between 0.1% and 5%, it affects women more often during the fifth or sixth decade of life and very rarely affects children. It may present different clinical forms and both white forms (papular, reticular and in the plaque) and red forms, (atrophic, erosive and bullous) can coexist (Figs. 12.4 and 12.5). The most commonly affected areas, regardless of their form, are the oral mucosa (bilateral, symmetric and asymptomatic), followed by the tongue and less frequently, the gum, lip and palate Symptoms are more frequent in the red forms and tend to be stinging, itching and ulcer. We have thought that maybe, symptoms are better summarized throughout the text.

OLP is a complex disease and, although its aetiology remains unknown, it is generally accepted that the lesions are produced in response to the T cells against the epithelial cells, and it is triggered by different stimuli such as psychological stress (Tvarijonaviciute et al. 2017, 2018), medicine consumption, hepatitis C virus infection and diabetes.

The annual malignancy of patients diagnosed with OLP is estimated between 0.5–1% (van der Waal et al. 2014). It is estimated that the annual incident rate for

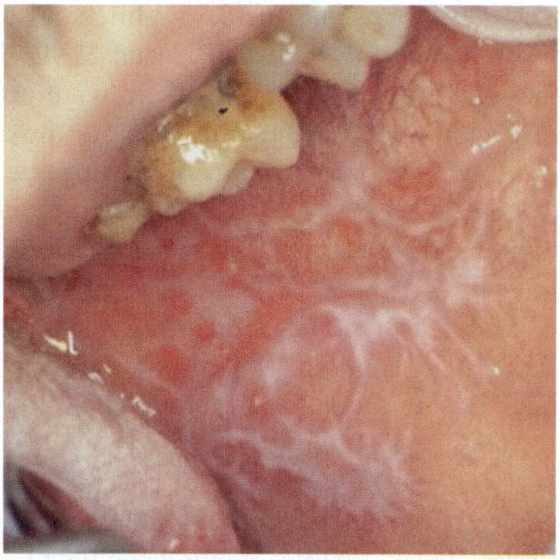

Fig. 12.4 Reticular OLP in left oral mucosa where reticular striae characteristic can be observed

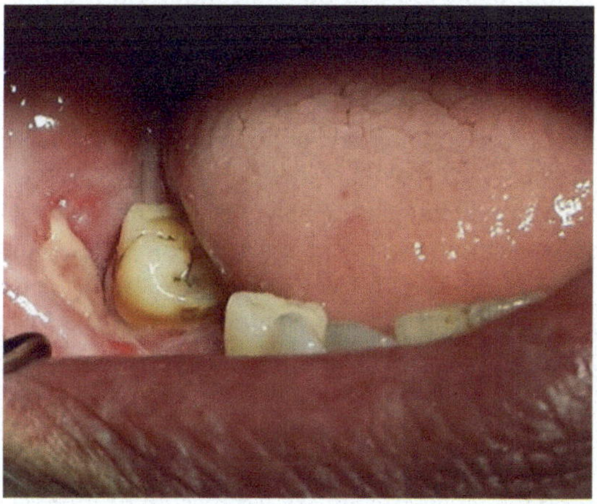

Fig. 12.5 Erosive OLP in right oral mucosa with painful ulcer

oral cancer is 5 per 100,000 inhabitants and the malignant transformation rate is 100 times higher in patients diagnosed with OLP (Lodi et al. 2005).

Rhodus et al. (2005) analysed the expression of the TNF-alpha cytokines, IL-1 alpha, IL-6 and IL-8 in the total non-stimulated saliva in healthy patients, patients diagnosed with OLP, with dysplasia and with oral squamous cell carcinoma (OSCC) to assess the change in concentration, searching for possible malignancy predictors. With regards to moderate dysplasia, the TNF-alpha and the IL-1alpha were significantly higher than in the control group, and there were no differences with respect to patients diagnosed with OSCC. With regards to severe dysplasia, the level of TNF-alpha was not significant with respect to the OSCC patients and the levels of IL-1alpha, IL-6 and IL-8 were significantly lower than in patients diagnosed with OSCC. The level of the four cytokines was not significantly different in the control and cases.

Nosratzehi et al. (2017) studied biomarkers that could indicate the OSCC malignancy from OLP lesions, analysing the MMPs (MMP-1, MMP-2, MMP-3 and

MMP-13) using the ELISA. While statistically significant differences were found when comparing the MMP-2 and MMP-13 concentrations, which were higher in the OSCC and OLP group, there were no statistically significant differences with respect to the OLP, the oral cancer or the control group samples.

However, apart from comparing patients diagnosed with dysplastic and non-dysplastic OLP or OSCC, long-term monitoring is necessary in order to establish malignancy markers. Long-term prospective studies with greater sampling sizes must be carried out.

Different hypothesis regarding the mechanisms involved in the OLP pathogenesis exist. Diagnosis is based on clinical and histopathological criteria and, in practice, the treatment is practically palliative. The most common OLP treatment includes corticosteroids, retinoids, cyclosporine, tacrolimus, phototherapy and surgery (Garcia-Pola et al. 2017). The studies that assessed the impact of steroid and anti-inflammatory therapy for OLP with regards to the concentration of cytokines in saliva suggest that this method can help to monitor the OLP response. Rhodus et al. (2006) observed that the concentration of cytokines after the topical application of dexamethasone 0.1% mouth wash reduces the levels of TNF-α, IL-1 α, IL-6 and IL-8 after 6 weeks of treatment. Ghallab et al. (2010) found a significant reduction in the salivary levels of IFN-, TNF- α and sTNFR-2 following treatment with prednisone (40–60 mg/day, no more than 60 days), suggesting that cytokines can help to monitor the OLP response.

More recent studies have used panel lectin microarray technology to search for salivary biomarkers in the OLP. Fang et al. compared the expressions in OLP cases and in healthy control groups and identified three lectins that increased, especially in the saliva of patients diagnosed with OLP: aleuria aurantia lectin (AAL); phytolacca americana (PWM); phaseolus vulgaris agglutinin (E + L) (Fang et al. 2018).

The salivary microRNAs differential expression studies in patients diagnosed with OLP reveal that microRNAs such as miR-4484 have a significantly higher expression in patients with OLP (Byun et al. 2015). Mehdipour et al. (2018) analysed the levels of salivary expression of miR-21, miR-125a, miR-31 and miR-200a in patients diagnosed with OLP, OSCC and healthy patients. They found a significantly higher expression of miR-21 in patients diagnosed with dysplastic OLP and OSCC in comparison to healthy patients. The miR-31 also presented a higher expression in patients diagnosed with dysplastic OLP and OSCC, but not in patients diagnosed with OLP without dysplasia nor in healthy patients.

The difficulty in finding a microRNA biomarker is that more than 2500 microRNAs have been identified so far in human beings, with different functions and with a feature that makes them difficultly exclusive which is the fact that one microRNA may have different targets, and one protein may be the target of different microRNAs. As Ma et al. state in their review of mRNAs-miRNAs and cytokines, it is observed that the TNF-alpha have numerous associated microRNAs and these microRNAs affect the expression of other cytokines (Ma et al. 2016).

Although the potential of saliva as a non-invasive OLP diagnostic and prognostic method is undeniable, it is also limited given that xerostomia and salivary gland hypofunction are also common in patients diagnosed with OLP (Larsen et al. 2017).

12.3 Salivary Markers in Oral Cancer

12.3.1 Proteomic and Peptidomic Analysis of Saliva in Oral Cancer

The main objective of the study of proteome in oral cancer is currently to facilitate an early diagnosis and to determine its stage, as well as to advise on the odds ratio of potentially malignant lesions becoming malignant. However, on account of the contradictory results demonstrated in the literature with regards to the cohorts and the small sample size used, it has been impossible to achieve these objectives (Spielmann and Wong 2011).

The latest advances in technologies related to proteomic and immunoassay techniques have allowed for the effective exploration of these analytes in saliva. Currently, the most commonly used techniques to analyse these proteins are: Enzyme-linked immunosorbent assay (ELISA), Radio-immunoassay, High performance liquid chromatography (HPLC), Two-dimensional gel electrophoresis followed by mass spectrometry (2DE-MS), 2DE and reverse-phase liquid chromatography (LC), followed by LC-tandem MS (2DE and LC-MS/MS), Matrix-assisted laser desorption/ionization-time-of-flight mass spectrometry (MALDI-TOF MS), and 2DE followed by MALDI-TOF MS (Schafer et al. 2014).

Nowadays, the most studied identified protein-based salivary biomarkers in relation to oral cancer by immunoassays are mainly cytokines (IL-6, IL-8, IL-1β), immunoglobulins, apoptosis inhibitors, glycoproteins involved in cell adhesion and molecules related to oxidative stress Battino et al. (2008). The biological nature of these proteins indicates the relevance of the inflammation and the immunological interactions of this pathology (Kaur et al. 2018).

Apart from these identifications, proteins can also be detected on a larger scale thanks to the development of techniques known as proteomics (i.e., MS-based approaches). The group led by Wong, identified, for the first time, the salivary proteome in pursuit of useful biomarkers in oral cancer diagnosis in a study of 40 cases and controls (1:1) using MALDI-TOF MS techniques. These authors found 46 differentially expressed peptides/proteins in a test Gallo et al. (2016). Subsequently the same group succeeded in generating a panel composed of 5 proteins (M2BP, MRP14, CD59, catalase, and profilin) which was capable of achieving a high specificity and sensitivity in the diagnosis of oral cancer (83% and 90% respectively) (Hu et al. 2008). More biomarkers were found by other groups using techniques based on mass spectrometry (de Jong et al. 2010; Wu et al. 2015; Dowling et al. 2008). Subsequently, other authors found more differentially expressed proteins using gel-based techniques such as the 2DE (Krapfenbauer et al. 2014; Jessie et al. 2013). Moreover, the most thorough study regarding salivary proteomic in oral cancer has been carried out in Taiwan. Aforementioned study included 460 patients, of which 96 were controls, 103 presented low-risk precancerous lesions, 130 presented high-risk precancerous lesions and 131 were diagnosed with oral cancer. This group identified a panel composed of four proteins MMP1, KNG1, ANXA2, and HSPA5, considered acceptably sensitive and specific in order to detect this pathology (Yu et al. 2016). Transferrin has also been studied and described as a promising biomarker (Jou et al. 2010). However, its usefulness has been questioned due to its restorative nature on a tissue level in relation to oral cancer, as this could facilitate the contamination of salivary samples with serum proteins such as transferrin (Kang and Kho 2018).

The application of liquid-based cytology specimens in the proteomic analysis of cells contained in the saliva, using techniques based on mass spectrometry has been a recent development (Yang et al. 2014). The proteomic analysis of these epithelial cells has been studied by means of cell sedimentation (Xie et al. 2008).

However, there are some obstacles that impede the effective introduction of these techniques. Primarily, the lack of standardisation and processing of samples, the presence of highly abundant proteins that complicate analysis, the reliance on bioinformatics databases to identify proteins and the constant discovery of genetic material involved in the transcription. The proteomic research remains in its early stages, as has been demonstrated by the two bases published thus far (Wilhelm et al. 2014).

Based on the discussed evidence, the authors believe in the progressive change from the tissue-based techniques to those based on liquid biological matrices such as saliva, which will be progressively introduced into modern medicine, and in fact it is believed that the study of the proteome of said matrix will prevail as one of the most promising OMICs.

12.3.2 Salivary miRNA Analysis in Oral Cancer

The specific characteristics of microRNAs make them attractive for the treatment and early diagnosis of many pathologies. In contrast to messenger RNAs which are unstable, microRNAs appear to be protected from salivary enzymes that can degrade the RNA by means of a vesicular system (Nylander et al. 2012; Liu et al. 2012a, b).

Studies of microRNAs in the saliva of patients diagnosed with oral cancer aim to find a microRNAs expression profile that allows for the early diagnosis of oral cancer and the prediction of its behaviour. Most microRNAs differential expression studies in the saliva of patients diagnosed with oral cancer firstly analyse the biopsy tissue samples by means of microarrays, and later validate its results with saliva and qRT-PCR (quantitative reverse transcriptase polymerase chain reaction).

Park et al. (2009) measured the expression with qRT-PCR of 314 microRNAs in the saliva of 50 patients diagnosed with OSCC and that of 50 healthy patients. The results revealed two miRNAs, the miR-125a and the miR-200a, significantly underexpressed in the saliva of patients diagnosed with OSCC.

Yang et al. (2013) studied the differential expression of microRNAs using microarray technology, capable of identifying 754 microRNAs using frozen tissue from progressive and non-progressive leucoplakia. Subsequently, they validated these results with saliva samples, obtaining a total of 25 differentially expressed microRNAs, 12 overexpressed microRNAs and 13 underexpressed microRNAs.

Another microRNAs study about saliva was the one carried out by Momen-Heravi et al. (2014), who used the NanoStringnCounter miRNA technology, which is capable of detecting 700 microRNAs probes. Samples of 34 patients were analysed, 9 patients diagnosed with OSCC before treatment, 8 patients diagnosed with OSCC in remission, 8 patients diagnosed with OLP and 9 healthy patients. A total of 13 microRNAs were determined as deregulated in the saliva of patients diagnosed with OSCC in comparison with healthy patients, 11 were underexpressed (miR-136, miR-147, meR-1250, miR-148a, miR-632, miR-668, miR-877, miR-503, miR-220a, miR-323-5p) and 2 were overexpressed (miR-24, miR27b). Although the sample was limited in size, they were able to determine the miR-27b as an effective biomarker in distinguishing between patients diagnosed with OSCC and healthy patients, patients diagnosed with OLP or patients in remission, with an 85.7% sensitivity and a 100% specificity.

Other studies use a more varied sample to analyse the differential expression. This is the case for the group led by Zahran et al. (2015), who studied three microRNAs in salivary samples, miR-184, miR-21 and miR-145 in 20 healthy patients, 40 patients suffering from pre-malignant lesions (20 diagnosed with dysplasia and 20 who were not suffering from it) who were monitored for at least 3 years, 20 patients diagnosed with OSCC and 20 patients diagnosed with recurrent aphthous stomatitis by means of qRT-PCR. These authors concluded that the miR-184 is a potential biomarker for identifying patients suffering from OSCC using saliva.

For the qRT-PCR technique, a control gene present in all the samples is needed. Some studies use other microRNAs as controls, such as miR-16 or miR-191 (Momen-Heravi et al. 2014; Liu et al. 2012a, b; Hung et al. 2016), and others use a small nucleolar RNA (such as RNU6 or SNORD68) (Shahidi et al. 2017; Yang et al. 2013; Zahran et al. 2015).

The heterogeneity of results can be associated with the technical differences between such studies. Apart from the heterogeneous nature that underlying molecular mechanisms have on cancer, different kits, platforms, controls and types of saliva are used to collect microRNAs, which can influence the final result. Park et al. (2009) found that there was twice as many isolated microRNAs in supernatant saliva compared to the total saliva.

According to Guerra et al. (2016), the ideal characteristics of an oral, head and neck cancer biomarker are high sensitivity and specificity, its presence in all affected patients and its cut-off value with a minimum overlap between states of normality and disease. For the time being, and bearing in mind the foregoing, this has not been possible with microRNAs.

12.3.3 Salivary Exosomes as Potential Biomarkers in Oral Cancer

The extracellular vesicles population is comprised of different types of small vesicles, including microvesicles (MVs) and exosomes. The two of these are vesicles which are surrounded by a membrane and that differ in their biogenesis and their biophysical properties, such as the size of the vesicles and the surface marker proteins, however, their biological roles, such as the intercellular communication, the maintenance of biological processes and the pathological processes to which they are closely linked to also differ, especially the cancerous process (Lee et al. 2012a).

Their size is only 50–150 nm diameter, and they are released into the extracellular environment and subsequently into different biological fluids such as saliva, urine and blood (Tkach and Théry 2016). The exosomes do not derive from the gemmation of the plasma membrane. They are secreted by peculiar cellular structures called microvesicular bodies (MVBs) that more than likely stem from a multiple cascade fusion of inner vesicles, including early endosomes, lysosomes and other structures that vary depending on their cellular origin (Fais et al. 2013).

Many types of cells can produce exosomes, such as the dendritic cells, the B and T cells, the mast cells and the epithelial cells. (O'Loughlin et al. 2012).

Tumour cells secrete large quantities of exosomes that promote tumour progression. The exosomes communicate with the surrounding cells by activating cell pathways (Guo et al. 2017) that enable tumour growth and development, the creation of new blood vessels that nurture it, the evasion of the immune response, and the creation of resistance to chemotherapeutic agents and metastasis, remotely preparing the pre-metastatic niche (King et al. 2012; Webber et al. 2010; Liu et al. 2006; Hood et al. 2011; Peinado et al. 2012). For this reason, exosomes are a relevant component of the tumour microenvironment and they are highly considered as one of the main contributors to tumour invasion and metastasis (Peinado et al. 2011).

The main component of the exosomes are the lipids, which compose their membrane (Record et al. 2014). In contrast to other microvesicles, the exosomes do not have nuclear DNA, but they may contain mitochondrial DNA (Sansone et al. 2017). Valadi et al. (2007a) was the first author to demonstrate the existence of messenger RNA and microRNA in its interior.

The exosomal protein composition includes an ubiquitin and a specific cellular protein and it carries out numerous biological functions. The exosomal proteins have been commonly identified as derived from the endocytic compartment or the plasma membrane and sometimes even from internal compartments such as the Golgi, nucleus, endoplasmic reticulum and mitochondria.

The exosomes constitutively express tetraspanins (CD63, CD9, CD81), endosomal and lysosomal markers (Rab5, LAMP) and heat shock proteins (HSP 70) (Andreu and Yáñez-Mó 2014).

By carrying out an adapted ELISA test which allows for the detection, characterisation and quantification of exosomes, it has been demonstrated that patients diagnosed with OSCC have significantly increased plasma levels of exosomes that express Cav1 in comparison with the plasma of healthy donors (Properzi et al. 2013) and also CD63 (Logozzi et al. 2009).

The discovery, 10 years ago, that the contents of the exosomes can be transferred to other cells by means of fusion in order to create phenotype alterations support intensive research in this field (Valadi et al. 2007b). The first author who reported the presence of exosomes in the saliva was Ogawa et al. (2008).

The collection of exosomes in saliva is easy and non-invasive. Furthermore, it contains less proteins than blood, which is why its identification and quantification is greatly simplified (Topkas et al. 2012; Schulz et al. 2013). It has been demonstrated that these can be stored at 4 °C without the need for deep freezing at −80 °C, greatly facilitating their use in clinical means (Kumeda et al. 2017).

In 2011, Ogawa et al. (2008) described two types of salivary exosomes that differed in their size and protein composition. The type I exosomes are larger and denser to the electrons than the type II exosomes, which are more similar to those present in other fluids and body environments. Both types contain proteomic markers such as CD63, Alix, Tsg101 and Hsp70, immunoglobulin A and immunoglobulin polymeric receptors, even though the specific protein composition is different. The human salivary exosomes participate in the catabolism of biopeptides and play an important role in the local immune response in the oral cavity. It is believed that the secretion of type I or type II exosomes depends, to a great extent, on the type of salivary gland that it is produced by (Ogawa et al. 2013). The Next-Generation Sequencing Technology has recently allowed for more thorough research regarding RNAs to be carried out, codifying long proteins. Said RNAs are composed of a high content of salivary exosomes that are capable of controlling tumour proteins translationally and which play a relevant role in the cell proliferation and cell death, as well as in the immune response (Nair et al. 2018). In 2016 Ogawa et al. demonstrated, in vitro, that the RNAs content and other pseudogenes included in the interior of the salivary exosomes are capable of being horizontally transferred to other cells, modulating the genetic expression of the receptors (Ogawa et al. 2016) and increasing their invasiveness and migration capability (Li et al. 2016).

Kim et al. (2017) checked the quantity and the type of extracellular vesicles in the plasma and saliva of mice with melanoma. The authors found that only 38.22% ± 18.55% of the vesicles found in the plasma appeared in saliva in contrast to the tumoral pathology, which is found in the oral cavity where the number of exosomes is very high due to the direct contact between the saliva and tumour.

Zlotogorski-Hurvitz et al. (2016) showed morphological and molecular differences in the salivary exosomes of patients diagnosed with OSCC with respect to healthy patients as a screening measure.

To conclude, we could state that the information available in this field is limited and extremely recent. Most of it is focused on the quantification and qualification of the content of the exosomes present in the saliva. However, we lack clear information about their biological role with respect to OSCC. More research is needed in this field, one that is gaining considerable ground due to its significance in the carcinogenic process.

12.3.4 DNA Salivary Methylation in Oral Cancer

The epigenetic term is understood as those heritable changes in the genetic expression that are not codified in the DNA sequencing (Hema et al. 2017; Singh et al. 2016). The most studied epigenetic modification in DNA is the cytosine methylation in the CpG dinucleotide (cytosine waste in the nucleotides). The other main group of epigenetic modifications includes post-translational modifications in the histones, mainly the changes by means of phosphorylation and deacetylation (Langie et al. 2017a, b; Lim et al. 2016a; Nagler 2009). The methylation silences the genetic expression by preventing the union of the transcription factors, without affecting the underlying sequence of the base pairs (Piyathilake et al. 2005). Numerous methylated genes in head and neck cancer have been described (Arantes et al. 2014; Arantes et al. 2015; Bryan et al. 2013; Carvalho et al. 2011; De et al. 2014; Galetzka et al. 2012; Gardner et al. 2015; Guerrero-Preston et al. 2011; Killian et al. 2014; Kim et al. 2012; Markopoulou et al. 2012; Matthews et al. 2013; Nagata et al. 2012; Ovchinnikov et al. 2012, 2014; Rettori et al. 2013a; Schussel et al. 2013; Simkin et al. 2012; Wu et al. 2015; Yang et al. 2013): p16INK4A, p15INK4B, p53, APC, E-cadherin, MGMT, MLH1, DAPK1, RASSF1A, RARβ as well as others that are developed below (Alvarez et al. 2018). All of these genes and molecules can be analysed not only in the tissues themselves, by means of a biopsy specimen, but they can also be measured throughout the free DNA assessment in the plasma/serum (Leon et al. 1977; Shapiro et al. 1983), saliva/oral rinses (Lopez et al. 2003) and urine.

The aberrant hypermethylation is common in head and neck tumours and it can be useful in saliva in order to monitor the disease in terms of recurrence and/or worsening. An important consideration is the validity of results obtained in saliva compared with the plasma. In terms of whole-genome methylation, Smith et al. (Smith et al. 2015) observed that the salivary methylome was positively correlated with the blood methylation in 85.5% of the CpG areas studied. Other studies (Langie et al. 2017a, b) confirm that these percentages reached 90% in the case of adults and teenagers, while other authors such as Godderis et al. (2015) were not able to confirm this.

A recent systematic review reveals that the studies that assess the DNA methylation show a high sensitivity for the diagnosis of head and neck cancer (HNC) and solid tissues (0.57) and a high specificity in saliva (0.89). The combination of multiple methylated genes was more sensitive for the diagnosis (Ji et al. 2001). A summary of all the studies and methylation markers used in the last 18 years can be observed in Table 12.2. Ji et al. (2016) have published a summary of the methylation markers with a high degree of sensitivity/specificity among the different types of tissues. By carrying out an individual analysis, it can be observed that the

Table 12.2 Study summary of methylation markers

Study	Methylated genes
Zhou et al. (2018)	CDKN2A
Cao et al. (2018)	mgmiR9-1, mgmiR124-1, mgmiR124-2, mgmiR124-3, mgmiR129-2, mgmiR137, and mgmiR148a
Ferlazzo et al. (2017)	p16 and O^6-methylguanine-DNA methyltransferase (MGMT)
Lim et al. (2016)	RASSF1α, p16 INK4a, TIMP3, PCQAP/MED15)
Germi et al. (2016)	ZRE2, ZRE3a, and ZRE3b
Liu et al. (2016)	S100A8
Nawaz et al. (2015)	EBNA1, LMP1, RASSF1A, DAPK, ITGA9, P16, WNT7A, CHFR, CYB5R2, WIF1, RIZ1, FSTL1
Arantes et al. (2015)	TIMP3, DCC, DAPK, CCNA1, AIM1, MGMT, CDH1, HIC1
Kis et al. (2014)	P16
Bhatia et al. (2014)	P16
Dang et al. (2013)	P16
Puttipanyalears et al. (2013)	ALU
Tian et al. (2013)	RASSF1A, CDKN2A, DLEC1, DAPK1, UCHL1
Rettori et al. (2013b)	CCNA1, DAPK, MGMT, SFRP1, TIMP3
You et al. (2013)	CDK10
Schussel et al. (2013)	DCC, EDNRB
Ovchinnikov et al. (2012)	RASSF1A, p16, DAPK1
Minor et al. (2012)	miR-9-1, miR-9-3
Nagata et al. (2012)	ECAD, TMEFF2, RARβ, MGMT, FHIT, WIF-1, DAPK, p16, HIN-1, TIMP3, p15, APC, SPARC
Zhang et al. (2012a)	EBNA1, LMP1, RASSF1A, DAPK
Demokan and Dalay (2011)	P16
Li et al. (2011)	P16, DAPK, MGMT, CDH1, RASSF1A
Weiss et al. (2011)	P16
Gyobu et al. (2011)	PAX6, ENST00000363328
Loyo et al. (2011)	AIM1, APC, CALCA, DCC, DLEC, DLC1, ESR, FHIT, KIF1A, PGP9.5, TIG1
Guerrero-Preston et al. (2011)	HOXA9, NID2, GATA4, KIF1A, EDNRB, DCC, MCAM, CALCA
Laytragoon et al. (2010)	P16
Pattani et al. (2010)	EDNRB
Kaur et al. (2010)	P16
Tawfik et al. (2011)	hMLH1
Su et al. (2010)	P16
Cao et al. (2009)	P16
Steinmann et al. (2009)	P16
Ghosh et al. (2009)	India

(continued)

Table 12.2 (continued)

Study	Methylated genes
Viet and Schmidt (2008)	GABRB3, IL11, INSR, NOTCH3, NTRK3, PXN, ERBB4, PTCH2, TMEFF1, TNFSF10, TWIST1, ADCYAP1, CEBPA, EPHA5, FGF3, HLF, AGTR1, BMP3, FGF8, NTRK3, FLT, IRAK3, KDR, NTRK, RASGRF1, WT1, ESR1, ETV1, GAS7, PKD2, WNT2, EPHA5, GALR1, KDR, p16, AGTR1, EYA4, IHH, NTRK3, NTRK3, TFPI2
Adams et al. (2008)	AHRR, p16, CBRP, CLDN3, MT1G, MGMT, RARβ, PGP9.5
Carvalho et al. (2008)	DCC, DAPK, ESR, CCNA1, CCND2, MINT1, MINT31, CDH1, AIM1, MGMT, p16, PGP9.5, RARβ, HIC1, RASSF1A, CALCA, TGFBR2, S100A2, RIZ1, RBM6
Righini et al. (2007)	P16
Franzmann et al. (2007)	CD44
Martone et al. (2007)	P16
Shaw et al. (2006)	P16
Maruya et al. (2004)	P16
Kulkarni and Saranath (2004)	P16
Weber et al. (2003)	P16
Wong et al. (2003)	P16
Tong et al. (2002)	EBV
Nakahara et al. (2001)	P16
Rosas et al. (2001)	P16
Sanchez et al. (2000)	P16

methylation markers for the saliva with scores above 90% in terms of sensitivity were: ECAD, DAPK/DCC/TIMP3, CCNA1/DAPK/DCC/TIMP3, DAPK/DCC AND CDH1. These results are consistent with those obtained by Arantes et al. (2014), who found that DAPK, DCC and TIMP3 were hypermethylated in nearly 90% of clinically T1 and T2 tumours.

The E-Chaderin is a cell adhesion molecule that has been widely studied in oral cancer (Chang et al. 2002; Maeda et al. 2007; Nakayama et al. 2001; Saito et al. 1998)(Nagata et al. 2012), the reduction in expression is associated with invasion and tumour destruction and its silencing due to methylation is associated with a reduction of this. Its hypermethylation is associated with the loss of cellular cohesiveness and the epithelium-mesenchymal transition. DAPK (Death-associated protein kinase) has been proposed as a tumour suppressor gene, and its silencing is associated to the development of malignant lesions. The hypermethylation is related to early events of the carcinogenesis (Liu et al. 2012a, b). The DCC hypermethylation (deleted in colorectal cancer) has also been related to premalignant and malignant oral lesions in oral rinses (Schussel et al. 2013). TIMP-3 or tissue inhibitor of metalloproteinase 3 relates to the activity inhibition of one of the extracellular matrix-destructive enzymes and its hypermethylation has been associated with later stages of the malignant disease and local recurrence (Rettori et al. 2013b). CCNA1 or Cyclin A1 is a molecule that binds to relevant regulators of the cellular cycle, such as proteins of the Rb family, transcription factor E2F-1 and proteins of the p21 family. Its hypermethylation, especially when combined with DAPK and TIMP3, has been associated with malignant lesions. CDH1 or cadherin-1 is a molecule which is also related to intercellular aggregation, its hypermethylation is associated to the loss of cellular cohesiveness and the epithelial-mesenchymal transition.

The methylation determination can also be useful in order to reveal the HPV positivity. Lim et al. (2016b), found that the methylation levels for the RASSF1α, p16 INK4a, TIMP3 and PCQAP/MED15 were superior in patients with negative HPV in comparison with a control group (71% sensitivity and 80% specificity). Conversely, the methylation levels were inferior in patients with positive HPV in comparison with the control groups (80% sensibility and 74% specificity).

We may conclude that the specific analysis of methylation in unique markers does not appear to be pathognomonic in the diagnosis of oral malignant lesions. However, the use of methylation panels with several genes provides a higher diagnostic sensitivity. It seems that the oral oncogenic process starts early with epigenetic silencing that can be determined in the saliva for the diagnosis, screening or evolutionary control in those patients diagnosed with oral cancer or patients at a greater risk of suffering from it.

12.3.5 Other Salivary Biomarkers in Oral Cancer

Salivary Metabolome in Oral Cancer

The Warburg effect refers to the dramatically altered use of nutrients in cancer cells, specifically by means of the aerobic glycolysis (Liberti and Locasale 2016). Recent advances in analytical chemistry have allowed for the exploration of this metabolome in different tissues and liquid biological matrices. In particular, saliva has been studied by means of capillary electrophoresis-mass spectrometry (CE-MS) and ultra-performance liquid chromatography (UPLC-MS) techniques.

Yan et al. were the first authors to study the metabolome in the saliva of patients diagnosed with oral cancer and those suffering from precancerous lesions (oral lichen planus and oral leucoplakia) using HPLC/MS analysis. These authors managed to generate an algorithm with a discriminatory capacity of 100% among lesions (Yan et al. 2008). Wei et al. validated these results subsequently by means of UPLC-MS, comparing these oral cancer cases with those of leucoplakia. In a ROC type analysis, they found a panel composed of three metabolites (i.e., Valine, lactic acid and phenylalanine) that presented a capacity as a biomarker (Wei et al. 2011). Other recently used techniques are the ultra-performance liquid chromatography-electrospray ionization-mass spectrometry (UPLC-ESI-MS) (Wang et al. 2014), nuclear magnetic resonance (NMR)-based metabolomics (Tiziani et al. 2009), and gas chromatography–mass spectrometry (GC-MS) (Shigeyama et al. 2018) in the study of oral cancer.

In spite of the promising results, there are no large-scale studies with a representative cohort of patients. Furthermore, its analysis has relevant limitations such as the difficulty of distinguishing metabolites derived from oral biofilm or oral tissues or the high variability that it can suffer due to physiological or environmental factors (Ishikawa et al. 2017).

Salivary Microbiome in Oral Cancer
The first evidences of the existence of bacteria with carcinogenic power were discovered in the 1930s (Vento and Tanko 2009). A body of evidences exists in relation to bacteria that may contribute to the development of oral cancer due to its capacity of generating carcinogenic metabolites such as the acetaldehyde or its capacity of maintaining chronic inflammation (Hooper et al. 2009).

Mager et al. (2005) reported for the first time the presence of different quantities of certain bacteria species in patients suffering from oral cancer in comparison with control patients. By means of checkerboard DNA-DNA hybridization (*Capnocytophagagingivalis, Prevotellamelaninogenica* and *Streptococcus mitis*) they found that they had a diagnostic capacity for this pathology of 80% sensitivity and 83% specificity. Another study carried out using 454 parallel sequencing of the 16S rRNA found certain clearly differentiated phyla between the control patients and those diagnosed with oral cancer, specifically *Firmicutes,* and Bacteroidetes (Pushalkar et al. 2011). Recent studies have focused on studying the salivary microbiome in the malignant transition of premalignant oral lesions (Lee et al. 2017; Hu et al. 2016). In this sense, the largest cohort of patients published to date has demonstrated by means of next-generation sequencing techniques that five genera are strongly altered (*Bacillus, Enterococcus, Parvimonas, Peptostreptococcus,* and *Slackia*) (Lee et al. 2017).

Circulating Tumour DNA and Other Genomic Biomarkers in the Saliva
These consist of DNA fragments of less than 160 bp that contain alterations on a genetic and epigenetic level. These fragments are generated by cells which are subject to different types of programmed cell deaths (Stewart and Tsui 2018). Certain literature has studied the presence of such biomarkers in saliva (van Ginkel et al. 2017). We will focus next on the alterations on a genetic level by the prior description of biomarkers derived from genome methylation.

Wang et al. demonstrated that saliva was a better option when studying tumour DNA in the case of oral cancer than serum, although serum was better for studying other types of cancer (Wang et al. 2015a, b).

Different researchers have specifically studied the instability of microsatellites and the loss of heterozygosity in the DNA of exfoliated cells from the saliva. Spafford et al. found mutations in a descriptive study in 92% of these types of cases. Subsequent literature confirmed such alterations (Okami et al. 2002). Minority alterations have been found in the mitochondrial DNA of patients diagnosed with head and neck cancer (Jiang et al. 2005).

Additional studies have also investigated the DNA derived from HPV on a salivary level, using mainly PCR techniques. It has been proven that the prognostic value of HPV-16 DNA on a salivary level in patients diagnosed with HPV positive head and neck cancers (Ahn et al. 2014; Chuang et al. 2008).

12.4 Salivary Markers in Non-oral Cancer

As it has been stated in the previous chapters, saliva is particularly useful for the diagnosis of potentially malignant oral and tumour lesions of the oral cavity (Khurshid et al. 2018). In addition to this, researchers have aimed at developing alternative methods for the diagnosis and monitoring of other tumours which are not related with the oral cavity, from major salivary gland cancer to tumours in further away organs such as the breasts or pancreas (Table 12.3).

With regards to parotid cancer, the alteration of miRNA has been studied in several types of cancer and 8 miRNAs were expressed in parotid cancer, concluding that the association of 2 biomarkers to detect parotid tumours was ideal (Matse et al.

Table 12.3 Study summary of cancer biomarkers from other cancers than oral cancer

Author	Type of cancer	Salivary biomarker
Matse et al. (2015)	Parotid cancer	5 miRNAs (hsa-miR-296-5p, hsa-miR-577, hsa-miR-1233, hsamiR-1267, and hsa-miR-1825) had a significantly higher expression level (lower ΔCt) in whole saliva samples from patients with a parotid gland neoplasm compared to their expression levels in whole saliva from healthy controls 3 miRNAs (hsa-miR-103a-3p, hsa-miR-211 and hsa-miR-425-5p) had a significantly higher expression level in whole saliva from controls compared to patients with a parotid gland neoplasm
Huang et al. (2012)	Parotid cancer	Fibroblast growth factor 2 (FGF2) and fibroblast growth factor receptor 1 (FGFR1)
Streckfus et al. (2000)	Breast cancer	Higher levels of c-erbB-2 (erb) and CA15-3 p53 levels were higher in control subjects
Tajmul et al. (2018)	Ovarian cancer	Lipocalin-2, indoleamine-2, 3-dioxygenase1 (IDO1) and S100A8
Lee et al. (2012b)	Ovarian Cancer	AGPAT1, B2M, BASP2, IER3, and IL1B
Zhang et al. (2012b)	Lung Cancer	CCNI, EGFR, FGF19, FRS2, and GREB1
Shiiki et al. (2011)	Prostate adenocarcinoma	Prostate specific antigen (PSA)
Humeau et al. (2015)	Pancreatic Cancer	KRAS, EMBD3L2, EACRV1, EDPM1
Ding et al. (2019)	Hepatocellular carcinoma	SOD2

2016). Breast cancer takes the lives of 500,000 women in the world each year. A new case of breast cancer is detected every 2 min in the world, indicating the importance of this disease in diagnostic medicine. Streckfus et al. (2000) have confirmed that the c-erB-2 tumour marker is present in the saliva and serum of ill women and not in healthy patients. Salivary measures of IL-6 and cortisol are used in these types of tumours in order to assess the stress level and apply Mindfulness-based stress reduction (MBSR) techniques (Lengacher et al. 2018). In terms of ovarian cancer, this is derived from the coelomic epithelium, the germ lines or the ovarian stroma, it corresponds to 6% of all cancers in women, 27% of gynaecological cancers, but it is the most lethal. The CA-125 tumour marker in saliva and serum has been determined, concluding that saliva has a higher specificity and a lower sensitivity for diagnosis than serum (Malathi et al. 2014). In relation to prostate cancer, it has been demonstrated that a positive correlation between the prostate-specific antigen (PSA) measured in the plasma with saliva exists (Shiiki et al. 2011). Pancreatic cancer is the 4th cause of death. The presence of extracellular RNA of pancreatic cancer in resectable stages has been determined. The multivariate logistic regression model that provided a better diagnosis rate was the one that combined 4 biomarkers (KRAS, EMBD3L2, EACRV1, EDPM1) (Humeau et al. 2015). Lung cancer has a high incidence and the mutations in EGFR are the specific biomarkers for the non-small cell carcinoma. Zhang et al. have developed a method to detect such mutations in saliva (Wei et al. 2011). Xiao et al. (2012) have found 16 candidate proteins in saliva to discriminate patients diagnosed with cancer from healthy patients with a high sensitivity and specificity.

12.5 Conclusion

From all this data and the evolution of analysis methods, we can predict that the use of saliva as a means of analysing body tumours will be a reality patent in hospital units.

References

Abdolsamadi H, Rafieian N, Goodarzi MT, Feradmal J, Davoodi P, Jazayeri M, Taghavi Z, Hoseyni SM, Ahmadi-Motamayel F (2014) Levels of salivary antioxidant vitamins and lipid peroxidation in patients with oral lichen planus and healthy individuals. Chonnam Med J 50(2):58–62

Adams L, Roth MJ, Abnet CC, Dawsey SP, Qiao YL, Wang GQ, Wei WQ, Lu N, Dawsey SM, Woodson K (2008) Promoter methylation in cytology specimens as an early detection marker for esophageal squamous dysplasia and early esophageal squamous cell carcinoma. Cancer Prev Res (Phila) 1(5):357–361

Ahn SM, Chan JY, Zhang Z, Wang H, Khan Z, Bishop JA, Westra W, Koch WM, Califano JA (2014) Saliva and plasma quantitative polymerase chain reaction-based detection and surveillance of human papillomavirus-related head and neck cancer. JAMA Otolaryngol Head Neck Surg 140(9):846–854

Ajona D, Pajares MJ, Chiara MD, Rodrigo JP, Jantus-Lewintre E, Camps C, Suarez C, Bagan JV, Montuenga LM, Pio R (2015) Complement activation product C4d in oral and oropharyngeal squamous cell carcinoma. Oral Dis 21(7):899–904

Alvarez PB, Sayans MP, Iruegas EP, Penaranda JMS, Alves MO, Carrion AB, Garcia AG (2018) State of methylation and epigenome studies in some types of head and neck cancer. Biomed Res 29(14). https://doi.org/10.4066/biomedicalresearch.29-18-104

Andreu Z, Yáñez-Mó M (2014) Tetraspanins in extracellular vesicle formation and function. Front Immunol 5:442

Arantes LM, de Carvalho AC, Melendez ME, Carvalho AL, Goloni-Bertollo EM (2014) Methylation as a biomarker for head and neck cancer. Oral Oncol 50(6):587–592

Arantes LM, de Carvalho AC, Melendez ME, Centrone CC, Góis-Filho JF, Toporcov TN, Caly DN, Tajara EH, Goloni-Bertollo EM, Carvalho AL, GENCAPO (2015) Validation of methylation markers for diagnosis of oral cavity cancer. Eur J Cancer 51(5):632–641

Ayinampudi BK, Narsimhan M (2012) Salivary copper and zinc levels in oral pre-malignant and malignant lesions. Journal of oral and maxillofacial pathology: JOMFP 16(2):178

Bagan JV, Scully C (2008) Recent advances in oral oncology 2007: epidemiology, aetiopathogenesis, diagnosis and prognostication. Oral Oncol 44(2):103–108

Battino M, Greabu M, Totan A, Bullon P, Bucur A, Tovaru S, Mohora M, Didilescu A, Parlatescu I, Spinu T, Totan C (2008) Oxidative stress markers in oral lichen planus. Biofactors (Oxford, England) 33(4):301–310

Baum BJ (1993) Principles of saliva secretion. Ann N Y Acad Sci 694(1):17–23

Bhatia V, Goel MM, Makker A, Tewari S, Yadu A, Shilpi P, Kumar S, Agarwal SP, Goel SK (2014) Promoter region hypermethylation and mRNA expression of MGMT and p16 genes in tissue and blood samples of human premalignant oral lesions and oral squamous cell carcinoma. Biomed Res Int 2014:248419

Brailo V, Vucicevic-Boras V, Cekic-Arambasin A, Alajbeg IZ, Milenovic A, Lukac J (2006) The significance of salivary interleukin 6 and tumor necrosis factor alpha in patients with oral leucoplakia. Oral Oncol 42(4):370–373

Brailo V, Vucicevic-boras V, Lukac J, Biocina-lukenda D, Zilic-alajbeg I, Milenovic A, Balija M (2012) Salivary and serum interleukin 1 beta, interleukin 6 and tumor necrosis factor alpha in patients with leukoplakia and oral cancer. Medicina oral, patologia oral y cirugia bucal 17(1):10

Brinkman BM, Wong DT (2006) Disease mechanism and biomarkers of oral squamous cell carcinoma. Curr Opin Oncol 18(3):228–233

Bryan AD, Magnan RE, Hooper AE, Harlaar N, Hutchison KE (2013) Physical activity and differential methylation of breast cancer genes assayed from saliva: a preliminary investigation. Ann Behav Med 45(1):89–98

Byun JS, Hong SH, Choi JK, Jung JK, Lee HJ (2015) Diagnostic profiling of salivary exosomal microRNAs in oral lichen planus patients. Oral Dis 21(8):987–993

Camisasca DR, da Rosgoncalves L, Soares MR, Sandim V, Nogueira FC, Garcia CH, Santana R, de Oliveira SP, Buexm LA, de Faria PA, Dias FL, Pereira DA, Zingali RB, Alves G, Lourenco SQ (2017) A proteomic approach to compare saliva from individuals with and without oral leucoplakia. J Proteome 151:43–52

Cao J, Zhou J, Gao Y, Gu L, Meng H, Liu H, Deng D (2009) Methylation of p16 CpG island associated with malignant progression of oral epithelial dysplasia: a prospective cohort study. Clin Cancer Res 15(16):5178–5183

Cao Y, Green K, Quattlebaum S, Milam B, Lu L, Gao D, He H, Li N, Gao L, Hall F, Whinery M, Handley E, Ma Y, Xu T, Jin F, Xiao J, Wei M, Smith D, Bornstein S, Gross N, Pyeon D, Song J, Lu SL (2018) Methylated genomic loci encoding microRNA as a biomarker panel in tissue and saliva for head and neck squamous cell carcinoma. Clin Epigenetics 10:43

Carvalho AL, Jeronimo C, Kim MM, Henrique R, Zhang Z, Hoque MO, Chang S, Brait M, Nayak CS, Jiang WW, Claybourne Q, Tokumaru Y, Lee J, Goldenberg D, Garrett-Mayer E, Goodman S, Moon CS, Koch W, Westra WH, Sidransky D, Califano JA (2008) Evaluation of promoter hypermethylation detection in body fluids as a screening/diagnosis tool for head and neck squamous cell carcinoma. Clin Cancer Res 14(1):97–107

Carvalho AL, Henrique R, Jeronimo C, Nayak CS, Reddy AN, Hoque MO, Califano JA (2011) Detection of promoter hypermethylation in salivary rinses as a biomarker for head and neck squamous cell carcinoma surveillance. Clin Cancer Res 17(14):4782–4789

Chang HW, Chow V, Lam KY, Wei WI, Yuen A (2002) Loss of E-cadherin expression resulting from promoter hypermethylation in oral tongue carcinoma and its prognostic significance. Cancer 94(2):386–392. https://doi.org/10.1002/cncr.10211

Chuang AY, Chuang TC, Chang S, Zhou S, Begum S, Westra WH, Ha PK, Koch WM, Califano JA (2008) Presence of HPV DNA in convalescent salivary rinses is an adverse prognostic marker in head and neck squamous cell carcinoma. Oral Oncol 44(10):915–919

Crowley E, Di Nicolantonio F, Loupakis F, Bardelli A (2013) Liquid biopsy: monitoring cancer-genetics in the blood. Nat Rev Clin Oncol 10(8):472

Cuevas-Cordoba B, Santiago-Garcia J (2014) Saliva: a fluid of study for OMICS. Omics: J Integr Biol 18(2):87–97

Dang J, Bian YQ, Sun JY, Chen F, Dong GY, Liu Q, Wang XW, Kjems J, Gao S, Wang QT (2013) MicroRNA-137 promoter methylation in oral lichen planus and oral squamous cell carcinoma. J Oral Pathol Med 42(4):315–321

De Jong EP, Xie H, Onsongo G, Stone MD, Chen XB, Kooren JA, Refsland EW, Griffin RJ, Ondrey FG, Wu B, Le CT, Rhodus NL, Carlis JV, Griffin TJ (2010) Quantitative proteomics reveals myosin and actin as promising saliva biomarkers for distinguishing pre-malignant and malignant oral lesions. PLoS One 5(6):E11148

De A, Sparreboom W, van den Berg A, Carlen ET (2014) Rapid microfluidic solid-phase extraction system for hyper-methylated DNA enrichment and epigenetic analysis. Biomicrofluidics 8(5):054119

Demokan S, Dalay N (2011) Role of DNA methylation in head and neck cancer. Clin Epigenetics 2(2):123–150. https://doi.org/10.1007/s13148-011-0045-3. Epub 2011 Jul 9

Ding F, Sun K, Sun N, Jiang Q, Cao M, Wu Z (2019) iTRAQ-based proteomics reveals SOD2 as a potential salivary biomarker in liver cancer. Int J Biol Markers 34(3):221–231

Dowling P, Wormald R, Meleady P, Henry M, Curran A, Clynes M (2008) Analysis of the saliva proteome from patients with head and neck squamous cell carcinoma reveals differences in abundance levels of proteins associated with tumour progression and metastasis. J Proteome 71(2):168–175

Epstein JB, Zhang L, Rosin M (2002) Advances in the diagnosis of oral premalignant and malignant lesions. J Can Dent Assoc 68(10):617–621

Fais S, Logozzi M, Lugini L, Federici C, Azzarito T, Zarovni N, Chiesi A (2013) Exosomes: the ideal nanovectors for biodelivery. Biol Chem 394(1):1–15

Fang L, Liu Q, He P, Wang X, Wang Y, Wei M, Chen L (2018) Alteration of salivary glycopatterns in oral lichen planus. Biomarkers 23(2):188–195

Ferlazzo N, Currò M, Zinellu A, Caccamo D, Isola G, Ventura V, Carru C, Matarese G, Ientile R (2017) Influence of MTHFR genetic background on p16 and MGMT methylation in oral squamous cell cancer. Int J Mol Sci 18(4)

Flores IL, Santos-Silva AR, Coletta RD, Leme AF, Lopes MA (2016) Low expression of angiotensinogen and dipeptidyl peptidase 1 in saliva of patients with proliferative verrucous leucoplakia. World J Clin Cases 4(11):356–363

Franzmann EJ, Reategui EP, Pedroso F, Pernas FG, Karakullukcu BM, Carraway KL, Hamilton K, Singal R, Goodwin WJ (2007) Soluble CD44 is a potential marker for the early detection of head and neck cancer. Cancer Epidemiol Biomark Prev 16:1348–1355

Galetzka D, Hansmann T, El Hajj N, Weis E, Irmscher B, Ludwig M et al (2012) Monozygotic twins discordant for constitutive BRCA1 promoter methylation, childhood cancer and secondary cancer. Epigenetics 7(1):47–54. https://doi.org/10.4161/epi.7.1.18814

Gallo C, Ciavarella D, Santarelli A, Ranieri E, Colella G, Lo Muzio L, Lo Russo L (2016) Potential salivary proteomic markers of oral squamous cell carcinoma. Cancer Genomics Proteomics 13(1):55–61

Gandara-Vila P, Pérez-Sayans M, Suárez-Peñaranda JM, Gallas-Torreira M, Somoza-Martín J, Reboiras-López M, Blanco-Carrión A, García-García A (2018) Survival study of leucoplakia malignant transformation in a region of northern Spain. Med Oral Patol Oral y Cir Bucal 23(4):E413

Garcia-Pola MJ, Gonzalez-Alvarez L, Garcia-Martin JM (2017) Treatment of oral lichen planus. Systematic review and therapeutic guide. Med Clin 149(8):351–362

Gardner KR, Sapienza C, Fisher JO (2015) Genetic and epigenetic associations to obesity-related appetite phenotypes among African-American children. Pediatr Obes 10(6):476–482

Germi R, Guigue N, Lupo J, Semenova T, Grossi L, Vermeulen O, Epaulard O, de Fraipont F, Morand P (2016) Methylation of Epstein-Barr virus Rta promoter in EBV primary infection, reactivation and lymphoproliferation. J Med Virol 88(10):1814–1820

Ghallab, N.A., El-Wakeel, N. And Shaker, O.G., 2010. Levels of salivary IFN-gamma, TNF-alfa, and TNF receptor-2 as prognostic markers in (erosive) oral lichen planus. Mediat Inflamm, 2010, pp. 847632

Ghosh A, Ghosh S, Maiti GP, Sabbir MG, Alam N, Sikdar N, Roy B, Roychoudhury S, Panda CK (2009) SH3GL2 and CDKN2A/2B loci are independently altered in early dysplastic lesions of head and neck: correlation with HPV infection and tobacco habit. J Pathol 217(3):408–419

Godderis L, Schouteden C, Tabish A, Poels K, Hoet P, Baccarelli AA, Van Landuyt K (2015) Global methylation and hydroxymethylation in DNA from blood and saliva in healthy volunteers. Biomed Res Int 2015:845041

Guerrero-Preston R, Soudry E, Acero J, Orera M, Moreno-López L, Macía-Colón G, Jaffe A, Berdasco M, Ili-Gangas C, Brebi-Mieville P, Fu Y, Engstrom C, Irizarry RA, Esteller M, Westra W, Koch W, Califano J, Sidransky D (2011) NID2 and HOXA9 promoter hypermethylation as biomarkers for prevention and early detection in oral cavity squamous cell carcinoma tissues and saliva. Cancer Prev Res (Phila) 4(7):1061–1072

Guerra ENS, Rêgo DF, Elias ST, Coletta RD, Mezzomo LAM, Gozal D, Canto GDL (2016) Diagnostic accuracy of serum biomarkers for head and neck cancer: a systematic review and meta-analysis. Crit Rev Oncol Hematol 101:93–118

Guo W, Gao Y, Li N, Shao F, Wang C, Wang P, Yang Z, Li R, He J (2017) Exosomes: new players in cancer (review). Oncol Rep 38(2):665–675

Gyobu K, Yamashita S, Matsuda Y, Igaki H, Niwa T, Oka D, Kushima R, Osugi H, Lee S, Suehiro S, Ushijima T (2011) Identification and validation of DNA methylation markers to predict lymph node metastasis of esophageal squamous cell carcinomas. Ann Surg Oncol 18(4):1185–1194

Hayes DF, Bast RC, Desch CE, Fritsche H Jr, Kemeny NE, Jessup JM et al (1996) Tumor marker utility grading system: a framework to evaluate clinical utility of tumor markers. JNCI: J Natl Cancer Inst 88(20):1456–1466

Hema KN, Smitha T, Sheethal HS, Mirnalini SA (2017) Epigenetics in oral squamous cell carcinoma. J Oral Maxillofac Pathol 21(2):252–259

Hoffmann RR, Yurgel LS, Campos MM (2011) Evaluation of salivary endothelin-1 levels in oral squamous cell carcinoma and oral leucoplakia. Regul Pept 166(1–3):55–58

Hood JL, San RS, Wickline SA (2011) Exosomes released by melanoma cells prepare sentinel lymph nodes for tumor metastasis. Cancer Res 71(11):3792–3801

Hooper SJ, Wilson MJ, Crean SJ (2009) Exploring the link between microorganisms and oral cancer: a systematic review of the literature. Head Neck 31(9):1228–1239

Hu S, Arellano M, Boontheung P, Wang J, Zhou H, Jiang J, Elashoff D, Wei R, Loo JA, Wong DT (2008) Salivary proteomics for oral cancer biomarker discovery. Clin Cancer Res 14(19):6246–6252

Hu X, Zhang Q, Hua H, Chen F (2016) Changes in the salivary microbiota of oral leucoplakia and oral cancer. Oral Oncol 56:E6–E8

Huang XF, Li W, Yang XD, Jiang WH, Chen XH, Hu QG (2012) Keratocystoma of the parotid gland: a clinicopathological study and literature review. Int J Oral Maxillofac Surg 41(2):256–260

Humeau M, Vignolle-Vidoni A, Sicard F, Martins F, Bournet B, Buscail L, Torrisani J, Cordelier P (2015) Salivary microRNA in pancreatic cancer patients. PLoS One 10(6):e0130996

Hung KF, Liu CJ, Chiu PC, Lin JS, Chang KW, Shih WY, Kao SY, Tu HF (2016) Microrna-31 upregulation predicts increased risk of progression of oral potentially malignant disorder. Oral Oncol 53:42–47

Ishikawa S, Sugimoto M, Kitabatake K, Tu M, Sugano A, Yamamori I, Iba A, Yusa K, Kaneko M, Ota S, Hiwatari K, Enomoto A, Masaru T, Iino M (2017) Effect of timing of collection of salivary metabolomic biomarkers on oral cancer detection. Amino Acids 49(4):761–770

Jaeger F, Assuncao AC, Caldeira PC, Queiroz-Junior CM, Bernardes VF, De Aguiar MC (2015) Is salivary epidermal growth factor a biomarker for oral leucoplakia? A preliminary study. Oral Surg Oral Med Oral Pathol Oral Radiol 119(4):451–458

Jessie K, Jayapalan JJ, Ong KC, Abdul Rahim ZH, Zain RM, Wong KT, Hashim OH (2013) Aberrant proteins in the saliva of patients with oral squamous cell carcinoma. Electrophoresis 34(17):2495–2502

Ji ZW, Oku N, Umeda M, Komori T (2001) Establishment of an oral squamous cell carcinoma cell line (NOS-1) exhibiting amplification of the erbb-1 oncogene and point mutation of p53 tumor suppressor gene: its biological characteristics and animal model of local invasion by orthotopic transplantation of the cell line. Oral Oncol 37(4):386–392

Ji X, Guan C, Jiang X, Li H (2016) Diagnostic accuracy of DNA methylation for head and neck cancer varies by sample type and number of markers tested. Oncotarget 7(48):80019–80032

Jiang WW, Masayesva B, Zahurak M, Carvalho AL, Rosenbaum E, Mambo E, Zhou S, Minhas K, Benoit N, Westra WH, Alberg A, Sidransky D, Koch W, Califano J (2005) Increased mitochondrial DNA content in saliva associated with head and neck cancer. Clin Cancer Res 11(7):2486–2491

Joseph BK (2002) Oral cancer: prevention and detection. Medical Principles and Practice : International Journal of the Kuwait University, Health Science Centre 11(Suppl 1):32–35

Jou YJ, Lin CD, Lai CH, Chen CH, Kao JY, Chen SY, Tsai MH, Huang SH, Lin CW (2010) Proteomic identification of salivary transferrin as a biomarker for early detection of oral cancer. Anal Chim Acta 681(1–2):41–48

Kang JH, Kho HS (2018) Blood contamination in salivary diagnostics: current methods and their limitations. Clin Chem Lab Med 57(8):1115–1124

Kaur J, Demokan S, Tripathi SC, Macha MA, Begum S, Califano JA, Ralhan R (2010) Promoter hypermethylation in Indian primary oral squamous cell carcinoma. Int J Cancer 127(10):2367–2373

Kaur J, Politis C, Jacobs R (2016) Salivary 8-hydroxy-2-deoxyguanosine, malondialdehyde, vitamin C, and vitamin E in oral pre-cancer and cancer: diagnostic value and free radical mechanism of action. Clin Oral Investig 20(2):315–319

Kaur J, Jacobs R, Huang Y, Salvo N, Politis C (2018) Salivary biomarkers for oral cancer and pre-cancer screening: a review. Clin Oral Investig 22(2):633–640

Khurshid Z, Zafar MS, Khan RS, Najeeb S, Slowey PD, Rehman IU (2018) Role of salivary biomarkers in oral cancer detection. Adv Clin Chem 86:23–70

Killian JK, Miettinen M, Walker RL, Wang Y, Zhu YJ, Waterfall JJ et al (2014) Recurrent epimutation of SDHC in gastrointestinal stromal tumors. Sci Transl Med 6(268):268ra177. https://doi.org/10.1126/scitranslmed.3009961

Kim BW, Esworthy RS, Hahn MA, Pfeifer GP, Chu FF (2012) Expression of lactoperoxidase in differentiated mouse colon epithelial cells. Free Radic Biol Med 52(9):1569–1576

Kim J, Shin H, Park J (2017) RNA in salivary extracellular vesicles as a possible tool for systemic disease diagnosis. J Dent Res 96(8):938–944

King HW, Michael MZ, Gleadle JM (2012) Hypoxic enhancement of exosome release by breast cancer cells. BMC Cancer 12(1):421

Kis A, Tatár TZ, Gáll T, Boda R, Tar I, Major T, Redl P, Gergely L, Pathol SK (2014) Frequency of genetic and epigenetic alterations of p14ARF and p16INK4A in head and neck cancer in a Hungarian population. Oncol Res 20(4):923–929

Kobayashi T, Maruyama S, Abe T, Cheng J, Takagi R, Saito C, Saku T (2012) Keratin 10-positive orthokeratotic dysplasia: a new leucoplakia-type precancerous entity of the oral mucosa. Histopathology 61(5):910–920

Krapfenbauer K, Drucker E, Thurnher D (2014) Identification of tumour-related proteins as potential screening markers by proteome analysis-protein profiles of human saliva as a predictive and prognostic tool. EPMA J 5(1), pp 20-5085-5-20. Ecollection 2014

Kulkarni V, Saranath D (2004) Concurrent hypermethylation of multiple regulatory genes in chewing tobacco associated oral squamous cell carcinomas and adjacent normal tissues. Oral Oncol 40:145–153

Kumeda N, Ogawa Y, Akimoto Y, Kawakami H, Tsujimoto M, Yanoshita R (2017) Characterization of membrane integrity and morphological stability of human salivary exosomes. Biol Pharm Bull 40(8):1183–1191

Langie SAS, Moisse M, Declerck K, Koppen G, Godderis L, Vanden Berghe W et al (2017a) Salivary DNA methylation profiling: aspects to consider for biomarker identification. Basic Clin Pharmacol Toxicol 121(Suppl 3):93–101

Langie SAS, SzarcVelSzic K, Declerck K, Traen S, Koppen G, Van Camp G et al (2017b) Correction: whole-genome saliva and blood DNA methylation profiling in individuals with a respiratory allergy. Plos One 12(8):e0183088

Larsen KR, Johansen JD, Reibel J, Zachariae C, Rosing K, Pedersen AML (2017) Oral symptoms and salivary findings in oral lichen planus, oral lichenoid lesions and stomatitis. BMC Oral Health 17(1):2

Laytragoon-Lewin N, Chen F, Castro J, Elmberger G, Rutqvist LE, Lewin F, Turesson I, Lundgren J (2010) DNA content and methylation of p16, DAPK and RASSF1A gene in tumour and distant, normal mucosal tissue of head and neck squamous cell carcinoma patients. Anticancer Res 30(11):4643–4648

Lee Y, El Andaloussi S, Wood MJA (2012a) Exosomes and microvesicles: extracellular vesicles for genetic information transfer and gene therapy. Hum Mol Genet 21(R1):125

Lee YH, Kim JH, Zhou H, Kim BW, Wong DT (2012b) Salivary transcriptomic biomarkers for detection of ovarian cancer: for serous papillary adenocarcinoma. J Mol Med (Berl) 90(4):427–434

Lee WH, Chen HM, Yang SF, Liang C, Peng CY, Lin FM, Tsai LL, Wu BC, Hsin CH, Chuang CY, Yang T, Yang TL, Ho SY, Chen WL, Ueng KC, Huang HD, Huang CN, Jong YJ (2017) Bacterial alterations in salivary microbiota and their association in oral cancer. Sci Rep 7(1):16540-017-16418-x

Lengacher CA, Reich RR, Paterson CL, Shelton M, Shivers S, Ramesar S et al (2018) A large randomized trial: effects of mindfulness-based stress reduction (MBSR) for breast cancer (BC) survivors on salivary cortisol and IL-6. Biol Res Nurs 21(1):39–49

Leon SA, Ehrlich GE, Shapiro B, Labbate VA (1977) Free DNA in the serum of rheumatoid arthritis patients. J Rheumatol 4(2):139–143

Li B, Wang B, Niu LJ, Jiang L, Qiu CC (2011) Hypermethylation of multiple tumor-related genes associated with DNMT3b up-regulation served as a biomarker for early diagnosis of esophageal squamous cell carcinoma. Epigenetics 6:307–316

Li L, Li C, Wang S, Wang Z, Jiang J, Wang W, Li X, Chen J, Liu K, Li C, Zhu G (2016) Exosomes derived from hypoxic oral squamous cell carcinoma cells deliver miR-21 to normoxic cells to elicit a prometastatic phenotype. Cancer Res 76(7):1770–1780

Liberti MV, Locasale JW (2016) The Warburg effect: how does it benefit cancer cells? Trends Biochem Sci 41(3):211–218

Lim Y, Sun CX, Tran P, Punyadeera C (2016a) Salivary epigenetic biomarkers in head and neck squamous cell carcinomas. Biomark Med 10(3):301–313. https://doi.org/10.2217/bmm.16.2

Lim Y, Wan Y, Vagenas D, Ovchinnikov D, Perry C, Davis M, Punyadeera C (2016b) Salivary DNA methylation panel to diagnose HPV-positive and HPV-negative head and neck cancers. BMC Cancer 16:749

Liu C, Yu S, Zinn K, Wang J, Zhang L, Jia Y, Kappes JC, Barnes S, Kimberly RP, Grizzle WE, Zhang HG (2006) Murine mammary carcinoma exosomes promote tumor growth by suppression of NK cell function. J Immunol (Baltimore, Md: 1950) 176(3):1375–1385

Liu CJ, Lin SC, Yang CC, CHENG HW, Chang KW (2012a) Exploiting salivary miR-31 as a clinical biomarker of oral squamous cell carcinoma. Head Neck 34(2):219–224

Liu Y, Zhou ZT, He QB, Jiang WW (2012b) DAPK promoter hypermethylation in tissues and body fluids of oral precancer patients. Medical Oncology (Northwood, London, England) 29(2):729–733

Liu K, Zhang Y, Zhang C, Zhang Q, Li J, Xiao F, Li Y, Zhang R, Dou D, Liang J, Qin J, Lin Z, Zhao D, Jiang M, Liang Z, Su J, Gupta VP, He M, Yang X (2016) Methylation of S100A8 is a promising diagnosis and prognostic marker in hepatocellular carcinoma. Oncotarget 7(35):56798–56810

Lodi G, Scully C, Carrozzo M, Griffiths M, Sugerman PB, Thongprasom K (2005) Current controversies in oral lichen planus: report of an international consensus meeting. Part 2. Clinical management and malignant transformation. Oral Surg Oral Med Oral Pathol Oral Radiol Endod 100(2):164–178

Logozzi M, De Milito A, Lugini L, Borghi M, Calabrò L, Spada M, Perdicchio M, Marino ML, Federici C, Iessi E, Brambilla D, Venturi G, Lozupone F, Santinami M, Huber V, Maio M, Rivoltini L, Fais S (2009) High levels of exosomes expressing CD63 and caveolin-1 in plasma of melanoma patients. PLoS One 4(4):E5219

Lopez M, Aguirre JM, Cuevas N, Anzola M, Videgain J, Aguirregaviria J, Martinez de Pancorbo M (2003) Gene promoter hypermethylation in oral rinses of leucoplakia patients–A diagnostic and/or prognostic tool? Eur J Cancer (Oxford, England: 1990) 39(16):2306–2309

Loyo M, Brait M, Kim MS, Ostrow KL, Jie CC, Chuang AY, Califano JA, Liégeois NJ, Begum S, Westra WH, Hoque MO, Tao Q, Sidransky D (2011) A survey of methylated candidate tumor suppressor genes in nasopharyngeal carcinoma. Int J Cancer 128(6):1393–1403

Ludwig JA, Weinstein JN (2005) Biomarkers in cancer staging, prognosis and treatment selection. Nat Rev Cancer 5(11):845

Ma H, Wu Y, Yang H, Liu J, Dan H, Zeng X, Zhou Y, Jiang L, Chen Q (2016) MicroRNAs in oral lichen planus and potential miRNA-mRNA pathogenesis with essential cytokines: a review. Oral Surg Oral Med Oral Pathol Oral Radiol 122(2):164–173

Maeda G, Chiba T, Aoba T, Imai K (2007) Epigenetic inactivation of E-cadherin by promoter hypermethylation in oral carcinoma cells. Odontology/the Society of the Nippon Dental University 95(1):24–29

Mager DL, Haffajee AD, Devlin PM, Norris CM, Posner MR, Goodson JM (2005) The salivary microbiota as a diagnostic indicator of oral cancer: a descriptive, non-randomized study of cancer-free and oral squamous cell carcinoma subjects. J Transl Med 3:27-5876-3-27

Malathi N, Mythili S, Vasanthi HR (2014) Salivary diagnostics: a brief review. ISRN Dent 2014

Markopoulou S, Nikolaidis G, Liloglou T (2012) DNA methylation biomarkers in biological fluids for early detection of respiratory tract cancer. Clin Chem Lab Med 50(10):1723–1731. https://doi.org/10.1515/cclm-2012-0124

Martone T, Gillio-Tos A, De Marco L, Fiano V, Maule M, Cavalot A, Garzaro M, Merletti F, Cortesina G (2007) Association between hypermethylated tumor and paired surgical margins in head and neck squamous cell carcinomas. Clin Cancer Res 13:5089–5094

Maruya S, Issa JP, Weber RS, Rosenthal DI, Haviland JC, Lotan R, El-Naggar AK (2004) Differential methylation status of tumor-associated genes in head and neck squamous carcinoma: incidence and potential implications. Clin Cancer Res 10:3825–3830

Matse JH, Yoshizawa J, Wang X, Elashoff D, Bolscher JG, Veerman EC, Leemans CR, Pegtel DM, Wong DT, Bloemena E (2015) Human salivary micro-RNA in patients with parotid salivary gland neoplasms. PLoS One 10(11):e0142264

Matse JH, Yoshizawa J, Wang X, Elashoff D, Bolscher JGM, Veerman ECI et al (2016) Correction: human salivary micro-RNA in patients with parotid salivary gland neoplasms. PLoS One 11(1):e0146701

Matthews AM, Kaur H, Dodd M, D'Souza J, Liloglou T, Shaw RJ, Risk JM (2013) Saliva collection methods for DNA biomarker analysis in oral cancer patients. Br J Oral Maxillofac Surg 51(5):394–398

Mcpherson RA, Pincus MR (2017) Henry's clinical diagnosis and management by laboratory methods E-book Elsevier Health Sciences

Mehdipour M, Shahidi M, Manifar S, Jafari S, Mashhadi Abbas F, Barati M, Mortazavi H, Shirkhoda M, Farzanegan A, ElmiRankohi Z (2018) Diagnostic and prognostic relevance of salivary microrna-21, −125a, −31 and -200a levels in patients with oral lichen planus – a short report. Cell Oncol (Dordr) 41(3):329–334

Michailidou E, Tzimagiorgis G, Chatzopoulou F, Vahtsevanos K, Antoniadis K, Kouidou S, Markopoulos A, Antoniades D (2016) Salivary mRNA markers having the potential to detect oral squamous cell carcinoma segregated from oral leucoplakia with dysplasia. Cancer Epidemiol 43:112–118

Minor J, Wang X, Zhang F, Song J, Jimeno A, Wang XJ, Lu X, Gross N, Kulesz-Martin M, Wang D, Lu SL (2012) Methylation of microRNA-9 is a specific and sensitive biomarker for oral and oropharyngeal squamous cell carcinomas. Oral Oncol 48(1):73–78

Momen-Heravi F, Trachtenberg AJ, Kuo WP, Cheng YS (2014) Genomewide study of salivary microRNAs for detection of oral cancer. J Dent Res 93(7 suppl):93S

Nagler RM (2009) Saliva as a tool for oral cancer diagnosis and prognosis. Oral Oncol 45(12):1006–1010

Nagata S, Hamada T, Yamada N, Yokoyama S, Kitamoto S, Kanmura Y, Nomura M, Kamikawa Y, Yonezawa S, Sugihara K (2012) Aberrant DNA methylation of tumor-related genes in oral rinse: a noninvasive method for detection of oral squamous cell carcinoma. Cancer 118(17):4298–4308

Nair S, Tang KD, Kenny L, Punyadeera C (2018) Salivary exosomes as potential biomarkers in cancer. Oral Oncol 84:31–40

Nakahara Y, Shintani S, Mihara M, Ueyama Y, Matsumura T (2001) High frequency of homozygous deletion and methylation of p16(INK4A) gene in oral squamous cell carcinomas. Cancer Lett 163:221–228

Nakayama S, Sasaki A, Mese H, Alcalde RE, Tsuji T, Matsumura T (2001) The E-cadherin gene is silenced by CpG methylation in human oral squamous cell carcinomas. Int J Cancer 93(5):667–673

Nawaz I, Moumad K, Martorelli D, Ennaji MM, Zhou X, Zhang Z, Dolcetti R, Khyatti M, Ernberg I, Hu LF (2015) Detection of nasopharyngeal carcinoma in Morocco (North Africa) using a multiplex methylation-specific PCR biomarker assay. Clin Epigenetics 7:89

Nosratzehi T, Alijani E, Moodi M (2017) Salivary MMP-1, MMP-2, MMP-3 and MMP-13 levels in patients with Oral lichen planus and squamous cell carcinoma. Asian Pac J Cancer Prev 18(7):1947–1951

Nylander E, Ebrahimi M, Wahlin YB, Boldrup L, Nylander K (2012) Changes in miRNA expression in sera and correlation to duration of disease in patients with multifocal mucosal lichen planus. J Oral Pathol Med 41(1):86–89

O'Loughlin AJ, Woffindale CA, Wood JAM (2012) Exosomes and the emerging field of exosome-based gene therapy. Curr Gene Ther 12(4):262–274

Ogawa Y, Kanai-Azuma M, Akimoto Y, Kawakami H, Yanoshita R (2008) Exosome-like vesicles with dipeptidyl peptidase IV in human saliva. Biol Pharm Bull 31(6):1059–1062

Ogawa Y, Taketomi Y, Murakami M, Tsujimoto M, Yanoshita R (2013) Small RNA transcriptomes of two types of exosomes in human whole saliva determined by next generation sequencing. Biol Pharm Bull 36(1):66–75

Ogawa Y, Tsujimoto M, Yanoshita R (2016) Next-generation sequencing of protein-coding and long non-protein-coding RNAs in two types of exosomes derived from human whole saliva. Biol Pharm Bull 39(9):1496–1507

Okami K, Imate Y, Hashimoto Y, Kamada T, Takahashi M (2002) Molecular detection of cancer cells in saliva from oral and pharyngeal cancer patients. Tokai J Exp Clin Med 27(3):85–89

Ovchinnikov DA, Cooper MA, Pandit P, Coman WB, Cooper-White JJ, Keith P, Wolvetang EJ, Slowey PD, Punyadeera C (2012) Tumor-suppressor gene promoter Hypermethylation in saliva of head and neck cancer patients. Transl Oncol 5(5):321–326. Epub 2012 Oct 1

Ovchinnikov DA, Wan Y, Coman WB, Pandit P, Cooper-White JJ, Herman JG, Punyadeera C (2014) DNA methylation at the novel CpG sites in the promoter of MED15/PCQAP gene as a biomarker for head and neck cancers. Biomark Insights 9:53–60. https://doi.org/10.4137/BMI.S16199

Panneer Selvam N, Sadaksharam J (2015) Salivary interleukin-6 in the detection of oral cancer and precancer. Asia Pac J Clin Oncol 11(3):236–241

Park NJ, Zhou H, Elashoff D, Henson BS, Kastratovic DA, Abemayor E, Wong DT (2009) Salivary microRNA: discovery, characterization, and clinical utility for oral cancer detection. Clin Cancer Res 15(17):5473–5477

Patel S, Metgud R (2015) Estimation of salivary lactate dehydrogenase in oral leukoplakia and oral squamous cell carcinoma: a biochemical study. J Cancer Res Ther 11(1):119–123

Pattani KM, Zhang Z, Demokan S, Glazer C, Loyo M, Goodman S, Sidransky D, Bermudez F, Jean-Charles G, McCaffrey T, Padhya T, Phelan J, Spivakovsky S, Bowne HY, Goldberg JD, Rolnitzky L, Robbins M, Kerr AR, Sirois D, Califano JA (2010) Endothelin receptor type B gene promoter hypermethylation in salivary rinses is independently associated with risk of oral cavity cancer and premalignancy. Cancer Prev Res (Phila) 3(9):1093–1103

Peinado H, Lavotshkin S, Lyden D (2011) The secreted factors responsible for pre-metastatic niche formation: old sayings and new thoughts. Semin Cancer Biol 21:139–146

Peinado H, Alečković M, Lavotshkin S, Matei I, Costa-Silva B, Moreno-Bueno G, Hergueta-Redondo M, Williams C, García-Santos G, Ghajar CM (2012) Melanoma exosomes educate bone marrow progenitor cells toward a pro-metastatic phenotype through MET. Nat Med 18(6):883–891

Piyathilake CJ, Bell WC, Jones J, Henao OL, Heimburger DC, Niveleau A, Grizzle WE (2005) Patterns of global DNA and histone methylation appear to be similar in normal, dysplastic and neoplastic oral epithelium of humans. Dis Markers 21(3):147–151

Properzi F, Logozzi M, Fais S (2013) Exosomes: the future of biomarkers in medicine. Biomarkers 7(5):769–778

Punyani SR, Sathawane RS (2013) Salivary level of interleukin-8 in oral precancer and oral squamous cell carcinoma. Clin Oral Investig 17(2):517–524

Pushalkar S, Mane SP, Ji X, Li Y, Evans C, Crasta OR, Morse D, Meagher R, Singh A, Saxena D (2011) Microbial diversity in saliva of oral squamous cell carcinoma. FEMS Immunol Med Microbiol 61(3):269–277

Puttipanyalears C, Subbalekha K, Mutirangura A, Kitkumthorn N (2013) Alu hypomethylation in smoke-exposed epithelia and oral squamous carcinoma. Asian Pac J Cancer Prev 14(9):5495–5501

Record M, Carayon K, Poirot M, Silvente-Poirot S (2014) Exosomes as new vesicular lipid transporters involved in cell-cell communication and various pathophysiologies. Biochim Biophys Acta 1841(1):108–120

Rettori MM, de Carvalho AC, Bomfim Longo AL, de Oliveira CZ, Kowalski LP, Carvalho AL, Vettore AL (2013a) Prognostic significance of TIMP3 hypermethylation in post-treatment salivary rinse from head and neck squamous cell carcinoma patients. Carcinogenesis 34(1):20–27

Rettori MM, de Carvalho AC, Longo AL, de Oliveira CZ, Kowalski LP, Carvalho AL, Vettore AL (2013b) TIMP3 and CCNA1 hypermethylation in HNSCC is associated with an increased incidence of second primary tumors. J Transl Med 11:316

Rhodus NL, Cheng B, Myers S, Bowles W, Ho V, Ondrey F (2005) A comparison of the pro-inflammatory, NF-kappab-dependent cytokines: TNF-alpha, IL-1-alpha, IL-6, and IL-8 in different oral fluids from oral lichen planus patients. Clin Immunol (Orlando, Fla) 114(3):278–283

Rhodus NL, Cheng B, Bowles W, Myers S, Miller L, Ondrey F (2006) Proinflammatory cytokine levels in saliva before and after treatment of (erosive) oral lichen planus with dexamethasone. Oral Dis 12(2):112–116

Righini CA, de Fraipont F, Timsit JF, Faure C, Brambilla E, Reyt E, Favrot MC (2007) Tumor-specific methylation in saliva: a promising biomarker for early detection of head and neck cancer recurrence. Clin Cancer Res 13:1179–1185

Robinson PN, Mickelson AR (2006) Early diagnosis of oral cavity cancers. Otolaryngol Clin N Am 39(2):295–306

Rosas SL, Koch W, da Costa Carvalho MG, Wu L, Califano J, Westra W, Jen J, Sidransky D (2001) Promoter hypermethylation patterns of p16, O6-methylguanine-DNA-methyltransferase, and death-associated protein kinase in tumors and saliva of head and neck cancer patients. Cancer Res 61:939–942

Saito Y, Takazawa H, Uzawa K, Tanzawa H, Sato K (1998) Reduced expression of E-cadherin in oral squamous cell carcinoma: relationship with DNA methylation of 5′ CpG island. Int J Oncol 12(2):293–298

Sanchez-Cespedes M, Esteller M, Wu L, Nawroz-Danish H, Yoo GH, Koch WM, Jen J, Herman JG, Sidransky D (2000) Gene promoter hypermethylation in tumors and serum of head and neck cancer patients. Cancer Res 60:892–895

Sansone P, Savini C, Kurelac I, Chang Q, Amato LB, Strillacci A, Stepanova A, Iommarini L, Mastroleo C, Daly L, Galkin A, Thakur BK, Soplop N, Uryu K, Hoshino A, Norton L, Bonafé M, Cricca M, Gasparre G, Lyden D, Bromberg J (2017) Packaging and transfer of mitochondrial DNA via exosomes regulate escape from dormancy in hormonal therapy-resistant breast cancer. Proc Natl Acad Sci U S A 114(43):E9075

Schafer CA, Schafer JJ, Yakob M, Lima P, Camargo P, Wong DT (2014) Saliva diagnostics: utilizing oral fluids to determine health status. Monogr Oral Sci 24:88–98

Schnipper LE, Davidson NE, Wollins DS, Tyne C, Blayney DW, Blum D, American Society of Clinical Oncology, Null (2015). American society of clinical oncology statement: a conceptual framework to assess the value of cancer treatment options. J Clin Oncol Off J Am Soc Clin Oncol 33(23):2563–2577

Schulz BL, Cooper-White J, Punyadeera CK (2013) Saliva proteome research: current status and future outlook. Crit Rev Biotechnol 33(3):246–259

Schussel J, Zhou XC, Zhang Z, Pattani K, Bermudez F, Jean-Charles G, McCaffrey T, Padhya T, Phelan J, Spivakovsky S, Brait M, Li R, Bowne HY, Goldberg JD, Rolnitzky L, Robbins M, Kerr AR, Sirois D, Califano JA (2013) EDNRB and DCC salivary rinse hypermethylation has a similar performance as expert clinical examination in discrimination of oral cancer/dysplasia versus benign lesions. Clin Cancer Res 19(12):3268–3275

Shahidi M, Jafari S, Barati M, Mahdipour M, Gholami MS (2017) Predictive value of salivary microrna-320a, vascular endothelial growth factor receptor 2, CRP and IL-6 in Oral lichen planus progression. Inflammopharmacology

Shapiro B, Chakrabarty M, Cohn EM, Leon SA (1983) Determination of circulating DNA levels in patients with benign or malignant gastrointestinal disease. Cancer 51(11):2116–2120

Sharma M, Bairy I, Pai K, Satyamoorthy K, Prasad S, Berkovitz B, Radhakrishnan R (2011) Salivary IL-6 levels in oral leucoplakia with dysplasia and its clinical relevance to tobacco habits and periodontitis. Clin Oral Investig 15(5):705–714

Shaw RJ, Liloglou T, Rogers SN, Brown JS, Vaughan ED, Lowe D, Field JK, Risk JM (2006) Promoter methylation of P16, RARbeta, E-cadherin, cyclin A1 and cytoglobin in oral cancer: quantitative evaluation using pyrosequencing. Br J Cancer 94:561–568

Shetty SR, Chadha R, Babu S, Kumari S, Bhat S, Achalli S (2012) Salivary lactate dehydrogenase levels in oral leucoplakia and oral squamous cell carcinoma: a biochemical and clinicopathological study. J Cancer Res Ther 8(Suppl 1):123

Shetty SR, Babu S, Kumari S, Shetty P, Hegde S, Karikal A (2015) Status of trace elements in saliva of oral precancer and oral cancer patients. J Cancer Res Ther 11(1):146–149

Shigeyama H, Wang T, Ichinose M, Ansai T, Lee SW (2018) Identification of volatile metabolites in human saliva from patients with oral squamous cell carcinoma via zeolite-based thin-film microextraction coupled with GC-MS. J Chromatogr B Analyt Technol Biomed Life Sci 1104:49–58

Shiiki N, Tokuyama S, Sato C, Kondo Y, Saruta J, Mori Y, Shiiki K, Miyoshi Y, Tsukinoki K (2011) Association between saliva PSA and serum PSA in conditions with prostate adenocarcinoma. Biomarkers 16(6):498–503

Siegel RL, Miller KD, Jemal A (2017) Cancer statistics, 2017. CA Cancer J Clin 67(1):7–30

Simkin M, Abdalla M, El-Mogy M, Haj-Ahmad Y (2012) Differences in the quantity of DNA found in the urine and saliva of smokers versus nonsmokers: implications for the timing of epigenetic events. Epigenomics 4(3):343–352

Singh NN, Peer A, Nair S, Chaturvedi RK (2016) Epigenetics: a possible answer to the undeciphered etiopathogenesis and behavior of oral lesions. J Oral Maxillof Pathol: JOMFP 20(1):122–128

Smith AK, Kilaru V, Klengel T, Mercer KB, Bradley B, Conneely KN, Binder EB (2015) DNA extracted from saliva for methylation studies of psychiatric traits: evidence tissue specificity and relatedness to brain. American journal of medical genetics. Part B, neuropsychiatric genetics: the official publication of the international society of. Psychiatr Genet 168B(1):36–44

Smith RA, Andrews KS, Brooks D, Fedewa SA, Manassaram-Baptiste D, Saslow D et al (2017) Cancer screening in the United States, 2017: a review of current American cancer society guidelines and current issues in cancer screening. CA Cancer J Clin 67(2):100–121

Spielmann N, Wong DT (2011) Saliva: diagnostics and therapeutic perspectives. Oral Dis 17(4):345–354

Steinmann K, Sandner A, Schagdarsurengin U, Dammann RH (2009) Frequent promoter hypermethylation of tumor-related genes in head and neck squamous cell carcinoma. Oncol Rep 22(6):1519–1526

Stewart CM, Tsui DWY (2018) Circulating cell-free DNA for non-invasive cancer management. Cancer Genet 228–229:169–179

Streckfus C, Bigler L, Dellinger T, Dai X, Kingman A, Thigpen JT (2000) The presence of soluble c-erbB-2 in saliva and serum among women with breast carcinoma: a preliminary study. Clin Cancer Res 6(6):2363–2370

Su PF, Huang WL, Wu HT, Wu CH, Liu TY, Kao SY (2010) p16(INK4A) promoter hypermethylation is associated with invasiveness and prognosis of oral squamous cell carcinoma in an age-dependent manner. Oral Oncol 46(10):734–739

Tajmul M, Parween F, Singh L, Mathur SR, Sharma JB, Kumar S, Sharma DN, Yadav S (2018) Identification and validation of salivary proteomic signatures for non-invasive detection of ovarian cancer. Int J Biol Macromol 108:503–514

Tawfik HM, El-Maqsoud NM, Hak BH, El-Sherbiny YM (2011) Head and neck squamous cell carcinoma: mismatch repair immunohistochemistry and promoter hypermethylation of hMLH1 gene. Am J Otolaryngol 32(6):528–536

Tian F, Yip SP, Kwong DL, Lin Z, Yang Z, Wu VW (2013) Promoter hypermethylation of tumor suppressor genes in serum as potential biomarker for the diagnosis of nasopharyngeal carcinoma. Cancer Epidemiol 37(5):708–713

Tiziani S, Lopes V, Gunther UL (2009) Early stage diagnosis of oral cancer using 1H NMR-based metabolomics. Neoplasia (New York, NY) 11(3): 269–276, 4p following 269

Tkach M, Théry C (2016) Communication by extracellular vesicles: where we are and where we need to go. Cell 164(6):1226–1232

Tong JH, Tsang RK, Lo KW, Woo JK, Kwong J, Chan MW, Chang AR, van Hasselt CA, Huang DP, To KF (2002) Quantitative Epstein-Barr virus DNA analysis and detection of gene promoter hypermethylation in nasopharyngeal (NP) brushing samples from patients with NP carcinoma. Clin Cancer Res 8:2612–2619

Topkas E, Keith P, Dimeski G, Cooper-White J, Punyadeera C (2012) Evaluation of saliva collection devices for the analysis of proteins. Clinica Chimica Acta; International Journal of Clinical Chemistry 413(13–14):1066–1070

Tvarijonaviciute A, Aznar-Cayuela C, Rubio CP, Ceron JJ, Lopez-Jornet P (2017) Evaluation of salivary oxidate stress biomarkers, nitric oxide and C-reactive protein in patients with oral lichen planus and burning mouth syndrome. J Oral Pathol Med 46(5):387–392

Tvarijonaviciute A, Aznar-Cayuela C, Rubio CP, Tecles F, Ceron JJ, Lopez-Jornet P (2018) Salivary antioxidant status in patients with Oral lichen planus: correlation with clinical signs and evolution during treatment with Chamaemelum nobile. Biomed Res Int 2018:5187549

van der Waal I (2014 July 01) Oral potentially malignant disorders: is malignant transformation predictable and preventable? Med Oral Patol Oral Cir Bucal 19(4):386

Valadi H, Ekström K, Bossios A, Sjöstrand M, Lee JJ, Lötvall JO (2007a) Exosome-mediated transfer of mRNAs and microRNAs is a novel mechanism of genetic exchange between cells. Nat Cell Biol 9(6):654–659

Valadi H, Ekström K, Bossios A, Sjöstrand M, Lee JJ, Lötvall JO (2007b) Exosome-mediated transfer of mRNAs and microRNAs is a novel mechanism of genetic exchange between cells. Nat Cell Biol 9(6):654

Van Ginkel JH, Slieker FJB, de Bree R, van Es RJJ, Van Cann EM, Willems SM (2017) Cell-free nucleic acids in body fluids as biomarkers for the prediction and early detection of recurrent head and neck cancer: a systematic review of the literature. Oral Oncol 75:8–15

Vento S, Tanko MN (2009) The bacterium that could cause cancer. Lancet Oncology 10(5):528–2045(09). 70038-5

Viet CT, Schmidt BL (2008) Methylation array analysis of preoperative and postoperative saliva DNA in oral cancer patients. Cancer Epidemiol Biomark Prev 17(12):3603–3611

Wang Q, Gao P, Cheng F, Wang X, Duan Y (2014) Measurement of salivary metabolite biomarkers for early monitoring of oral cancer with ultra performance liquid chromatography-mass spectrometry. Talanta 119:299–305

Wang Q, Yu Q, Lin Q, Duan Y (2015a) Emerging salivary biomarkers by mass spectrometry. Clinicachimica Acta Int J Clin Chem 438:214–221

Wang Y, Springer S, Mulvey CL, Silliman N, Schaefer J, Sausen M, James N, Rettig EM, Guo T, Pickering CR, Bishop JA, Chung CH, Califano JA, Eisele DW, Fakhry C, Gourin CG, Ha PK, Kang H, Kiess A, Koch WM, Myers JN, Quon H, Richmon JD, Sidransky D, Tufano RP, Westra WH, Bettegowda C, Diaz LA Jr, Papadopoulos N, Kinzler KW, Vogelstein B, Agrawal N (2015b) Detection of somatic mutations and HPV in the saliva and plasma of patients with head and neck squamous cell carcinomas. Sci Transl Med 7(293):293ra104

Warnakulasuriya S, Johnson NW, Van Der Waal I (2007) Nomenclature and classification of potentially malignant disorders of the oral mucosa. J Oral Pathol Med 36(10):575–580

Warnakulasuriya S, Kovacevic T, Madden P, Coupland VH, Sperandio M, Odell E, Moller H (2011) Factors predicting malignant transformation in oral potentially malignant disorders among patients accrued over a 10-year period in south East England. J Oral Pathol Med 40(9):677–683

Weber A, Wittekind C, Tannapfel A (2003) Genetic and epigenetic alterations of 9p21 gene products in benign and malignant tumors of the head and neck. Pathol Res Pract 199:391–397

Webber J, Steadman R, Mason MD, Tabi Z, Clayton A (2010) Cancer exosomes trigger fibroblast to myofibroblast differentiation. Cancer Res 70(23):9621–9630

Wei J, Xie G, Zhou Z, Shi P, Qiu Y, Zheng X, Chen T, Su M, Zhao A, Jia W (2011) Salivary metabolite signatures of oral cancer and leucoplakia. Int J Cancer 129(9):2207–2217

Weiss D, Basel T, Sachse F, Braeuninger A, Rudack C (2011) Promoter methylation of cyclin A1 is associated with human papillomavirus 16 induced head and neck squamous cell carcinoma independently of p53 mutation. Mol Carcinog 50(9):680–688

Wilhelm M, Schlegl J, Hahne H, Gholami AM, Lieberenz M, Savitski MM, Ziegler E, Butzmann L, Gessulat S, Marx H, Mathieson T, Lemeer S, Schnatbaum K, Reimer U, Wenschuh H, Mollenhauer M, Slotta-Huspenina J, Boese JH, Bantscheff M, Gerstmair A, Faerber F, Kuster B (2014) Mass-spectrometry-based draft of the human proteome. Nature 509(7502):582–587

Wong TS, Man MW, Lam AK, Wei WI, Kwong YL, Yuen AP (2003) The study of p16 and p15 gene methylation in head and neck squamous cell carcinoma and their quantitative evaluation in plasma by real-time PCR. Eur J Cancer 39:1881–1887

Wu CC, Chu HW, Hsu CW, Chang KP, Liu HP (2015) Saliva proteome profiling reveals potential salivary biomarkers for detection of oral cavity squamous cell carcinoma. Proteomics 15(19):3394–3404

Xiao H, Zhang L, Zhou H, Lee JM, Garon EB, Wong DTW (2012) Proteomic analysis of human saliva from lung Cancer patients using two-dimensional difference gel electrophoresis and mass spectrometry. Mol Cell Proteomics 11(2):M111.012112

Xie H, Onsongo G, Popko J, de Jong EP, Cao J, Carlis JV, Griffin RJ, Rhodus NL, Griffin TJ (2008) Proteomics analysis of cells in whole saliva from oral cancer patients via value-added three-dimensional peptide fractionation and tandem mass spectrometry. Mol Cell Proteomics 7(3):486–498

Yan SK, Wei BJ, Lin ZY, Yang Y, Zhou ZT, Zhang WD (2008) A metabonomic approach to the diagnosis of oral squamous cell carcinoma, oral lichen planus and oral leucoplakia. Oral Oncol 44(5):477–483

Yang, Y., Li, Y.X., Yang, X., Jiang, L., Zhou, Z.J. And Zhu, Y.Q., 2013. Progress risk assessment of oral premalignant lesions with saliva miRNA analysis. BMC Cancer, 13, pp. 129
Yang Y, Rhodus NL, Ondrey FG, Wuertz BR, Chen X, Zhu Y, Griffin TJ (2014) Quantitative proteomic analysis of oral brush biopsies identifies secretory leukocyte protease inhibitor as a promising, mechanism-based oral cancer biomarker. PLoS One 9(4):E95389
You Y, Yang W, Wang Z, Zhu H, Li H, Lin C, Ran Y (2013) Promoter hypermethylation contributes to the frequent suppression of the CDK10 gene in human nasopharyngeal carcinomas. Cell Oncol (Dordr) 36(4):323–331
Yu JS, Chen YT, Chiang WF, Hsiao YC, Chu LJ, See LC, Wu CS, Tu HT, Chen HW, Chen CC, Liao WC, Chang YT, Wu CC, Lin CY, Liu SY, Chiou ST, Chia SL, Chang KP, Chien CY, Chang SW, Chang CJ, Young JD, Pao CC, Chang YS, Hartwell LH (2016) Saliva protein biomarkers to detect oral squamous cell carcinoma in a high-risk population in Taiwan. Proc Natl Acad Sci USA 113(41):11549–11554
Zahran F, Ghalwash D, Shaker O, Al-Johani K, Scully C (2015) Salivary micrornas in oral cancer. Oral Dis 21(6):739–747
Zhang Z, Sun D, Hutajulu SH, Nawaz I, Nguyen Van D, Huang G, Haryana SM, Middeldorp JM, Ernberg I, Hu LF (2012a) Development of a non-invasive method, multiplex methylation specific PCR (MMSP), for early diagnosis of nasopharyngeal carcinoma. PLoS One 7(11):e45908
Zhang L, Xiao H, Zhou H, Santiago S, Lee JM, Garon EB, Yang J, Brinkmann O, Yan X, Akin D, Chia D, Elashoff D, Park NH, Wong DTW (2012b) Development of transcriptomic biomarker signature in human saliva to detect lung cancer. Cell Mol Life Sci 69(19):3341–3350
Zhou C, Shen Z, Ye D, Li Q, Deng H, Liu H, Li J (2018) The association and clinical significance of CDKN2A promoter methylation in head and neck squamous cell carcinoma: a meta-analysis. Cell Physiol Biochem 50(3):868–882
Zlotogorski-Hurvitz A, Dayan D, Chaushu G, Salo T, Vered M (2016) Morphological and molecular features of oral fluid-derived exosomes: oral cancer patients versus healthy individuals. J Cancer Res Clin Oncol 142(1):101–110

Chapter 13
Saliva in Sport Sciences

Jose J. Cerón

13.1 Introduction and General Concepts

Saliva is a biological sample that due to its easy and simple collection can have a wide potential application in sport sciences to evaluate different biomarkers. There are previous excellent reviews in which the possible applications of saliva in sport sciences have been indicated (Lindsay and Costello 2017; Papacosta and Nassis 2011). In this chapter we will review the main biomarkers of potential use in sport that can be measured in saliva grouped in biomarkers of stress, immune system and inflammation, muscle damage, anaerobic metabolism and oxidative status (Fig. 13.1). For each biomarker we will provided general ideas and updated information that can help to gain knowledge about its interpretation and possible applications in sport sciences.

It is important to indicate that as Lindsay and Costello (2017) pointed out, there are four main forms to express the results of salivary biomarkers in exercise studies:

1. Absolute concentration (i.e. g or mmol/L) or activities in case of enzymes (i.e. IU/L)
2. Corrected by Flow rate (i.e. g or mmol/min), that is obtained by multiplying the unit of point 1 by the volume obtained by unit of time, for example during 1 min of sampling. By this correction the changes in flow rate are explored, that can be due to psychological causes and related to parasympathetic withdrawal. As will be later mentioned, this has been used in case of IgA, in which a study reported a 50% of decrease in secretion rate but no changes in absolute concentration after a 160 km run.

J. J. Cerón (✉)
Interdisciplinary Laboratory of Clinical Analysis (Interlab-UMU), Regional Campus of International Excellence Campus Mare Nostrum, University of Murcia, Murcia, Spain
e-mail: jjceron@um.es

A. Tvarijonaviciute et al. (eds.), *Saliva in Health and Disease*,
https://doi.org/10.1007/978-3-030-37681-9_13

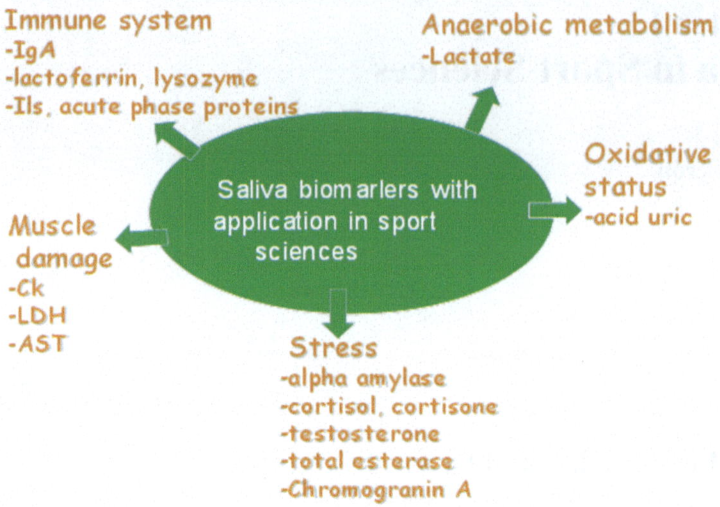

Fig. 13.1 Main biomarkers that can be measured in saliva with application in sport sciences, grouped by the biological information that can provide

3. Corrected by protein content (i.e. g or mmol/mg protein). This has been used in some reports but there are authors who indicated that it is not adequate to correct the values by concentration (Blannin et al. 1998)
4. Concentration relative to saliva osmolality (mg/mOsm). This has been recommended for IgA measurements (Sari-Sarraf et al. 2007), although further studies should be made about the real utility of this correction.

13.2 Biomarkers of Stress

13.2.1 Alpha Amylase

Salivary alpha-amylase (sAA; EC 3.2.1.1) is secreted by the parotid gland in response to adrenergic activity and increases in psychological and physical stress situations. Although it can be measured by concentration, usually the measurement of sAA activity is preferred for the evaluation of sAA in exercise. The activity is expressed without any correction by total protein or flow, since this can lead to errors in its interpretation (Contreras Aguilar et al. 2017).

Koubuchi and Suzuki (2014) reviewed the situations in sport science that can increase amylase and indicate that sAA increases tended to be more pronounced at exercise intensities >70% VO_2max in healthy young individuals. Another evidence of the relation between sAA and exercise intensity is the fact that amylase in saliva might be used as method for determining the anaerobic threshold in exercise. Being the anaerobic threshold defined as the oxygen consumption above which aerobic energy production is supplemented by anaerobic mechanisms, causing a sustained increase in lactate and metabolic acidosis (Calvo et al. 1997).

Although further research about possible factors that can influence sAA in exercise and also possible applications of sAA should be performed in the future, this enzyme was higher in response to competition in experienced compared to novice athletes, and was positively associated with performance and interest in team-bonding (Kivlighan and Granger 2006). However, increases in alpha-amylase correlated with a decrease in physical activity has been described in older adults (Strahler et al. 2010). Regarding possible new applications of this enzyme, recently the increases in sAA has been found after administration of probiotic supplement in rugby players suggesting a possible role as a host defence peptide (Pumpa et al. 2019).

13.2.2 Cortisol

Cortisol is secreted by the adrenals after activation of the hypothalamus-hypophysis due to any stressor stimulus. Cortisol in saliva is only in its free form, whereas in serum it is in its free form and also bound to proteins. Higher responses to exercise in cortisol in saliva than in serum have been described. This is because an exponential increase in free cortisol occurs when the binding capacity of the corticosteroid-binding globulin, which is the protein that binds cortisol in serum, is exceeded (Del Corral et al. 2016).

Usually it is described that salivary cortisol increases after exercise. However, in some sports, such as soccer, conflicting results have been obtained. Since some authors indicated an increase of salivary cortisol of 78% after a soccer game compared to basal values taken before, (Torphe and Sunderland 2012), whereas other did not find significant changes (Moreira et al. 2009). These differences could be possible related with different performances or training of the individuals of the study. Also the period of time in which the saliva is collected can influence, since for example it has been described that immediately after a rugby match there is an increase in cortisol, that returns to normal values in 4 h (Elloumi et al. 2003).

It is also important to consider the circadian influence of the cortisol. Therefore, in trials, it is recommended to do comparisons of cortisol with a control day, in which the same population is sampled at the same time without doing the exercise. However, in short duration studies (less than 60 min) cortisol can be compared with baseline (Del Corral et al. 2016).

In general cortisol in saliva can be used for:

- *Evaluate training.* Usually intensive training produces a short-term effect with increase in cortisol in saliva (Li et al. 2012). However, in situations of overtraining cortisol can be decreased (Filaire et al. 2013). For this purpose, it is very interesting the evaluation of alpha-amylase and cortisol together. For example, in cases of overtraining in tennis players, there is an asymmetry between the two markers, with a rise in the alpha-amylase awakening response and a higher alpha-amylase activity output, but a decrease in the overall output of salivary cortisol. Therefore, in these cases the ratio of sAA over cortisol increases, being this ratio in correlation with a decrease in the performance (Filaire et al. 2013). The increase in alpha-amylase in correlation with a decrease in physical activity found in older adults would be in line with these findings (Strahler et al. 2010).

- *Establish the optimal recovery time after an intense exercise.* This could be established by the time in which cortisol reaches its basal values before the exercise. For example, after a rugby match there is an increase in salivary cortisol that returned to basal values after 4 h and reach values lower than the basal concentrations, even until the day 5 after the match. It is possible that this long time is required to restore the break-down of homeostasis induced by the very hard mental and physical strain associated with a rugby match (Elloumi et al. 2003). In general, it could be indicated that decreased values of cortisol in comparison with basal levels would be indicative of overtraining or that the body is recovering from intensive efforts.
- As other possible applications of the cortisol, interestingly, a higher difference between values of cortisol from saliva collected at the morning, and the values of cortisol in saliva collected after a training test, is related with winning games in an international rugby union competition (Crewther et al. 2018).

In veterinary science, in horses that participated in exercise competitions, increases of cortisol in saliva were detected, possible due to the exposure to a novel environment compared to home. However, there was no consistent relationship between baseline salivary cortisol concentrations and competition scores (Munk et al. 2017).

13.2.3 Others Biomarkers of Stress

13.2.3.1 Cortisone

It is a metabolite of the cortisol being at higher concentration than cortisol in saliva and also increases more than cortisol after a physical exercise. In addition, compared to cortisol in saliva, it is more correlated with plasma total and plasma free cortisol (Del Corral et al. 2016).

13.2.3.2 Testosterone

In the section of cortisol, it was indicated that cortisol decreases until the day 5 after the rugby match were observed. However, for salivary testosterone measured for the same period of time, only slight decreases were observed just after the game, with further increases that resulted in values higher than the basal ones, at day 5 (Elloumi et al. 2003). Since in this trial cortisol was decreased until day 5, a high ration testosterone/cortisol could reflect a situation of recovery of the body.

13.2.3.3 Total Esterase

An assay for total esterase, being carbonic-anhydrase-VI the esterase that contribute more to the activity of the assay, has been developed. Total esterase in saliva increased after an indoor football match and positively correlated with salivary alpha-amylase (Tecles et al. 2016).

13.2.3.4 Chromogranin A

It has been described increased in chromogranin A, measured by western blott, after an incremental maximal exercise in swimmers, being this increase highly correlated with alpha-amylase and salivary lactate (Bocanegra et al. 2012). In another study in swimmers chromogranin A correlated positively with adrenalin (and therefore with the sympathetic activity), total proteins in saliva and with the intensity during a 21 week training season (Diaz-Gomez et al. 2013).

13.3 Biomarkers of the Immune System and Inflammation

13.3.1 Immunoglobulin A (IgA)

IgA is involved in the defense of the mucosal surfaces. There are two main controversies about its relation with exercise because:

- some authors did not find decreases in IgA after an acute exercise, for example a football game (Torphe and Sunderland 2012) and in general, low-moderate intensity exercises does not produce changes in IgA (Walsh et al. 1999). However, high intensity exercise such as a marathon, 50 km ski race or more than 2 h of cycling (Nieman et al. 2002; Tomasi et al. 1982; Walsh et al. 2002) produce decreases in salivary IgA, although these decreases are of lower magnitude in well trained individuals (Mackinnon and Hooper 1994). This data could indicate that IgA seems to decrease only in cases in which exercise is excessive or the athlete is not enough trained.
- another controversy is if the IgA decrease produced after exercise can be related or not with certain immunosuppression and decrease in performance. This is because some authors indicate that decreases in IgA are associated with an increase risk of infections (Svendsen et al. 2016). Maybe the cause of the controversy could be explained because that although an excessive exercise can

decrease IgA, this decrease can represent a risk of infections if: (1) the excessive exercise is maintained, (2) the athlete is prone to infection due to predisposition to a pro-inflammatory response and a dysregulated anti-inflammatory cytokine response to intense exercise.

IgA has been described to be affected by the secretion rate, since in an experiment consisting in running 160 km, there were no changes in total concentration in salivary IgA after the run, but a decrease of 50% was observed when IgA was expressed by secretion rate (amount/minute) (Nieman et al. 2003). Although it is interesting to consider the secretion rate, it is not recommended to correct IgA by total protein to take in consideration the possible effect of dehydration, since the changes in total protein are not related to evaporative loss of saliva water and they just increase after exercise (Walsh et al. 1999). Also it can be important in the interpretation of IgA to consider the time after exercise in which it is measured, since it has been described decreases in IgA few hours after exercise that are no longer observed after 2 days (Canto et al. 2018).

In general there are two major potential applications of IgA in exercise:

- *To identify situations of excessive training*, fact that could help to trainers to decrease or adjust the intensity of the training performed (Shephard and Shek 1998). Overall it could be stated that a lack of decrease in IgA would be a marker of adequate training sessions. Even there are some intensive training systems that can increase IgA in saliva and therefore augment the oral mucosal immunity (Antualpa et al. 2018).
- *To detect possible situations of risks of infection*, possible due to excessive exercise or due to the lack of IgA by other causes such as genetic condition. For example, risk of upper respiratory tract infection (URS) has been found to occur due to decrease of IgA concentrations after an intensive endurance training. In these situations, dietary supplementation with bovine colostrum, probiotics and selected antioxidants can reduce the incidence or severity of URS in some athletes (Gleeson and Pyne 2016). In this line, marathon runners who did not receive a polysaccharide-based multiingredient supplement for 15 days prior a race, had a decrease of salivary IgA corrected by total protein after the race, while runners with this supplementation did not show changes in salivary IgA and had a better immunoregulation (Roca et al. 2019)

13.3.2 Lactoferrin and Lysozyme

Both are antimicrobial compounds and would have a similar role than salivary IgA, in protecting mucosal surfaces from external pathogens.

For both analytes controversial results have been published. In the case of lactoferrin increases in serum have been reported after high and middle intensity exercise (5000 steps running at 180 and 130 steps/min respectively) (Inoue et al. 2004). But no changes in lactoferrin in saliva were found after a marathon. Interestingly, a correlation between salivary lactoferrin and blood lymphocyte counts were found in this study before the race (Canto et al. 2018).

Also there is some controversy regarding the response of lysozyme after exercise. Decreases in lysozyme have been reported in athletes over a training (Papacosta and Nassis 2011) or after a marathon (Canto et al. 2018). However increases in this protein have been found after exercise with moderate intensity, being further increased after exercise with high intensity (Ligtenberg et al. 2015). Higher levels of lysozyme 2 days after a race were detected in marathon runners who developed a lower respiratory tract infection in the 2 weeks after the race in comparison with those without infection, this could be related with the protective function of this enzyme (Canto et al. 2018).

13.3.3 Interleukines (IL) and Acute Phase Proteins

Both are biomarkers of inflammation. Although it could be interesting to monitor inflammation associated to exercise, there are relatively few studies about that. In one of these studies, it was found that serum and saliva IL-6 showed different patters after an acute strenuous exercise. Serum IL-6 significantly rose in the late recovery after the exercise, while the salivary levels showed a modest and not significant increase immediately after the termination of the test. In this report a relation between IL-6 and lactate in saliva was found and it was postulated that IL-6 can be produced predominantly by type II fibers. These authors indicated the need of studies to clarify possible interferences with ILs measurements due to different sample collection methods and the importance of using validated assays (Minetto et al. 2005a, b). Also, the acute phase protein haptoglobin has been described to increase in saliva after exercise, opening a new interesting field about the response of the acute phase proteins to exercise and its possible practical applications as biomarkers of inflammation (Mateo et al. 2019).

In the future other markers of immune system such as adenosine deaminase could be evaluated in exercise and their possible correlations with other biomarkers studied.

13.4 Biomarkers of Muscle Damage

13.4.1 Creatine Kinase (CK), Lactate Dehydrogenase (LDH) and Aspartate Aminotransferase (AST)

These are enzymes that are inside muscle fibers and that increase in serum after a muscle damage. CK and LDH can increase in saliva after an indoor football game. However, there are some limitations in the use of these enzymes:

- A high interindividual variability of results, with some individuals showing no increases in these analytes (Barranco et al. 2018).
- A poor correlation between saliva and plasma. This low correlation could be due to a delay in the pass from blood to saliva. In addition, for CK, a threshold level in the movement from plasma to saliva is possible, since this protein appears in lower concentration in saliva compared with serum. Curiously, in animals such as horses or dogs, a correlation of CK in serum and plasma has been reported (Tvarijonaviciute et al. 2017; Contreras Aguilar et al. 2019).
- Lack of specificity since these enzymes are increased in saliva in periodontal disease, that in case of AST can increase in this condition until five fold (Todorovic et al. 2006). Also salivary LDH is increased in oral cancers (Shetty 2012). In addition the CK total and CK-MB are increased in saliva of patients with myocardial infarction or heart muscle damage (Mirzah-Dizgah and Jafari-Sebet 2011; Mirzaii-Dizgah et al. 2012), although this fact make possible that CK, together with troponin in saliva, could be used to detect cardiac damage in athletes.

Overall more studies are needed in order to assess and clarify if measurements of CK and LDH in saliva could be potentially used to evaluate possible muscle stress or damage in cases of intensive exercise. Also it would be interesting not only to evaluate the changes in these enzymes in acute muscle stress, but also in situations of chronic muscle stress or true muscle damage.

13.5 Biomarkers of Anaerobic Metabolism

The measurement of lactate in saliva corrected by total protein has been studied for the assessment of fatigue induced during repeated explosive effort sequences.

A correlation between lactate in saliva and blood has been described after intensive exercise. However the values of this correlation differ between authors. In a study ranged between 0.5 in athletes to 0.38 in non-athletes (Tekus et al. 2012). Other studies reported similar correlations but only in non-trained and when results were corrected by total protein (Franco-Martinez et al. 2019). On the other hand, higher correlations (around 0.9) were described by Segura et al. (1996). These authors indicated that the stimulation with citric acid produced higher correlations between lactate in serum and saliva. Interestingly, blood and saliva lactate are highly correlated with the total protein in saliva, therefore it has been suggested that total protein in saliva could be potentially used a marker of salivary anaerobic threshold, which is the phase in which a subject pass from the aerobic to anaerobic metabolism (Bortolini et al. 2009; Justino et al. 2018).

13.6 Biomarkers of Oxidative Status

Saliva has an antioxidant system integrated by various metabolites and enzymes. Uric acid (UA) is one of the most important antioxidant molecules in saliva, contributing around 70% of the total salivary antioxidant capacity (Battino et al. 2002) and being able to chelate transition metals and to react with biological oxidants. Also UA is the biomarker more sensitive to oxidative damage. Since in a previous report UA was the only analyte showing a significant change when a panel of oxidative stress biomarkers in saliva was measured after an acute session of resistance exercise, it was concluded that uric acid should be one of the most important biomarkers to be analyzed in saliva in order to evaluate oxidative stress (Deminice et al. 2010).

13.7 Concluding Remarks

- Coaches or practitioners can have advantages of the measurement of certain analytes in saliva in order to get data from their athletes, specially to evaluate and detect possible situations of stress or immunodeficiency that could predispose to a lower performance or even lesions.
- There is a high variability in the components of saliva between different subjects (Gleeson et al. 2011). Therefore individual player monitoring is highly recommended, and ideally resting values of the same individual should be used as reference for comparative purposes.
- Some analytes in saliva such as IL-6 or muscle enzymes can have different response than in serum. For example, after an exercise, serum IL-6 significantly rose in the late recovery while saliva concentration showed a modest and not significant increase immediately after the termination of the test. This could suggest, in the case of IL-6 that there are specific mechanisms responsible for IL-6 release in saliva, which are independent of the systemic compartment and the circulating pools, being unlikely that this molecule diffuses though the acinar cells of the salivary glands from the bloodstream in to saliva, and that serum levels can influence salivary concentration as occurs for steroid hormones (Minetto et al. 2005a, b).
- Further studies in order to detect new biomarkers by "omics" techniques will be of interest in order to increase the range of analytes in saliva that can be applied for evaluation of the physiological responses to sport.
- It would be recommended in the future to clarify various aspects in the different analytes, such as: their origin, how the results should be given, possible gender influence, if the individual should be used as control instead the use of reference ranges, influence of sampling material and circadian rhythms, haemolysis effect or ideal storage conditions.

References

Antualpa K, Aoki MS, Moreira A (2018) Intensified training period increases salivary IgA responses but does not affect the severity of upper respiratory tract infection symptoms in prepuberal rhythmic gymnast. Pediatr Exerc Sci 30(2):189–197

Barranco T, Tvarijonaviciute A, Tecles F, Carrillo JM, Sánchez-Resalt C, Jimenez-Reyes P, Rubio M, García-Balletbó M, Cerón JJ, Cugat R (2018) Changes in creatine kinase, lactate dehydrogenase and aspartate aminotransferase in saliva samples after an intense exercise: a pilot study. J Sports Med Phys Fitness 58(6):910–916

Battino M, Ferreiro MS, Gallardo I, Newman HN, Bullon P (2002) The antioxidant capacity of saliva. J Clin Periodontol 29:189–194

Blannin AK, Robson PJ, Walsh NP, Clark AM, Glennon L, Gleeson M (1998) The effect of exercising to exhaustion at different intensities on saliva immunoglobulin A protein and electrolyte secretion. Int J Sports Med 19:547–552

Bocanegra OL, Diaz MM, Teixeira RR, Soares SS, Espindola FS (2012) Determination of the lactate threshold by means of salivary biomarkers: chromogranin A as novel marker of exercise intensity. Eur J Appl Physiol 112(9):3195–3203

Bortolini MJ, De Agostini GG, Reis IT, Lamounier RP, Blumberg JB, Espindola FS (2009) Total protein of whole saliva as a biomarker of anaerobic threshold. Res Q Exerc Sport 80(3):604–610

Calvo F, Chicharro JL, Bandrés F, Lucía A, Pérez M, Alvarez J, Mojares LL, Vaquero AF, Legido JC (1997) Anaerobic threshold determination with analysis of salivary amylase. Can J Appl Physiol 22(6):553–561

Cantó E, Roca E, Perea L, Rodrigo-Troyano A, Suarez-Cuartin G, Giner J, Feliu A, Soria JM, Nescolarde L, Vidal S, Sibila O (2018) Salivary immunity and lower respiratory tract infections in non-elite marathon runners. PLoS One 13:11

Contreras-Aguilar MD, Escribano D, Martínez-Subiela S, Martínez-Miró S, Rubio M, Tvarijonaviciute A, Tecles F, Cerón JJ (2017) Influence of the way of reporting alpha-Amylase values in saliva in different naturalistic situations: a pilot study. PLoS One 12(6):e0180100

Contreras-Aguilar MD, Escribano D, Martínez-Subiela S, Martín-Cuervo M, Lamy E, Tecles F, Cerón JJ (2019) Changes in saliva analytes in equine acute abdominal disease: a sialochemistry approach. BMC Vet Res 15(1):187

Crewther BT, Potts N, Kilduff LP, Drawer S, Cook CJ (2018) Can salivary testosterone and cortisol reactivity to a mid-week stress test discriminate a match outcome during international rugby union competition? J Sci Med Sport 21(3):312–316

Del Corral P, Schurman RC, Kinza SS, Fitzgerald MJ, Kordick CA, Rusch JL, Nadolski JB (2016) Salivary but not plasma cortisone tracks the plasma cortisol response to exercise: effect of time of day. J Endocrinol Investig 39(3):315–322

Deminice R, Sicchieri T, Payão PO, Jordão AA (2010) Blood and salivary oxidative stress biomarkers following an acute session of resistance exercise in humans. Int J Sports Med 31(9):599–603

Díaz Gómez MM, Bocanegra Jaramillo OL, Teixeira RR, Espindola FS (2013) Salivary surrogates of plasma nitrite and catecholamines during a 21-week training season in swimmers. PLoS One 21(8):5

Elloumi M, Maso F, Michaux O, Robert A, Lac G (2003) Behaviour of saliva cortisol [C], testosterone [T] and the T/C ratio during a rugby match and during the post-competition recovery days. Eur J Appl Physiol 90(1–2):23–28

Filaire E, Ferreira JP, Oliveira M, Massart A (2013) Diurnal patterns of salivary alpha-amylase and cortisol secretion in female adolescent tennis players after 16 weeks of training. Psychoneuroendocrinology 38(7):1122–1132

Franco-Martínez L, Tvarijonaviciute A, Martínez-Subiela S, Márquez G, Martínez Díaz N, Cugat R, Cerón JJ, Jiménez-Reyes P (2019) Changes in lactate, ferritin, and uric acid in saliva after repeated explosive effort sequences. J Sports Med Phys Fitness 59(6):902–909

Gleeson M, Pyne DB (2016) Respiratory inflammation and infections in high-performance athletes. Immunol Cell Biol 94(2):124–131

Gleeson M, Bishop N, Oliveira M, McCauley T (2011) Tauler PSex differences in immune variables and respiratory infection incidence in an athletic population. Exerc Immunolo Rev 17:122–134

Inoue H, Sakai M, Kaida Y, Kaibara K (2004) Blood lactoferrin reléase induced by running exercise in normal volunteers: antibacterial activity. Clin Chim Acta 341:165–172

Justino AB, Neto AM, Bortolini MJS, Peixoto LG, Prado ES, Garcia-Hortal M, Botelho FV, Espindola FS (2018) Salivary total protein as a biomarker of anaerobic threshold in elderly athletes subjected to an incremental physical test. Biol Sport 35(1):57–65

Kivlighan KT, Granger DA (2006) Salivary alpha-amylase response to competition: relation to gender, previous experience, and attitudes. Psychoneuroendocrinol 31(6):703–714

Koibuchi E, Suzuki Y (2014) Exercise upregulates salivary amylase in humans (review). Exp Ther Med 7(4):773–777

Li TL, Lin HC, Ko MH, Chang CK, Fang SH (2012) Effects of prolonged intensive training on the resting levels of salivary immunoglobulin A and cortisol in adolescent volleyball players. J Sports Med Phys Fitness 52(5):569–573

Ligtenberg AJ, Brand HS, van den Keijbus PA, Veerman EC (2015) The effect of physical exercise on salivary secretion of MUC5B, amylase and lysozyme. Arch Oral Biol 60:1639–1644

Lindsay A, Costello JT (2017) Realising the potential of urine and saliva as diagnostic tools in sport and exercise medicine. Sports Med 47(1):11–31

Mackinnon LT, Hooper S (1994) Mucosal (secretory) immune system responses to exercise of varying intensity and during overtraining. Int J Sports Med 15(Suppl 3):S179–S183

Mateo SV, Contreras Aguilar MD, López-Jornet P, Jimenez-Reyes P, Ceron JJ, Tvarijonaviciute A, Martinez-Subiela S (2019) Development and evaluation of a rapid and sensitive homogeneous assay for haptoglobin measurements in saliva. Microchem J 150:104159

Minetto M, Rainoldi A, Gazzoni M, Terzolo M, Borrione P, Termine A, Saba L, Dovio A, Angeli A, Paccotti P (2005a) Differential responses of serum and salivary interleukin-6 to acute strenuous exercise. Eur J Appl Physiol 93:679–686

Minetto M, Rainoldi A, Gazzoni M, Terzolo M, Borrione P, Termine A, Saba L, Dovio A, Angeli A, Paccotti P (2005b) Differential responses of serum and salivary interleukin-6 to acute strenuous exercise. Eur J Appl Physiol 93(5–6):679–686

Mirzah-Dizgah I, Jafari-Sabet M (2011) Unstimulated whole saliva creatine phosphokinase in acute myocardial infarction. Oral Dis 17:597–600

Mirzaii-Dizgah I, Hejazi SF, Riahi E, Salehi MM (2012) Saliva-based creatine kinase MB measurement as a potential point-of-care testing for detection of myocardial infarctation. Clin Oral Investig 16:775–779

Moreira A, Arsati F, de Oliveira Lima Arsati YB, da Silva DA, de Araújo VC (2009) Salivary cortisol in top-level professional soccer players. Eur J Appl Physiol 106(1):25–30

Munk R, Jensen RB, Palme R, Munksgaard L, Christensen JW (2017) An exploratory study of competition scores and salivary cortisol concentrations in Warmblood horses. Domest Anim Endocrinol 61:108–116

Nieman DC, Henson DA, Fagoaga OR, Utter AC, Vinci DM, Davis JM, Nehlsen-Cannarella SL (2002 Jan) Change in salivary IgA following a competitive marathon race. Int J Sports Med 23(1):69–75

Nieman DC, Dumke CI, Henson DA, McAnulty SR, McAnulty LS, Lind RH, Morrow JD (2003) Immune and oxidative changes during and following the Western States Endurance run. Int J Sports Med 24(7):541–547

Papacosta E, Nassis GP (2011) Saliva as a tool for monitoring steroid, peptide and immune markers in sport and exercise science. J Sci Med Sport 14(5):424–434

Pumpa KL, McKune AJ, Harnett J (2019) A novel role of probiotics in improving host defence of elite rugby union athlete: a double blind randomized controlled trial. J Sci Med Sport 22(8):876–881

Roca E, Cantó E, Nescolarde L, Perea L, Bayes-Genis A, Sibila O, Vidal S (2019) Effects of a polysaccharide-based multi-ingredient supplement on salivary immunity in non-elite marathon runners. J Int Soc Sports Nutr 16(1):14
Sari-Sarraf V, Reilly T, Doran DA, Atkinson G (2007) The effects of single and repeated bouts of soccer-specific exercise on salivary IgA. Arch Or Biol 52:526–532
Segura R, Javierre C, Ventura JL, Lizarraga MA, Campos B, Garrido E (1996) A new approach to the assessment of anaerobic metabolism: measurement of lactate in saliva. Br J Sports Med 30(4):305–309
Shephard RJ, Shek PN (1998) Acute and chronic over-exertion: do depressed immune responses provide useful markers? Int J Sports Med 19(3):159–171
Shetti S (2012) Salivary lactate dehydrogenase levels in oral leukoplakia and oral squamous cell carcinoma; a biochemical and clinicopathological study. J Cancer Res 8:123–125
Strahler J, Berndt C, Kirschbaum C, Rohleder N (2010) Aging diurnal rhythms and chronic stress: distinct alteration of diurnal rhythmicity of salivary alpha-amylase and cortisol. Biol Psychol 84(2):248–256
Svendsen IS, Hem E, Gleeson M (2016) Effect of acute exercise and hypoxia on markers of systemic and mucosal immunity. Eur J Appl Physiol 116(6):1219–1229
Tecles F, Tvarijonaviciute A, De Torre C, Carrillo JM, Rubio M, García M, Cugat R, Cerón JJ (2016) Total esterase activity in human saliva: validation of an automated assay, characterization and behaviour after physical stress. Scand J Clin Lab Invest 76(4):324–330
Tékus E, Kaj M, Szabó E, Szénási NL, Kerepesi I, Figler M, Gábriel R, Wilhelm M (2012) Comparison of blood and saliva lactate level after maximum intensity exercise. Acta Biol Hung 63(Suppl 1):89–98
Thorpe R, Sunderland C (2012) Muscle damage, endocrine, and immune marker response to a soccer match. J Strength Cond Res 26(10):2783–2790
Todorovic T, Dozic I, Vicente-Barrero M, Ljuskovic B, Pejovic J, Marjanovic M, Knezevic M (2006) Salivary enzymes and periodontal disease. Med Oral Patol Oral Cir Bucal 1:115–119
Tomasi TB, Trudeau FB, Czerwinski D, Erredge S (1982) Immune parameters in athletes before and after strenuous exercise. J Clin Immunol 2(3):173–178
Tvarijonaviciute A, Barranco T, Rubio M, Carrillo JM, Martinez-Subiela S, Tecles F, Carrillo JD, Ceron JJ (2017) Measurement of Creatine kinase and aspartate aminotransferase in saliva of dogs: a pilot study. BMC Vet Res 13:168
Walsh NP, Blannin AK, Clark AM, Cook L, Robson PJ, Gleeson M (1999) The effects of high-intensity intermittent exercise on saliva IgA, total protein and alpha-amylase. J Sports Sci 17:129–134
Walsh NP, Bishop NC, Blackwell J, Wierzbicki SG, Montague JC (2002) Salivary IgA response to prolonged exercise in a cold environment in trained cyclists. Med Sci Sports Exerc 34:1632–1637

Chapter 14
Salivary Biomarkers in Welfare Studies

Damián Escribano and Fernando Tecles

14.1 Introduction

Welfare is a concept with an increasing interest in both humans and animals. The human welfare can be understood in three general ways: (1) the individual experiences of happiness or feels satisfied; (2) satisfaction of desires; and (3) enjoyment of certain objective goods, such as health, education, personal relationships and recreation (Parfit 1984; Griffin 1986; Nussbaum 2000; Posner 2008). All together it suggests that human welfare comprises a state in which basic and secondary necessities are satisfied. In this context, the health together with the absence of physical and/or emotional pain and stress are mandatory to consider a state of welfare.

Meanwhile, animal welfare was defined by five major pillars which remark that welfare is achieved when an animal is free of (1) hunger and thirst; (2) discomfort; (3) pain, injury and disease; (4) fear and distress; and is (5) free to express its normal behavior (council directive 98/58/EC). These pillars are called the 'five freedoms' and, therefore, welfare is understood as the absence of negative experiences (negative welfare). However, this acceptation of animal welfare has been less and less used over time, since it represents an ideal situation that is rarely achieved. In addition, the complete absence of freedoms could not be quantified (Green and Mellor 2011). Therefore, the concept of animal welfare evolved to that situation in which the animal had positive experiences and a positive affective state (positive welfare) (Duncan 2005). Both welfare acceptations could be considered as being highly correlated since the presence of a positive state necessarily implies the absence of its negative counterpart. Therefore, they can hardly vary independently (Russell and Carroll 1999). This is the reason why some authors describe welfare as a relative concept (Spruijt et al. 2001), in which for example health implies necessarily the

D. Escribano (✉) · F. Tecles
Interdisciplinary Laboratory of Clinical Analysis (Interlab-UMU), Regional Campus of International Excellence Campus Mare Nostrum, University of Murcia, Murcia, Spain
e-mail: det20165@um.es

A. Tvarijonaviciute et al. (eds.), *Saliva in Health and Disease*,
https://doi.org/10.1007/978-3-030-37681-9_14

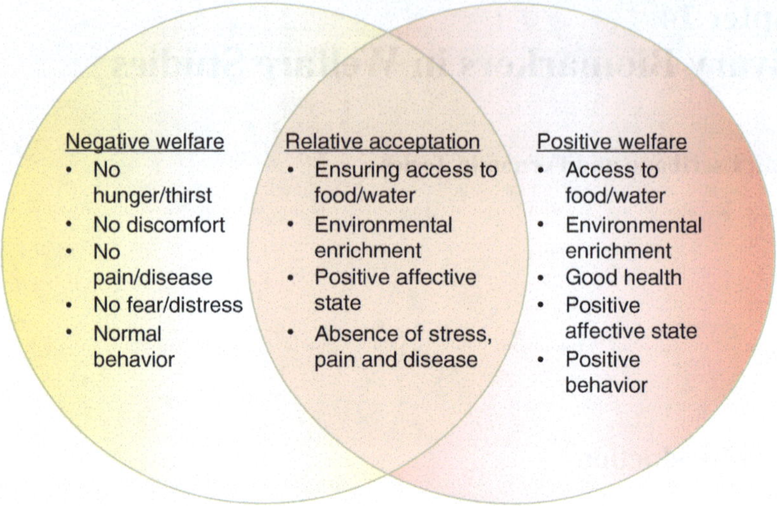

Fig. 14.1 Different acceptations of animal welfare

absence of disease, stress or pain (Fig. 14.1). Furthermore, the positive welfare could be better assessed and quantified, assessing space allowance, food and environmental enrichment, behavior (activities that are more liable to be performed when animals are in positive affective states) or physiological markers including salivary biomarkers (Yeates and Main 2008).

Currently, different substances in saliva have been postulated as biomarkers of welfare since their levels are altered in presence of stress and/or pain and reflect the activity of axes or systems involved in the homeostasis regulation. The two principals are the sympathetic nervous system (SNS) and the Hypothalamic-pituitary-adrenal axis (HPA). However, other systems are also interrelated including the immune system or Hypothalamic-pituitary-gonadal axis (HPG). In this chapter, salivary biomarkers used to evaluate welfare in humans and animals are reviewed being organized in relation with the major system or axis to which these have been associated.

14.2 Salivary Biomarkers of Sympathetic Nervous System (SNS)

Catecholamines (epinephrine and norepinephrine) are released when the SNS is activated and these substances are the major representatives of this system. However, these biomarkers have showed variability, instability and are difficult to measure in saliva (Schwab et al. 1992; Kennedy et al. 2001; Moya et al. 2006). Therefore, salivary biomarkers such as alpha amylase and chromogranin A (CgA), which reflect SNS activity, have been used in both veterinary and human medicine.

14.2.1 Salivary Alpha Amylase Activity (sAA)

sAA is one of the major enzymes in the oral cavity and its primary function, the hydrolysis of starch and glycogen, is involved in defence against bacteria. A low sAA activity has been related to a higher risk of oral infection (Scannapieco et al. 1993). In addition, it is secreted in response to adrenergic activity by parotid gland and it has been considered as a biomarker for evaluation of both physical and psychological stress in humans due to sympathetic activation (Nater and Rohleder 2009; Contreras-Aguilar et al. 2017a), being considered as a marker of acute stress. However, this has been recently discussed because the parasympathetic nervous system (PNS) also takes part in sAA release, since sAA can be secreted in response to mainly or only parasympathic innervation. In addition, since PNS also plays a role in salivary flow rate and protein secretion, the reliability of sAA as a marker of SNS is under discussion. Some methodological aspects also contribute to this fact, such as reporting results ignoring the flow rate effect, and the lack of a standardized procedure for saliva collection since the different ways could also affect stimulation of the autonomous nervous system affecting sAA release (Bosch et al. 2011). In addition, sAA has been reported to be a protein potentialy affected by usual type of diet. For example, Mandel et al. (2010) found different copy numbers of sAA according to the richness of starch in population diets. In spite of this, sAA levels have been correlated positively with SNS stress response in adults (Guglielminotti et al. 2012). Even so, few human studies have reported significant changes in response to stressful stimuli (Morrison et al. 2003; Schäffer et al. 2008; Hill-Soderlund et al. 2008). In a prospective study involving 76 healthy subjects who elected removal of their third molars, the sAA levels were significantly lower during surgery with respect with the pre-surgery consult, and remained decreased during the post-surgery follow-up visits (Robles et al. 2012). This salivary biomarker has been also studied in a group of 73 children with cancer subjected to three different emotion regulatory strategies (distraction, reappraisal or reassurance) before and after cold pressor pain induced task. The results showed that sAA rose in the less effective strategy (reassurance), mainly in those children that showed lower levels of positive affect. Concluding that although sAA increased due to pain, this effect could be minimized by positive affect (Jenkins et al. 2018), and therefore it could be more related with perception of pain rather than pain itself. In a study performed with 45 women suffering tension-headache and 45 healthy volunteers, sAA was significantly higher in patients with headache, and there was significant correlation between sAA activity and the perception of pain assessed by a questionnaire (Vahedi et al. 2018).

In veterinary medicine, sAA has been also used as possible SNS biomarker in pigs (Fuentes et al. 2011), sheep (Fuentes-Rubio et al. 2016; Contreras-Aguilar et al. 2019b), dogs (Contreras-Aguilar et al. 2017b) and horse (Contreras-Aguilar et al. 2018a). In horse, which looks like one of the species where the use of sAA gave better results, its levels have been proposed as a biomarker of pain-induced stress with an acute abdominal disease. In this study involving horses with acute abdominal disease, sAA activity was significantly higher in diseased animals rather than in healthy ones, and it was significantly correlated with the Equine Acute Abdominal Pain scales (version 1), indicating that it could be considered as a marker of pain and welfare (Contreras-Aguilar et al. 2018a) and the risk of death (Contreras-Aguilar et al. 2019c). However, there are contradictory results and no always significant increases were found after application of stress stimulus. For example, in pigs sAA increase after immobilization with nose snare has been observed (Fuentes et al. 2011) but not after weaning in piglets (Escribano et al. 2019b). Therefore, different dynamics after stress induction have been described. In addition, although the stress is a complex reaction that varies according to each individual idiosyncrasy, in the case of sAA activity all studies showed a higher inter-individual variability. This high variability could be an important disadvantage for its use as salivary stress biomarker. A possible explication is the fast-reacting biological response of this enzyme or its low durability after a stress stimulus (Nater and Rohleder 2009). For example, in human Groza et al. (1971) described a peak of sAA only one to three minutes after a mental stress and van Stegeren et al. (2008) found that the elevation of sAA was only about 10 min. In the same form, in veterinary medicine the higher levels after a stress model application seem to be obtained during the first minutes and would behave into account that different isoform of sAA can be involved in both animals and humans (Contreras-Aguilar et al. 2018b, 2019e). A great advantage to using sAA is that its liberation appears as independent of saliva flow rate and also the age has no impact in their levels (Rohleder et al. 2006; van Stegeren et al. 2008). However, Contreras-Aguilar et al. (2017a) described that is important to take into account the different ways of expressing sAA (enzymatic vs. concentration) and the factor involved for an objective interpretation of results.

14.2.2 Chromogranin A (CgA)

CgA is an acidic glycoprotein of 49 KDa that is stored and co-released together with catecholamines, fundamentally in nerves, chromaffin cells and adrenal medulla (Dimsdale et al. 1992; Takiyyuddin et al. 1994). CgA was shown to be stored in the acinar cell and produced in the salivary gland of human and animals (Sato et al. 2002; Saruta et al. 2005) and its liberation is induced by catecholamine secretion (Kanno et al. 1999). For this reason, salivary CgA has been used as a sensitive alternative of catecholamine for monitoring the SNS. In human medicine, salivary CgA has been measured in stress studies against different type of stress stimulus including social (Lee et al. 2006), psychosomatic (Toda et al. 2007) or mental stress (Rai and Kaur 2011; Obayashi 2013). In veterinary medicine, this salivary biomarker has been employed in stress and welfare studies in cows (Ninomiya and Sato 2011) or pigs (Escribano et al. 2014b, 2015, 2019b; Fàbrega et al. 2019). In those stressful situations, increases in salivary concentrations of CgA, with approximately mean increase of 50%, both human as animals have been described. Furthermore, in pigs the increase of CgA has also been related to chronic stress, being correlated with cortisol levels in hair (Casal et al. 2017).

Unlike catecholamines (stability of up to 2 h; Kennedy et al. 2001), salivary CgA is stable up to 1 year when stored at −20 °C (Escribano et al. 2014c). As limitations in human, its levels appear to follow a circadian rhythm, similarly to salivary cortisol, with a peak after awakening, and significant changes have been associated with symptoms of oral dryness (Den et al. 2007). However, in animal species such as pigs no circadian pattern was detected for salivary CgA in either season and there were no significant effects of gender and age (Escribano et al. 2014c).

14.3 Salivary Biomarker of Hypothalamic-Pituitary-Adrenal (HPA) and Gonadal (HPG) Axes.

14.3.1 Cortisol

The salivary cortisol is the most important representative and traditionally used biomarker of the HPA axis. It is important to indicate that only the fraction of free cortisol that is not bound to serum proteins is quantified in saliva, which is considered the biologically active form. The corticotrophin-releasing hormone (CRH) is released from hypothalamus followed by induction of the release of the peptide adrenocorticotrophic hormone (ACTH) from the anterior pituitary into the systemic circulation. ACTH finally induces liberation of the 'stress' hormone cortisol that is produced and released by the adrenal cortex. Cortisol exerts diverse systemic effects that help alleviate the inciting stress, including but not limited to metabolic, cardiovascular, and immunomodulatory effects (Hart 2012). It is the most studied and traditionally used salivary biomarker in literature. Cortisol can be measured in a reliable and accurate way in saliva, both in humans (Woolston et al. 1983; Obmiński et al. 1997) as well as in animal species (Cook et al. 1996; Peeters et al. 2011). It has been demonstrated that salivary cortisol concentrations are valid indicators of circulating cortisol levels and, therefore, represent an adequate way to evaluate the HPA axis response. In addition, salivary cortisol has good stability and saliva may be stored at 5 °C for up to 3 months or at −20 °C or −80 °C for at least 1 year (Naumova et al. 2012) for cortisol measurement. Since it is the most studied and used salivary biomarker, there are a lot of results against different types of stressors. For example, in species such as porcine, it has been used as salivary stress and welfare biomarker for three decades in multiple studies against different stimulus (Table 14.1). In humans, it has also been used in several studies in relationship with loss of welfare, stress or evaluation of pain. For example, Müller (2011) evaluated the salivary cortisol response in 64 male volunteers receiving painful electric skin stimuli. An increase in cortisol concentration was reported, which also correlated with the perceived pain intensity and subjective helplessness ratings. In a similar approach in which 46 healthy men were stressed by performing a medical test, increased cortisol

Table 14.1 Stressors related to salivary cortisol levels alterations in pigs

Stressor	References
Transport and accommodation in the slaughterhouse	Brown et al. (2005) and Soler et al. (2013)
Castration	Schönreiter et al. (1999)
Immobilization	Muneta et al. (2010), Fuentes et al. (2011) and Huang et al. (2017)
Social stress	Jarvis et al. (2006), Smulders et al. (2006) and van der Staay et al. (2008)
Food or water Deprivation	Parrott and Misson (1989) and Ott et al. (2014)
Ear tagging, Electric shocks, blood sampling	Merlot et al. (2011)
Disease stress	Koopmans et al. (2012)
Weaning	Escribano et al. (2019b)
Stress due to acute and chronic administration of ACTH.	Bushong et al. (2000).

was negatively correlated with pain thresholds after electrical stimulation, concluding that increased cortisol levels increase pain and reduce tolerance to pain (Choi et al. 2012). Pain induced by cold pressor significantly increased salivary cortisol in healthy individuals, showing a positive correlation with pain ratings (Goodin et al. 2012). In a study involving 31 women with osteoarthritis, those women with higher pain score had higher cortisol throughout the day (Carlesso et al. 2016).

Although cortisol can be considered as a reference biomarker, important factors must be taken into account. The salivary levels of free cortisol are characterized by circadian fluctuation and its secretion does not occur at a constant and consistent level in all mammals; concentrations in the morning are significantly higher than those in the evening and a peak occurs upon awakening both in human as animals species (Benton and Yates 1990; Ruis et al. 1997; Russell et al. 2010; Sjörs et al. 2012). On the other hand, much variability has been found in results depending on the stressor stimulus and the different reports. In horses with colic, cortisol was significantly higher than in healthy ones, but did not correlate with the Equine Acute Abdominal Pain Score 1 as an indicator of pain (Contreras-Aguilar et al. 2018a). For the effects produced by castration in salivary cortisol, some apparently inconsistent results have been obtained. In calves, no change in cortisol was observed between non-castrated calves and calves castrated by two different methods (band or knife) (Marti et al. 2017). In a similar approach conducted in 150 calves, salivary cortisol was higher in surgically castrated calves than in non-castrated calves, and a trend to cortisol reduction was observed after analgesic administration (Moya et al. 2014), suggesting pain as a factor increasing cortisol levels. However, these authors did not find increased cortisol in the calves castrated by band, contrarily to other authors that reported increased salivary cortisol after band castration (Pang et al. 2006, 2011) but with similar values than non-castrated ones when an analgesic treatment was administered (González et al. 2010). In pigs, a group of sows that was restrained with a soft rope snare around the maxilla and an intravenous catheter was inserted through the ear vein showed an increase in salivary cortisol that remained for 10 min compared with another group that was only restrained in the same way without being catheterized. It could be postulated that the pain and stress produced by the catheter could be responsible for the increased cortisol, although the catheterized group showed also a higher frequency of stress related behaviour such as head shaking and trembling (Yun et al. 2017).

Additionally to what was stated above, recent studies reported that there are no variations in cortisol after 24 and 48 h of transport or during periods of fasting or isolation in pigs (Ott et al. 2014; Escribano et al. 2014b, 2015), indicating that different types of stressors elicited different physiological stress responses, and therefore including various salivary biomarkers in stress evaluation seems essential.

14.3.2 Cortisone

Cortisone is also a steroid hormone and is considered the inactive form of cortisol. Cortisol has been seen to pass into cortisone in various tissues such as the salivary gland and hair, through the action of the enzyme 11β-hydroxysteroid dehydrogenase (11β-HSD). Recent studies in human found that salivary cortisone could act as a stress biomarker with high discriminatory power and significant correlations with subjective and autonomic stress measures (Bae et al. 2019). The ratio cortisol/cortisone has been also suggested as a possible biomarker of utility in response to acute stress in pregnant women (La Marca-Ghaemmaghami et al. 2013). Further studies about the utility of measurement of cortisone or even the ratio cortisol/cortisone are needed to know its possible usefulness in stress and wellness studies.

14.3.3 Testosterone

Gonadotropin releasing hormone is released from hypothalamus and finally testosterone is produced from androgens in the testes and ovaries. A small amount is also produced in the adrenal gland. Its salivary levels are strongly correlated with free serum testosterone levels in humans and it has been related with the modulation of pain sensation (Choi et al. 2012). In their study, Choi et al. (2012) reported a decreased salivary testosterone level when individuals were stressed, and the decreased testosterone positively correlated with pain thresholds after the electrical stimulation, concluding that a decreased salivary testosterone associates to pain and reduces tolerance to pain. In a study involving women, those with premenstrual dysphoric disorder had lower salivary testosterone levels and correlated with pain threshold, indicating a lower pain tolerance (Bartley et al. 2015). In addition, this salivary biomarker has also been related with depression and anxiety disorders (Giltay et al. 2012), stress by exercise (Gatti and De Palo 2011), as well as with the dominance and aggressive behaviour (Mehta and Josephs 2010; Montoya et al. 2012;

Romero-Martínez et al. 2013). Increases of salivary testosterone that are not accompanied by cortisol have been reported in human studies in situations that could indicate a predisposition to violence (Romero-Martínez et al. 2013). In pigs, it has been also associated with inflammatory stimulus (Moya et al. 2006) and similarly to humans, significantly higher levels of salivary testosterone together with decreased levels in salivary cortisol have been observed in an experimental model of transport and accommodation in the slaughterhouse in pigs that were predisposed to violence after mixing with unfamiliar animals (Escribano et al. 2014b). Therefore, this salivary biomarker could be useful in this type of stressful situations. In addition, its levels do not seem to be influenced by a circadian pattern at least in pigs (Escribano et al. 2014c). Also, salivary testosterone levels have been related to stress rather pain. For example, in an experiment in which a group of pigs was isolated as a source of stress, testosterone levels increased after regrouping, similarly as cortisol did (Escribano et al. 2015), where animals showed lesions due to hierarchical fights.

14.4 Salivary Biomarkers of the Immune System

14.4.1 Immunoglobulin A (IgA)

Salivary immunoglobulin A (IgA) is an immunoglobulin that is released in saliva and oral mucosal membrane and constitutes the main specific immune defence in saliva. Its function includes the blockade of adherence and penetration of microorganisms (Walker et al. 2004). Salivary IgA can be increased by both PNS and SNS activation (Carpenter et al. 2000; Allgrove et al. 2008). On the contrary, pain may reduce the salivary levels of this immunoglobulin. Levels of IgA were studied in a group of patients with galvanic pain and they were significantly lower than in asymptomatic controls (Syrjänen et al. 1984). The correlation between the salivary levels of IgA and oral pain has been studied in 20 patients after an orthodontic treatment. Although salivary IgA levels and oral pain intensity showed a tendency to a negative correlation, the results were not statistically significant, although the authors concluded that the reduction of the IgA concentrations after orthodontic treatment could have something to do with the pain perceived by the patient (da Silva Campos et al. 2010). In humans, salivary IgA has been also reported as a potential psychosomatic stress biomarker, moreover related to mental stress (Takatsuji et al. 2008; Ulmer-Yaniv et al. 2018). However, some studies have suggested a negative correlation between its levels and mental stress (Ng et al. 1999; Wetherell et al. 2004; Engeland et al. 2016). In veterinary medicine, the IgA expression has been related with stress in the pig saliva after restraint or isolation (Muneta et al. 2010; Escribano et al. 2015) and also it has been described as a useful marker of stress in dog (Kikkawa et al. 2003) or even as possible biomarker of well-being in elephant (Edwards et al. 2019). A great disadvantage of this salivary biomarker is that the flow rate of saliva and also oral contamination affects IgA levels (Spiropoulos et al. 1993; Klentrou et al. 2002). As an advantage, it is a stable biomarker that can be stored frozen for up to three months (Ng et al. 1999).

14.4.2 Acute Phase Proteins (APPs)

The acute phase response is a complex non-specific reaction of the organism that occurs rapidly after any tissue damage. One of the main characteristics of this acute phase response is the appearance of changes in the concentration of plasma proteins called acute phase proteins (APPs) (Eckersall 1995). In general, there is a controversy over if APPs can serve as markers of stress, since there are studies, fundamentally in pigs, where no changes in C-reactive protein (CRP) and haptoglobin (Hp) due to psychosocial stress was observed (Escribano et al. 2015). However, in other stress models such as food deprivation or after restraint, changes in Hp have been found (Ott et al. 2014; Huang et al. 2017). It has also been described that serum amyloid A can be a stress biomarker in models of isolation and transport (Soler et al. 2013), showing a more prolonged response than cortisol. In human, evidences demonstrating APPs reliable respond to acute stress are absent (Slavish et al. 2015). There are authors who indicate that there is a connection between APPs and stress, since the activation of the SNS and the release of catecholamines induces the release of pro-inflammatory cytokines such as interleukin 6 (IL-6) that induces the production of liver enzymes. However, more studies should be carried out to clarify the relationship between acute and chronic stress situations and APPs.

14.4.3 Interleukins

Although the literature is limited, several inflammatory markers (including IL-1β, TNF-α, and IL-6) have been reliably determined from saliva and have increased significantly in response to stress across multiple studies, with effect sizes ranging from very small to very large (Slavish et al. 2015). However, these interleukins usually appear in lower concentrations and usually increase in smaller amounts compared to acute phase proteins. Since the affective state could affect immunity, studies in humans have shown increases in some interleukins (IL-2 and IL-3) in saliva, as well as decreases in IL-6 and TNF-α, in individuals in a positive affective state (Pressman and Cohen 2005). The opposite is found when an individual suffers pain, showing increases in the salivary levels of interleukins. For example, salivary levels of IL-4, 6, 8 and 10 have been studied in humans suffering experimental pain, increasing due to cold pressor task, moreover in older adults rather than younger adults (Cruz-Almeida et al. 2017). Unstimulated (passively collected) saliva was collected in non-verbal children with and without pain and the pain group showed higher levels of IL-1α and 8 (Symons et al. 2015). Salivary IL-6 was increased after pressure pain in patients with fibromyalgia, and the increase was higher as the severity of the disease was higher (Geiss et al. 2012). These results are indicating immune system up-regulation in individuals with pain (Shubayev et al. 2010). The soluble tumor necrosis factor-α receptor II (sTNFαRII) reflects TNF-α activity (Aukrust et al. 1994) and it is more stable and easier to measure than TNF-α in the laboratory (Diez-Ruiz et al. 1995). The salivary levels of this biomarker have been studied in individuals after inducing pain by different modalities such as cold pressor, hot water or ischemia, but in one report decreased levels of this biomarker after pain induction were found (Goodin et al. 2012), contrarily to previous research that reported increase in salivary pro-inflammatory cytokine due to pain. The authors postulated that one possible reason for this discrepancy was the increased salivary cortisol levels reported in their research, which could have down-regulated the expression of pro-inflammatory cytokines. In veterinary medicine, also it has been described that some interleukins such as IL-18 can vary in stressful situations in animals (Muneta et al. 2011).

14.5 Other Salivary Biomarkers Related to Stress and Welfare

The biomarkers described above are those in which more studies have been carried out and that have been more related to one of the main systems or axes. However, as more studies are carried out, and the sensitivity of the techniques used in their measurement improves, more substances are measured and suggested as possible salivary biomarkers of welfare. In the next part of the chapter, we will describe some of these 'new' substances in which good results have been obtained in welfare studies.

14.5.1 Salivary Oxidative Biomarkers

Oxidative stress is a term used to indicate an imbalance between the production of reactive oxygen species (ROS) in the organism and the ability of the antioxidant molecules to neutralize them (Sordillo and Aitken 2009). Chronic psychological stress appears to accelerate biological aging, and oxidative damage is an important potential mediator of this process. However, the mechanisms by which psychological stress promotes oxidative damage are poorly understood (Aschbacher et al. 2013). In general, for the evaluation of the oxide-reducing balance, both the oxidizing and antioxidant molecules can be determined. For example, hydrogen peroxide, as well as the different enzymatic and non-enzymatic antioxidant systems, such as the total antioxidant capacity (TAC) of an organic fluid. Although most studies have used serum to assess the presence of oxidative stress, others have used saliva for this purpose, namely associated to exercise (Gonzalez et al. 2008; Deminice et al. 2010). However, recent studies performed in animal species such as pig (Rubio et al. 2019a) or sheep (Rubio et al. 2019b) reported changes in both antioxidant and oxidant state markers associated with exposure to a stressful stimulus. It must be taken into account that there are buccal pathologies such as periodontitis, caries, oral precancerosis, and other local oral pathologies that are associated with oxidative stress and therefore they could also induce changes in the salivary levels of those biomarkers (Tóthová et al. 2015). In any way, it would be of interest to perform further studies about how oxidative salivary biomarkers change in saliva in different physiological conditions or situations of stress or loss welfare.

14.5.2 Total Esterase Activity (TEA)

Total Esterase Activity (TEA) includes the sum of the activities of various esterases present in saliva such as carbonic anhydrase, cholinesterase and lipase, among others. It is considered a possible marker of the SNS, since it is correlated with CgA, and is elevated in humans by physical stress (Tecles et al. 2016) and also significant increases after application of a model of acute of stress in animals have been described (Tecles et al. 2017). One advantage of TEA as a salivary biomarker is that it is easy to measure and does not require immunological reagents such as CgA, cortisol or some acute phase proteins. However, its stability is limited (even at −80 °C), so it is recommended to measure it as soon as possible (Barranco et al. 2019). Further studies should be performed to compare salivary total esterase and the enzymes involved in this activity with these other salivary analytes in order to assess if they could be more sensitive to evaluate stress.

14.5.3 Lipase Activity

The lipase activity of the saliva seems to come from the serous glands of the tongue, and its secretion appears to be regulated by the SNS, so it could also be considered as a marker of the activity of the autonomic nervous system. Saliva lipase is found in very small amounts in human saliva, but in other animal species a higher level of activity has been found that increases after a stressful stimulus, such as in pigs (Tecles et al. 2017) or sheep where lipase showed the greatest percentage of increase after inducing stress by facing a dog and after shearing than with other biomarkers (Contreras-Aguilar et al. 2019b).

14.5.4 Lactate Dehydrogenase (LDH)

LDH is a ubiquitous enzyme that catalyzes the reaction of lactate production via pyruvate reduction during anaerobic glycolysis and its extracellular presence is related to cell necrosis and tissue breakdown (Lokesh et al. 2016). In relation to stress studies, Radaković et al. (2018) showed that adrenaline caused an increase in the total activity of LDH, LDH1 and LDH2 isoenzymes in rats. Drouet et al. (2015) showed significant increases in plasma levels of LDH in rats after a model of acute restraint stress in relation to the control group. In a recent article, Escribano et al. (2019a) found that activity of LDH could potentially be used as a salivary biomarker for both short-term acute stress and long-term chronic stress of lameness. An advantage for its use is that LDH can be measured by spectrophotometric assays that are cheaper, faster and easier to perform that the immunoassays required for measurement of other proteins.

14.5.5 Neuropeptides

Neuropeptides have been also measured in human saliva as possible welfare biomarkers. Nerve growth factor (NGF) plays roles in neuronal protection and regeneration. Its concentration increases during inflammation and high salivary levels have been reported in response to painful stimuli (Nam et al. 2007). Calcitonin gene-related peptide (CGRP) and brain derived neurotropic factor (BDNF) are neuropeptides abundant in nervous tissue that have been implicated in migraine and headache. Both peptides increase their concentration in serum and saliva during an active pain (Ashina et al. 2000; Jang et al. 2011; Zidverc-Trajkovic et al. 2009; Fischer et al. 2012). Substance P (SP) can be also measured in both serum and saliva and a high level of this neuropeptide is associated with chronic pain (Parris et al. 1990; Jang et al. 2011; Greco et al. 2008), although controversy exists due to recent research (Kallman et al. 2018). Regarding β-endorphin (β-End), it seems to have local production and/or release in salivary glands, but in a recent research no difference was seen in its salivary concentrations between chronic neuropathic patients and healthy subjects (Kallman et al. 2018). Glutamate is a neurotransmitter with a significant role in nociceptive processes caused by inflammation or nerve injury (Bleakman et al. 2006), and its receptors have been found on nociceptive nerves in skin and muscles (Wozniak et al. 2012; Gerdle et al. 2014). A recent research has reported that glutamate can be measured in human saliva (Jasim et al. 2018), although its usefulness as welfare biomarker has not been evaluated yet. In addition, although the neuropeptides are susceptible to be measured in human saliva, their levels could be significantly affected by the different saliva collection methods, flow rate or even the salivary gland that secreted the saliva (Jasim et al. 2018).

Regarding neuropeptides in veterinary medicine, the presence of SP has been proven in saliva of rats (Goedert et al. 1982) but it has not been evaluated as a pain biomarker. However, SP has been studied in calves just after castration and the levels were compared with those found in non-castrated calves. No difference between non-castrated and castrated calves in salivary substance P was observed (Marti et al. 2017).

14.5.6 *Oxytocin*

Oxytocin is a neuropeptide hormone composed of nine amino acids (Carter et al. 2007). In animal models, oxytocin is activated by stress and anxiety-inducing stimuli, and thereby modulates and inhibits both SNS and HPA stress activity (Amico et al. 2004; Ring et al. 2006; Windle et al. 2004). This biomarker has been usually employed as an indicator of a good emotional state in plasma samples of humans (Miller et al. 2009) or dogs (Odendaal and Meintjes 2003; Okamoto et al. 2009). In dogs, it can be measured in urine, increasing after positive emotion such as eating, exercising and stroking (Mitsui et al. 2011). Therefore, contrarily to most biomarkers used for evaluating welfare, such as cortisol, catecholamines or alpha-amylase that usually are associated with stress and negative situations, oxytocin has the particularity of being associated with positive experiences. Initially, due to the very low concentration in saliva, some authors did not recommend its measurement (Horvat-Gordon et al. 2005). However, due to the development of more sensitive assays, different studies performed in last years demonstrated that salivary concentrations of oxytocin change in stressful situations being a reliable substitute for wellbeing evaluation in humans (Holt-Lunstad et al. 2011; Blagrove et al. 2012). In addition, it has been reported the presence of oxytocin receptor in the human salivary gland (Forsyth and Margaret 2009). Therefore, saliva is being increasingly used for oxytocin measurement in humans in welfare studies (Carter et al. 2007; White-Traut et al. 2009; Weisman et al. 2012), especially in those about relaxation (Ooishi et al. 2017).

Recently, new sensitive salivary oxytocin assays have been developed and validated in dogs (MacLean et al. 2018) and sows (López-Arjona et al. 2019). In dogs, concurrent elevations in salivary oxytocin during nursing were detected (MacLean et al. 2018). In pigs, a significant decrease in salivary oxytocin was observed after 9 days of lactation (López-Arjona et al. 2019).

Overall, evidences suggest oxytocin to be a promising salivary biomarker for welfare studies, although further investigations are needed to confirm its utility in both human and veterinary medicine.

14.6 Considerations About the Use of Salivary Markers of Welfare and Pain

Salivary biomarkers have been widely used for welfare or pain assessment in both humans and animal species (Table 14.2). Furthermore, saliva is considered the sample of election in this type of studies mainly due to its non-invasiveness. However, apart from the specific problems inherent to each specific biomarker, some general aspects should be taken into account for its use.

A specific aspect to consider for the proper use of this type of sample is the way how the biomarkers enter the saliva, since this fact could highly condition their

Table 14.2 Response of the mostly employed salivary biomarkers in the stress studies in humans and different veterinary species

Salivary biomarker	Species	Response and kind of stimulus	References
Alpha Amylase	Human	No change due to noise level and a large variability	Morrison et al. (2003)
		↑ after 10 min of mental stress	van Stegeren et al. (2008)
		↑ When pregnant women are taken to the operating room	Guglielminotti et al. (2012)
		No ↑ during the surgery.↑ levels in anticipation of surgery	Robles et al. (2012)
		↑ at 15 min after an academic activity in veterinary students	Contreras-Aguilar et al. (2017a)
		↑due to pain in children with cancer	Jenkins et al. (2018)
		↑ in patient with Frequent Episodic Tension-Type Headache	Vahedi et al. (2018)
		↑ just after psychological and physical stress models	Contreras-Aguilar et al. (2019e)
	Pig	No increase during a restraint stress	Muneta et al. (2010)
		↑ at 30 and at 60 min after nose-snare application	Fuentes et al. (2011)
		No increase after weaning in piglets	Escribano et al. (2019b)
		No correlation with pain score system of lameness and prolapses	Contreras-Aguilar et al. (2019a)
	Sheep	↑ after immobilization and confrontation with a sheepdog	Fuentes-Rubio et al. (2016)
		↑ by facing a dog and ↑ after shearing	Contreras-Aguilar et al. (2019b)
	Horse	↑ in horses with acute abdomen than in healthy horses correlation with score in Equine Acute Abdominal Pain scales	Contreras-Aguilar et al. (2018a) and Contreras-Aguilar et al. (2019c)
	Dog	↑ after ejaculation	Contreras-Aguilar et al. (2017b)
		No differences between dogs with pyometra and healthy dogs	Tecles et al. (2018)
Chromogranin A	Human	↑ after venepuncture in children	Lee et al. (2006)
		↑ at 1, 2 and 3 weeks after stress due simulated microgravity	Rai and Kaur (2011)
		↑ levels in isolation stress in major competition of swimmers	Chennaoui et al. (2016)
		Inhibition of increase after arithmetic tasks due to black tea.	Yoto et al. (2018)

(continued)

Table 14.2 (continued)

Salivary biomarker	Species	Response and kind of stimulus	References
	Pig	↑ at 15 min after nose-snare application	Escribano et al. (2013)
		No significant change after I.V. administration of lipopolysaccharide	Escribano et al. (2014a)
		↑ at 30 and 60 min during lairage at the slaughterhouse	Escribano et al. (2014b)
		↑ at 30 min after isolation and at 30 min after regrouping	Escribano et al. (2015)
		↓ at 1 and 2 months of environmental enrichment	Casal et al. (2017)
		↑concentration in postpartum dysgalactia syndrome sows	Kaiser et al. (2018)
		↑ at 1 and 2 days post-weaning in piglets	Escribano et al. (2019b)
	Cow	↑ levels in isolation stress	Ninomiya and Sato (2011)
Cortisol	Human	↑ due to painful electric skin stimuli	Müller (2011)
		↑ reduces tolerance to pain due to painful electric skin stimuli	Choi et al. (2012)
		↑ throughout the day in women with ↑Osteoarthritis pain scores	Carlesso et al. (2016)
		↑ after an acute laboratory stressor	Masih et al. (2019)
		No change to short-term broadband noise and sounds of nature	Aydin and Searchfield (2019)
		↑ after exposure to noise	Jafari et al. (2019)
	Pig	↑ at 15 min after nose-snare application	Escribano et al. (2013)
		↑ after the first administration of lipopolysaccharide	Escribano et al. (2014a)
		↑ immediately after unloading at the slaughterhouse	Escribano et al. (2014b)
		No increase after isolation and ↑after regrouping	Escribano et al. (2015)
		No ↓ due to environmental enrichment and herbal compound	Casal et al. (2017)
		↑ concentration in postpartum dysgalactia syndrome sows	Kaiser et al. (2018)
		↑ at 1 day post-weaning in piglets	Escribano et al. (2019b)
	Sheep	↑ after immobilization and confrontation with a sheepdog	Fuentes-Rubio et al. (2016)
		↑ following transport and sham shearing	Coulon et al. (2014)
		↑ one day post-surgery in a sheep model (unilateral tibia osteotomy)	Häger et al. (2017)
		↑ by facing a dog and ↑ after shearing in relation to baseline, not between the stress and the control group.	Contreras-Aguilar et al. (2019b)

(continued)

Table 14.2 (continued)

Salivary biomarker	Species	Response and kind of stimulus	References
	Cow	↑ after social separation in dairy cattle	Hernandez et al. (2014)
		↑ in band-castrated	Marti et al. (2015)
		↑ after transrectal examination in reactivity dairy cows	Kovács et al. (2016)
		↑ during long-distance transport recently weaned beef calves	Marti et al. (2017)
		↑ in cows with mastitis	Contreras-Aguilar et al. (2019d)
	Horse	↑ after the transrectal ultrasound examination in nonlactating mares	Schönbom et al. (2015)
		↑ after weaning method	Mach et al. (2017)
		↑ in horses with acute abdomen than in healthy horses	Contreras-Aguilar et al. (2019c)
	Dog	↑in response to stress in German shepherd dogs	Svobodová et al. (2014)
		↓ after olfactory and auditory stimulation on separation anxiety	Shin and Shin (2016)
		↓ concentration in sociable dogs	Shin and Shin (2017)
Testosterone	Human	Its imbalance in relation to cortisol is predictive for dominance	Mehta and Josephs (2010)
		↓ after pain induced by electrical stimulation	Choi et al. (2012)
		↓ in female patients with a depressive disorder and anxiety	Giltay et al. (2012)
		↑ may increase proneness to violence	Romero-Martínez et al. (2013)
		↓ indicated a lower pain tolerance	Bartley et al. (2015)
	Pig	↑ at 15 min after nose-snare application	Escribano et al. (2013)
		↑ after unloading, at 30 and 60 min after slaughterhouse transport	Escribano et al. (2014b)
		↑ at 30 min after regrouping. No increase after isolation.	Escribano et al. (2015)
IgA	Human	Correlated inversely with self-reported levels of stress in nurse	Ng et al. (1999)
		The response to acute and cumulative acute multitasking stress depends of the task, but also upon individual perceptions.	Wetherell et al. (2004)
		↑ increased immediately after the examination in nursing student	Takatsuji et al. (2008)
		Psychological stress was negatively associated with immunity	Engeland et al. (2016)
		↑ in depressed mothers	Ulmer-Yaniv et al. (2018)

(continued)

Table 14.2 (continued)

Salivary biomarker	Species	Response and kind of stimulus	References
	Pig	↑ at 10 and 20 min during restraint	Muneta et al. (2010)
		↑ at 1 and 5 days after of isolation. No increase after regrouping	Escribano et al. (2015)
	Dog	↓ immediately after and 30 min after the noise stress	Kikkawa et al. (2003)
		No clear response to stress in German shepherd dogs	Svobodová et al. (2014)

response and, therefore, their utility assessing welfare or pain. The possible effect of the different saliva collection methods, as well as flow rate or even the salivary gland that secreted the saliva can have on the biomarkers should be further studied, since this could affect not only results but also the way how they should be interpreted (Contreras-Aguilar et al. 2017a; Jasim et al. 2018).

In addition, general aspects, not directly associated with the sample should also be considered. These aspects include:

1. Adaptive response against stress or pain. This response fundamentally depends on the type of stimulus or disturbance that is affecting the animal and cause the loss of welfare. For example, in the case of using isolation as stress stimulus in pigs (a more psychological factor), increase of CgA or IgA associated to SNS were found whereas after regrouping (a more physical factor) increase of biomarkers associated to HPA and HPG systems such as cortisol and testosterone were observed (Escribano et al. 2015). Therefore, these biomarkers must be selected according to the type of stimulus that has caused loss of welfare. Due to this fact, as well as because normally more than one factor is involved in stress, the inclusion of various salivary biomarkers in welfare studies is recommended (Martínez-Miró et al. 2019; Paszynska et al. 2016).
2. Intra-individual and inter-individual variability along with the external variables that can have influence the results make it difficult to compare data within and among studies. Cobb et al. (2016) in a systematic review and meta-analysis about salivary cortisol measurement in domestic canines highlighted the importance of carefully controlling experimental design to compare samples within and between individual dogs, as well as establishing and using best practices for saliva collection. Standing out that the results could be the reflection of a plethora of factors. In the same line, in other species such as porcine, where less variability is found due to same living and management conditions, also response of different magnitudes against the same stressor stimulus were reported (Rey-Salgueiro et al. 2018). Just for instance, Contreras-Aguilar et al. (2019b) reported increases of salivary cortisol in sheep after facing a dog and after shearing in relation to its basal levels, however, not increase in relation to the control group were found. Therefore, an additional and fundamental factor in stress studies also is to be able to know the basal or normal levels of the individuals or animals used.

14.7 Conclusions

Saliva can be considered a very useful sample with a high potential to assess the degree of pain, stress and, thus, welfare since saliva contains a great number of biomarkers that provide information about the main physiological systems involved in the stress response. Nevertheless, three main points should be acknowledged when using saliva in welfare studies: (1) A panel of salivary biomarkers involved in different physiological systems should be used, since there is no one single ideal biomarker for welfare; (2) The salivary biomarkers must be selected according to the type of stimulus that has caused the loss of welfare; (3) Basal or normal levels of the salivary biomarkers for each of studied individual should be known. However, further studies would be recommended in order to clarify some aspects of this field as if there are biomarkers that could differentiate between acute and chronic stress or if there are specific biomarkers that can be more useful in specific stressful stimuli.

Acknowledgements Damian Escribano has a post-doctoral fellowshisp (IJC2018-035105-I) "Juan de la Cierva Incorporación" supported by the "Ministerio de Economía y Competitividad", Spain.

References

Allgrove JE, Gomes E, Hough J, Gleeson M (2008) Effects of exercise intensity on salivary antimicrobial proteins and markers of stress in active men. J Sports Sci 6:653–661

Amico JA, Mantella RC, Vollmer RR, Li X (2004) Anxiety and stress responses in female oxytocin deficient mice. J Neuroendocrinol 16:319–324

Aschbacher K, O'Donovan A, Wolkowitz OM, Dhabhar FS, Su Y, Epel E (2013) Good stress, bad stress and oxidative stress: insights from anticipatory cortisol reactivity. Psychoneuroendocrinology 38:1698–1708

Ashina M, Bendtsen L, Jensen R, Sakai F, Olesen J (2000) Possible mechanisms of glyceryl-trinitrate-induced immediate headache in patients with chronic tension-type headache. Cephalalgia 20:919–924

Aukrust P, Frøland SS, Liabakk NB, Müller F, Nordøy I, Haug C, Espevik T (1994) Release of cytokines, soluble cytokine receptors, and interleukin-1 receptor antagonist after intravenous immunoglobulin administration in vivo. Blood 84:2136–2143

Aydin N, Searchfield GD (2019) Changes in tinnitus and physiological biomarkers of stress in response to short-term broadband noise and sounds of nature. Complement Ther Med 46:62–68

Bae YJ, Reinelt J, Netto J, Uhlig M, Willenberg A, Ceglarek U, Villringer A, Thiery J, Gaebler M, Kratzsch J (2019) Salivary cortisone, as a biomarker for psychosocial stress, is associated with state anxiety and heart rate. Psychoneuroendocrinology 101:35–41

Barranco T, Rubio CP, Tvarijonaviciute A, Rubio M, Damia E, Lamy E, Cugat R, Cerón JJ, Tecles F, Escribano D (2019) Changes of salivary biomarkers under different storage conditions: effects of temperature and length of storage. Biochemia Med 29(1):010706

Bartley EJ, Palit S, Kuhn BL, Kerr KL, Terry EL, DelVentura JL, Rhudy JL (2015) Natural variation in testosterone is associated with hypoalgesia in healthy women. Clin J Pain 31:730–739

Benton LA, Yates FE (1990) Ultradian adrenocortical and circulatory oscillations in conscious dogs. Am J Phys 258:R578–R590

Blagrove M, Fouquet NC, Baird AL, Pace-Schott EF, Davies AC, Neuschaffer JC, Henley-Einion JA, Weidemann CT, Thome J, McNamara P, Turnbull OH (2012) Association of Salivary-Assessed Oxytocin and Cortisol Levels with Time of Night and Sleep Stage. J Neural Transm 119:1223–1232

Bleakman D, Alt A, Nisenbaum ES (2006) Glutamate receptors and pain. Semin Cell Dev Biol 17:592–604

Bosch JA, Veerman EC, de Geus EJ, Proctor GB (2011) α-Amylase as a reliable and convenient measure of sympathetic activity: don't start salivating just yet! Psychoneuroendocrinology 36:449–453

Brown SN, Knowles TG, Wilkins LJ, Chadd SA, Warriss PD (2005) The response of pigs to being loaded or unloaded onto commercial animal transporters using three systems. Vet J 170:91–100

Bushong DM, Friend TH, Knabe DA (2000) Salivary and plasma cortisol response to adrenocorticotropin administration in pigs. Lab Anim 34:171–181

Carlesso LC, Sturgeon JA, Zautra AJ (2016) Exploring the relationship between disease-related pain and cortisol levels in women with osteoarthritis. Osteoarthr Cartil 24:2048–2054

Carpenter GH, Proctor GB, Anderson LC, Zhang XS, Garrett JR (2000) Immunoglobulin A secretion into saliva during dual sympathetic and parasympathetic nerve stimulation of rat submandibular glands. Exp Physiol 85:281–286

Carter C, Pournajafi-Nazarloo SH, Kramer KM, Ziegler TE, White-Traut R, Bello D, Schwertz D (2007) Oxytocin: behavioral associations and potential as a salivary biomarker. Ann N Y Acad Sci 1098:312–322

Casal N, Manteca X, Escribano D, Cerón JJ, Fàbrega E (2017) Effect of environmental enrichment and herbal compound supplementation on physiological stress indicators (chromogranin A, cortisol and tumour necrosis factor-α) in growing pigs. Animal 11:1228–1236

Chennaoui M, Bougard C, Drogou C, Langrume C, Miller C, Gomez-Merino D, Vergnoux F (2016) Stress biomarkers, mood states, and sleep during a major competition: "success" and "failure" athlete's profile of high-level swimmers. Front Physiol 14:7–94

Choi JC, Chung MI, Lee YD (2012) Modulation of pain sensation by stress-related testosterone and cortisol. Anaesthesia 67:1146–1151

Cobb ML, Iskandarani K, Chinchilli VM, Dreschel NA (2016) A systematic review and meta-analysis of salivary cortisol measurement in domestic canines. Domest Anim Endocrinol 57:31–42

Contreras-Aguilar MD, Escribano D, Martínez-Subiela S, Martínez-Miró S, Rubio M, Tvarijonaviciute A, Tecles F, Cerón JJ (2017a) Influence of the way of reporting alpha-Amylase values in saliva in different naturalistic situations: a pilot study. PLoS One 12(6):e0180100

Contreras-Aguilar MD, Tecles F, Martínez-Subiela S, Escribano D, Bernal LJ, Cerón JJ (2017b) Detection and measurement of alpha-amylase in canine saliva and changes after an experimentally induced sympathetic activation. BMC Vet Res 13:266

Contreras-Aguilar MD, Escribano D, Martín-Cuervo M, Tecles F, Cerón JJ (2018a) Salivary alpha-amylase activity and cortisol in horses with acute abdominal disease: a pilot study. BMC Vet Res 14:156

Contreras-Aguilar MD, Escribano D, Martínez-Subiela S, Martínez-Miró S, Cerón JJ, Tecles F (2018b) Changes in alpha-amylase activity, concentration and isoforms in pigs after an experimental acute stress model: an exploratory study. BMC Vet Res 14:256

Contreras-Aguilar MD, Escribano D, Martínez-Miró S, López-Arjona M, Rubio CP, Martínez-Subiela S, Cerón JJ, Tecles F (2019a) Application of a score for evaluation of pain, distress and discomfort in pigs with lameness and prolapses: correlation with saliva biomarkers and severity of the disease. Res Vet Sci 6(126):155–163

Contreras-Aguilar MD, Escribano D, Quiles A, López-Arjona M, Cerón JJ, Martínez-Subiela S, Hevia ML, Tecles F (2019b) Evaluation of new biomarkers of stress in saliva of sheep. Animal 13:1278–1286

Contreras-Aguilar MD, Martínez-Subiela S, Cerón JJ, Martín-Cuervo M, Tecles F, Escribano D (2019c) Salivary alpha-amylase activity and concentration in horses with acute abdominal disease: association with outcome. Equine Vet J 51:569–574

Contreras-Aguilar MD, Monkeviciene I, Ceron JJ, Silinskas I, Vallejo-Mateo PJ, Tecles F, Martinez-Subiela S, Tvarijonaviciute A, Zelvyte R (2019d) Biochemical changes in saliva of cows with inflammation: a pilot study. Res Vet Sci 124:383–386

Contreras-Aguilar MD, Vialaret J, de Périère DD, Escribano D, Lehmann S, Tecles F, Cerón JJ, Hirtz C (2019e) Variation of human salivary alpha-amylase proteoforms in three stimulation models. Clin Oral Investig. https://doi.org/10.1007/s00784-019-03021-9

Cook NJ, Schaefer AL, Lepage P, Morgan Jones SD (1996) Salivary vs. serum cortisol for the assessment of adrenal activity in swine. Can J Anim Sci 76:329–335

Coulon M, Lévy F, Ravel C, Nowak R, Boissy A (2014) Mild effects of gestational stress and social reactivity on the onset of mother-young interactions and bonding in sheep. Stress 17:460–470

Cruz-Almeida Y, Aguirre M, Sorenson H, Tighe P, Wallet SM, Riley JL (2017) 3rd Age differences in salivary markers of inflammation in response to experimental pain: does venipuncture matter? J Pain Res 10:2365–2372

da Silva Campos MJ, Souza Alves CC, Barbosa Raposo NR, Ferreira AP, Farinazzo Vitral RW (2010) Influence of salivary secretory immunoglobulin A level on the pain experienced by orthodontic patients. Med Sci Monit 16:CR405–CR409

Deminice R, Sicchieri T, Payão PO, Jordão AA (2010) Blood and salivary oxidative stress biomarkers following an acute session of resistance exercise in humans. Int J Sports Med 31:599–603

Den R, Toda M, Nagasawa S (2007) Circadian rhythm of human salivary chromogranin A. Biomed Res 28:57–60

Diez-Ruiz A, Tilz GP, Zangerle R, Baier-Bitterlich G, Wachter H, Fuchs D (1995) Soluble receptors for tumour necrosis factor in clinical laboratory diagnosis. Eur J Haematol 54:1–8

Dimsdale JE, O'Connor DT, Ziegler M, Mills P (1992) Chromogranin A correlates with norepinephrine release rate. Life Sci 51:519–525

Drouet JB, Fauvelle F, Maunoir-Regimbal S, Fidier N, Maury R, Peinnequin A, Denis J, Buguet A, Canini F (2015) Differences in prefrontal cortex GABA/glutamate ratio after acute restraint stress in rats are associated with specific behavioral and neurobiological patterns. Neuroscience 29:155–165

Duncan IJH (2005) Science-based assessment of animal welfare: farm animals. Sci Tech Rev 24:483–492

Eckersall PD (1995) Acute phase proteins as markers of inflammatory lesions. Comp Haematol Int 5:93–97

Edwards KL, Bansiddhi P, Paris S, Galloway M, Brown JL (2019) The development of an immunoassay to measure immunoglobulin A in Asian elephant feces, saliva, urine and serum as a potential biomarker of well-being. Conserv Physiol 7:coy077

Engeland CG, Hugo FN, Hilgert JB, Nascimento GG, Junges R, Lim HJ, Marucha PT, Bosch JA (2016) Psychological distress and salivary secretory immunity. Brain Behav Immun 52:11–17

Escribano D, Soler L, Gutiérrez AM, Martínez-Subiela S, Cerón JJ (2013) Measurement of chromogranin A in porcine saliva: validation of a time-resolved immunofluorometric assay and evaluation of its application as a marker of acute stress. Animal 7(4):640–647

Escribano D, Campos PH, Gutiérrez AM, Le Floc'h N, Cerón JJ, Merlot E (2014a) Effect of repeated administration of lipopolysaccharide on inflammatory and stress markers in saliva of growing pigs. Vet J 200:393–397

Escribano D, Fuentes-Rubio M, Cerón JJ (2014b) Salivary testosterone measurements in growing pigs: validation of an automated chemiluminescent immunoassay and its possible use as an acute stress marker. Res Vet Sci 97:20–25

Escribano D, Gutiérrez AM, Fuentes-Rubio M, Cerón JJ (2014c) Saliva chromogranin A in growing pigs: a study of circadian patterns during daytime and stability under different storage conditions. Vet J 199(3):355–359

Escribano D, Gutiérrez AM, Tecles F, Cerón JJ (2015) Changes in saliva biomarkers of stress and immunity in domestic pigs exposed to a psychosocial stressor. Res Vet Sci 102:38–44

Escribano D, Horvatić A, Contreras-Aguilar MD, Guillemin N, Cerón JJ, Tecles F, Martinez-Miró S, Eckersall PD, Manteca X, Mrljak V (2019a) Changes in saliva proteins in two conditions of compromised welfare in pigs: an experimental induced stress by nose snaring and lameness. Res Vet Sci 125:227–234

Escribano D, Ko HL, Chong Q, Llonch L, Manteca X, Llonch P (2019b) Salivary biomarkers to monitor stress due to aggression after weaning in piglets. Res Vet Sci 123:178–183

Fàbrega E, Marcet-Rius M, Vidal R, Escribano D, Cerón JJ, Manteca X, Velarde A (2019) The effects of environmental enrichment on the physiology, behaviour, productivity and meat quality of pigs raised in a hot climate. Animals (Basel) 9:E235

Fischer M, Wille G, Klien S, Shanib H, Holle D, Gaul C, Broessner G (2012) Brain-derived neurotrophic factor in primary headaches. J Headache Pain 13:469–475

Forsyth IA, Margaret CN (2009) Introduction: the myoepithelial cell and milk letdown; entrance to the multifunctional role of oxytocin. J Mammary Gland Biol Neoplasia 14:221–222

Fuentes M, Tecles F, Gutiérrez A, Otal J, Martínez-Subiela S, Cerón JJ (2011) Validation of an automated method for salivary alpha-amylase measurements in pigs (Sus scrofa domesticus) and its application as a stress biomarker. J Vet Diagn Investig 23:282–287

Fuentes-Rubio M, Fuentes F, Otal J, Quiles A, Luisa HM (2016) Validation of an assay for quantification of alpha-amylase in saliva of sheep. Can J Vet Res 80:197–202

Gatti R, De Palo EF (2011) An update: salivary hormones and physical exercise. Scand J Med Sci Sports 21:157–169

Geiss A, Rohleder N, Anton F (2012) Evidence for an association between an enhanced reactivity of interleukin-6 levels and reduced glucocorticoid sensitivity in patients with fibromyalgia. Psychoneuroendocrinology 37:671–684

Gerdle B, Ghafouri B, Ernberg M, Larsson B (2014) Chronic musculoskeletal pain: review of mechanisms and biochemical biomarkers as assessed by the microdialysis technique. J Pain Res 7:313–326

Giltay EJ, Enter D, Zitman FG, Penninx BW, van Pelt J, Spinhoven P, Roelofs K (2012) Salivary testosterone: associations with depression, anxiety disorders, and antidepressant use in a large cohort study. J Psychosom Res 72:205–213

Goedert M, Nagy JI, Emson PC (1982) The origin of substance P in the rat submandibular gland and its major duct. Brain Res 252:327–333

Gonzalez D, Marquina R, Rondon N, Rodriguez-Malaver A, Reyes R (2008) Effects of aerobic exercise on uric acid, total antioxidant activity, oxidative stress, and nitric oxide in human saliva. Res Sport Med 16:128–137

González LA, Schwartzkopf-Genswein KS, Caulkett NA, Janzen E, McAllister TA, Fierheller E, Schaefer AL, Haley DB, Stookey JM, Hendrick S (2010) Pain mitigation after band castration of beef calves and its effects on performance, behavior, Escherichia coli, and salivary cortisol. J Anim Sci 88:802–810

Goodin BR, Smith MT, Quinn NB, King CD, McGuire L (2012) Poor sleep quality and exaggerated salivary cortisol reactivity to the cold pressor task predict greater acute pain severity in a non-clinical sample. Biol Psychol 91:36–41

Greco R, Tassorelli C, Sandrini G, Di Bella P, Buscone S, Nappi G (2008) Role of calcitonin gene-related peptide and substance P in different models of pain. Cephalalgia: an international journal of. Headache 28:114–126

Green TC, Mellor DJ (2011) Extending ideas about animal welfare assessment to include 'quality of life' and related concepts. N Z Vet J 59:263–271

Griffin J (1986) Well-being: its meaning, measurement, and moral importance. Clarendon paperbacks, pp 10–16

Groza P, Zamfir V, Lungu D (1971) Postoperative salivary amylase changes in children. Rev Roum Physiol 8:307–312

Guglielminotti J, Dehoux M, Mentré F, Bedairia E, Montravers P, Desmonts JM, Longrois D (2012) Assessment of salivary amylase as a stress biomarker in pregnant patients. Int J Obstet Anesth 21:35–39

Häger C, Biernot S, Buettner M, Glage S, Keubler LM, Held N, Bleich EM, Otto K, Müller CW, Decker S, Talbot SR, Bleich A (2017) The Sheep Grimace Scale as an indicator of post-operative distress and pain in laboratory sheep. PLoS One 12(4):e0175839

Hart KA (2012) The use of cortisol for the objective assessment of stress in animals: pros and cons. Vet J 192:137–139
Hernandez CE, Thierfelder T, Svennersten-Sjaunja K, Berg C, Orihuela A, Lidfors L (2014) Time lag between peak concentrations of plasma and salivary cortisol following a stressful procedure in dairy cattle. Acta Vet Scand 9:56–61
Hill-Soderlund AL, Mills-Koonce WR, Propper C, Calkins SD, Granger DA, Moore GA, Gariepy JL, Cox MJ (2008) Parasympathetic and sympathetic responses to the strange situation in infants and mothers from avoidant and securely attached dyads. Dev Psychobiol 50:361–376
Holt-Lunstad J, Birmingham W, Light KC (2011) The influence of depressive symptomatology and perceived stress on plasma and salivary oxytocin before, during and after a support enhancement intervention. Psychoneuroendocrinology 36:1249–1256
Horvat-Gordon M, Granger DA, Schwartz EB, Nelson VJ, Kivlighan KT (2005) Oxytocin is not a valid biomarker when measured in saliva by immunoassay. Physiol Behav 84:445–448
Huang Y, Liu Z, Liu W, Yin C, Ci L, Zhao R, Yang X (2017) Short communication: salivary haptoglobin and chromogranin A as non-invasive markers during restraint stress in pigs. Res Vet Sci 114:27–30
Jafari MJ, Khosrowabadi R, Khodakarim S, Khodagholi F, Mohammadian F (2019) The effects of combined exposure to noise and heat on human salivary cortisol and blood pressure. Int J Occup Saf Ergon 22:1–22
Jang MU, Park JW, Kho HS, Chung SC, Chung JW (2011) Plasma and saliva levels of nerve growth factor and neuropeptides in chronic migraine patients. Oral Dis 17:187–193
Jarvis S, Moinard C, Robson SK, Baxter E, Ormandy E, Douglas AJ, Seckl JR, Russell JA, Lawrence AB (2006) Programming the offspring of the pig by prenatal social stress: neuroendocrine activity and behaviour. Horm Behav 49:68–80
Jasim H, Carlsson A, Hedenberg-Magnusson B, Ghafouri B, Ernberg M (2018) Saliva as a medium to detect and measure biomarkers related to pain. Sci Rep 8(1):3220
Jenkins BN, Granger DA, Roemer RJ, Martinez A, Torres TK, Fortier MA (2018) Emotion regulation and positive affect in the context of salivary alpha-amylase response to pain in children with cancer. Pediatr Blood Cancer 65(6):e26973
Kaiser M, Jacobsen S, Andersen PH, Bækbo P, Cerón JJ, Dahl J, Escribano D, Theil PK, Jacobson M (2018) Hormonal and metabolic indicators before and after farrowing in sows affected with postpartum dysgalactia syndrome. BMC Vet Res 14(1):334
Kallman TF, Ghafouri B, Bäckryd E (2018) Salivary beta-endorphin and substance P are not biomarkers of neuropathic chronic painpropensity. Heliyon 4(8):e00718
Kanno T, Asada N, Yanase H, Iwanaga T, Ozaki T, Nishikawa Y, Iguchi K, Mochizuki T, Hoshino M, Yanaihara N (1999) Salivary secretion of highly concentrated chromogranin A in response to noradrenaline and acetylcholine in isolated and perfused rat submandibular glands. Exp Physiol 84:1073–1083
Kennedy B, Dillon E, Mills PJ, Ziegler MG (2001) Catecholamines in human saliva. Life Sci 69:87–99
Kikkawa A, Uchida Y, Nakade T, Taguchi K (2003) Salivary secretory IgA concentrations in beagle dogs. Vet Med Sci 65:689–693
Klentrou P, Cieslak T, MacNeil M, Vintinner A, Plyley M (2002) Effect of moderate exercise on salivary immunoglobulin A and infection risk in humans. Eur J Appl Physiol 87:153–158
Koopmans SJ, van der Staay FJ, Le Floc'h N, Dekker R, van Diepen JT, Jansman AJ (2012) Effects of surplus dietary L-tryptophan on stress, immunology, behavior, and nitrogen retention in endotoxemic pigs. J Anim Sci 90:241–251
Kovács L, Kézér FL, Kulcsár-Huszenicza M, Ruff F, Szenci O, Jurkovich V (2016) Hypothalamic-pituitary-adrenal and cardiac autonomic responses to transrectal examination differ with behavioral reactivity in dairy cows. J Dairy Sci 99:7444–7457
La Marca-Ghaemmaghami P, La Marca R, Dainese SM, Haller M, Zimmermann R, Ehlert U (2013) The association between perceived emotional support, maternal mood, salivary cortisol,

salivary cortisone, and the ratio between the two compounds in response to acute stress in second trimester pregnant women. J Psychosom Res 75:314–320
Lee T, Shimizu T, Iijima M, Obinata K, Yamashiro Y, Nagasawa S (2006) Evaluation of psychosomatic stress in children by measuring salivary chromogranin A. Acta Paediatr 95:935–939
Lokesh K, Kannabiran J, Rao MD (2016) Salivary lactate dehydrogenase (LDH)- a novel technique in oral cancer detection and diagnosis. J Clin Diagn Res 10(2):ZC34–ZC37
López-Arjona M, Mateo SV, Manteca X, Escribano D, Cerón JJ, Martínez-Subiela S (2019) Oxytocin in saliva of pigs: an assay for its measurement and changes after farrowing. Domest Anim Endocrinol 70:106384
Mach N, Foury A, Kittelmann S, Reigner F, Moroldo M, Ballester M, Esquerré D, Rivière J, Sallé G, Gérard P, Moisan MP, Lansade L (2017) The effects of weaning methods on gut microbiota composition and horse physiology. Front Physiol 8:535
MacLean EL, Gesquiere LR, Gee N, Levy K, Martin WL, Carter CS (2018) Validation of salivary oxytocin and vasopressin as biomarkers in domestic dogs. J Neurosci Methods 293:67–76
Mandel AL, Peyrot d, Gachons C, Plank KL, Alarcon S, Breslin PA (2010) Individual differences in AMY1 gene copy number, salivary α-amylase levels, and the perception of oral starch. PLoS One 5:e13352
Marti S, Devant M, Amatayakul-Chantler S, Jackson JA, Lopez E, Janzen ED, Schwartzkopf-Genswein KS (2015) Effect of anti-gonadotropin-releasing factor vaccine and band castration on indicators of welfare in beef cattle. J Anim Sci 93:1581–1591
Marti S, Meléndez DM, Pajor EA, Moya D, Heuston CEM, Gellatly D, Janzen ED, Schwartzkopf-Genswein KS (2017) Effect of band and knife castration of beef calves on welfare indicators of pain at three relevant industry ages: II. Chronic pain. J Anim Sci 95:4367–4380
Martínez-Miró S, Tecles F, Ramón M, Escribano D, Hernández F, Madrid J, Orengo J, Martínez-Subiela S, Manteca X, Cerón JJ (2019) Causes, consequences and biomarkers of stress in swine: an update. BMC Vet Res 125:227–234
Masih T, Dimmock JA, Guelfi KJ (2019) The effect of a single, brief practice of progressive muscle relaxation after exposure to an acute stressor on subsequent energy intake. Stress Health. https://doi.org/10.1002/smi.2891. [In press]
Mehta PH, Josephs RA (2010) Testosterone and cortisol jointly regulate dominance: evidence for a dual-hormone hypothesis. Horm Behav 58:898–906
Merlot E, Mounier AM, Prunier A (2011) Endocrine response of gilts to various common stressors: a comparison of indicators and methods of analysis. Physiol Behav 102:259–265
Miller SC, Kennedy C, DeVoe D, Hickey M, Nelson T, Kogan L (2009) An examination of changes in oxytocin levels in men and women before and after interaction with a bonded dog. Anthrozoös 22(2009):31–42
Mitsui S, Yamamoto M, Nagasawa M, Mogi K, Kikusui T, Ohtani N, Ohta M (2011) Urinary oxytocin as a noninvasive biomarker of positive emotion in dogs. Horm Behav 60:239–243
Montoya ER, Terburg D, Bos PA, van Honk J (2012) Testosterone, cortisol, and serotonin as key regulators of social aggression: a review and theoretical perspective. Motiv Emot 36:65–73
Morrison WE, Haas EC, Shaffner DH, Garrett ES, Fackler JC (2003) Noise, stress, and annoyancein a pediatric intensive care unit. Crit Care Med 31:113
Moya SL, Boyle L, Arkins S, Lynch PB (2006) Effects of husbandry and low-dose lipopolysaccharide challenge on the acute phase response of young pigs, pp 24–30. https://t-stor.teagasc.ie/handle/11019/982. Accessed 13 July 19.
Moya D, González LA, Janzen E, Caulkett NA, Fireheller E, Schwartzkopf-Genswein KS (2014) Effects of castration method and frequency of intramuscular injections of ketoprofen on behavioral and physiological indicators of pain in beef cattle. J Anim Sci 92:1686–1697
Müller MJ (2011) Helplessness and perceived pain intensity: relations to cortisol concentrations after electrocutaneous stimulation in healthy young men. Biopsychosocial Med 30:5–8
Muneta Y, Yoshikawa T, Minagawa Y, Shibahara T, Maeda R, Omata Y (2010) Salivary IgA as a useful non-invasive marker for restraint stress in pigs. Vet Med Sci 72:1295–1300

Muneta Y, Minagawa Y, Nakane T, Shibahara T, Yoshikawa T, Omata Y (2011) Interleukin-18 expression in pig salivary glands and salivary content changes during acute immobilization stress. Stress 14:549–556
Nam JW, Chung JW, Kho HS, Chung SC, Kim YK (2007) Nerve growth factor concentration in human saliva. Oral Dis 13:187–192
Nater UM, Rohleder N (2009) Salivary alpha-amylase as a non-invasive biomarker for the sympathetic nervous system: current state of research. Psychoneuroendocrinology 34:486–496
Naumova EA, Sandulescu T, Al Khatib P, Thie M, Lee WK, Zimmer S, Arnold WH (2012) Acute short-term mental stress does not influence salivary flow rate dynamics. PLoS One 7:e51323
Ng V, Koh D, Ong HY, Chia SE, Ong CN (1999) Are salivary immunoglobulin A and lysozyme biomarkers of stress among nurses? J Occup Environ Med 41:920–927
Ninomiya S, Sato S (2011) The assessment of the effect of presenting a companion's face picture on social isolation stress using saliva sampling in cow. Anim Sci J 82:787–791
Nussbaum MC (2000) Women and human development: the capabilities approach. Cambridge University Press, New York, pp 78–80
Obayashi K (2013) Salivary mental stress proteins. Clin Chim Acta 425:196–201
Obmiński Z, Wojtkowiak M, Stupnicki R, Golec L, Hackney AC (1997) Effect of acceleration stress on salivary cortisol and plasma cortisol and testosterone levels in cadet pilots. J Physiol Pharmacol 48:193–200
Odendaal J, Meintjes RA (2003) Neurophysiological correlates of affiliative behaviour between humans and dogs. Vet J 165:296–301
Okamoto Y, Ohtani N, Uchiyama H, Ohta M (2009) The feeding behavior of dogs correlates with their responses to commands. J Vet Med Sci 71:1617–1621
Ooishi Y, Mukai H, Watanabe K, Kawato S, Kashino M (2017) Increase in salivary oxytocin and decrease in salivary cortisol after listening to relaxing slow-tempo and exciting fast-tempo music. PLoS One 12(12):e0189075
Ott S, Soler L, Moons CP, Kashiha MA, Bahr C, Vandermeulen J, Janssens S, Gutiérrez AM, Escribano D, Cerón JJ, Berckmans D, Tuyttens FA, Niewold TA (2014) Different stressors elicit different responses in the salivary biomarkers cortisol, haptoglobin, and chromogranin A in pigs. Res Vet Sci 97:124–128
Pang WY, Earley B, Sweeney T, Crowe MA (2006) Effect of carprofen administration during banding or burdizzo castration of bulls on plasma cortisol, in vitro interferon-gamma production, acute-phase proteins, feed intake, and growth. J Anim Sci 84:351–359
Pang WY, Earley B, Murray M, Sweeney T, Gath V, Crowe MA (2011) Banding or Burdizzo castration and carprofen administration on peripheral leukocyte inflammatory cytokine transcripts. Res Vet Sci 90:127–132
Parfit D (1984) Reasons and persons. Oxford University Press, Oxford, pp 493–502
Parris WC, Kambam JR, Naukam RJ, Rama Sastry BV (1990) Immunoreactive substance P is decreased in saliva of patients with chronic back pain syndromes. Anesth Analg 70:63–67
Parrott RF, Misson BH (1989) Changes in pig salivary cortisol in response to transport simulation, food and water deprivation, and mixing. Br Vet J 145:501–505
Paszynska E, Dmitrzak-Weglarz M, Tyszkiewicz-Nwafor M, Slopien A (2016) Salivary alpha-amylase, secretory IgA and free cortisol as neurobiological components of the stress response in the acute phase of anorexia nervosa. World J Biol Psychiatry 17(4):266–273
Peeters M, Sulon J, Beckers J, Ledoux D, Vandeheede M (2011) Comparison between blood serum and salivary cortisol concentrations in horses using an adrenocorticotropic hormone challenge. Equine Vet J 43:487–493
Posner EA (2008) Human Welfare, Not Human Rights. Columbia Law Rev 108:1758–1802
Pressman SD, Cohen S (2005) Does positive affect influence health? Psychol Bull 131:925–971
Radaković M, Borozan S, Djelić N, Ivanović S, Miladinović DĆ, Ristanić M, Spremo-Potparević B, Stanimirović Z (2018) Nitroso-oxidative stress, acute phase response, and cytogenetic damage in wistar rats treated with adrenaline. Oxidative Med Cell Longev 21:1805354. eCollection 2018

Rai B, Kaur J (2011) Salivary stress markers and psychological stress in simulated microgravity: 21 days in 6° head-down tilt. J Oral Sci 53:103–107

Rey-Salgueiro L, Martinez-Carballo E, Fajardo P, Chapela MJ, Espiñeira M, Simal-Gandara J (2018) Meat quality in relation to swine well-being after transport and during lairage at the slaughterhouse. Meat Sci 142:38–43

Ring RH, Malberg JE, Potestio L, Ping J, Boikess S, Luo B et al (2006) Anxiolytic-like activity of oxytocin in male mice: behavioral and autonomic evidence, therapeutic implications. Psychopharmacology 185(2):218–225

Robles TF, Sharma R, Park KS, Harrell L, Yamaguchi M, Shetty V (2012) Utility of a salivary biosensor for objective assessment of surgery-related stress. J Oral Maxillofac Surg 70:2256–2263

Rohleder N, Wolf JM, Maldonado EF, Kirschbaum C (2006) The psychosocial stress-induced increase in salivary alpha-amylase is independent of saliva flow rate. Psychophysiology 43:645–652

Romero-Martínez A, González-Bono E, Lila M, Moya-Albiol L (2013) Testosterone/ cortisol ratio in response to acute stress: a possible marker of risk for marital violence. Soc Neurosci 8:240–247

Rubio CP, Mainau E, Cerón JJ, Contreras-Aguilar MD, Martínez-Subiela S, Navarro E, Tecles F, Manteca X, Escribano D (2019a) Biomarkers of oxidative stress in saliva in pigs: analytical validation and changes in lactation. BMC Vet Res 15:144

Rubio CP, Contreras-Aguilar MD, Quiles A, López-Arjona M, Cerón JJ, Martínez-Subiela S, Hevia ML, Escribano D, Tecles F (2019b) Biomarkers of oxidative stress in saliva of sheep: analytical performance and changes after an experimentally induced stress. Res Vet Sci 123:71–76

Ruis MA, Te Brake JH, Engel B, Ekkel ED, Buist WG, Blokhuis HJ, Koolhaas JM (1997) The circadian rhythm of salivary cortisol in growing pigs: effects of age, gender, and stress. Physiol Behav 62:623–630

Russell JA, Carroll JM (1999) On the bipolarity of positive and negative affect. Psychol Bull 125:3–30

Russell G, Henley D, Leendertz J, Douthwaite J, Wood S, Stevens A, Woltersdorf W, Peeters B, Ruigt G, White A, Velhuis J, Lightman S (2010) Rapid glucocorticoid receptor-mediated inhibition of hypothalamic–pituitary– adrenal ultradian activity in healthy males. J Neurosci 30:6106–6115

Saruta J, Tsukinoki K, Sasaguri K, Ishii H, Yasuda M, Osamura YR, Watanabe Y, Sato S (2005) Expression and localization of chromogranin A gene and protein in human submandibular gland. Cells Tissues Organs 180:237–244

Sato F, Kanno T, Nagasawa S, Yanaihara N, Ishida N, Hasegawa T, Iwanaga T (2002) Immunohistochemical localization of chromogranin A in the acinar cells of equine salivary glands contrasts with rodent glands. Cells Tissues Organs 172:29–36

Scannapieco FA, Torres G, Levine MJ (1993) Salivary alpha-amylase: role in dental plaque and caries formation. Crit Rev Oral Biol Med 4:301–307

Schäffer L, Burkhardt T, Müller-Vizentini D, Rauh M, Tomaske M, Mieth RA, Bauersfeld U, Beinder E (2008) Cardiac autonomic balance in small-for-gestational-age neonates. Am J Physiol Heart Circ Physiol 294:H884–H890

Schönbom H, Kassens A, Hopster-Iversen C, Klewitz J, Piechotta M, Martinsson G, Kißler A, Burger D, Sieme H (2015) Influence of transrectal and transabdominal ultrasound examination on salivary cortisol, heart rate, and heart rate variability in mares. Theriogenology 83:749–756

Schönreiter S, Huber H, Lohmüller V, Zanella AJ, Unshelm J, Henke J, Erhardt W (1999) [Salivary cortisol as a stress parameter in piglets]. [Article in German]. Tierarztl Prax Ausg G Grosstiere Nutztiere 27:175–179

Schwab KO, Heubel G, Bartels H (1992) Free epinephrine, norepinephrine and dopamine in saliva and plasma of healthy adults. Eur J Clin Chem Clin Biochem 30:541–544

Shin YJ, Shin NS (2016) Evaluation of effects of olfactory and auditory stimulation on separation anxiety by salivary cortisol measurement in dogs. J Vet Sci 17:153–158

Shin YJ, Shin NS (2017) Relationship between sociability toward humans and physiological stress in dogs. J Vet Med Sci 79:1278–1283
Shubayev V, Kato K, Myers RR (2010) Cytokines in pain. In: Kruger L, Light AR (eds) Source translational pain research: from mouse to man, Chapter 8. Frontiers in Neuroscience. CRC Press/Taylor & Francis, Boca Raton
Sjörs A, Ljung T, Jonsdottir IH (2012) Long-term follow-up of cortisol awakening response in patients treated for stressrelated exhaustion. Br Med J Open 2:e001091
Slavish DC, Graham-Engeland JE, Smyth JM, Engeland CG (2015) Salivary markers of inflammation in response to acute stress. Brain Behav Immun 44:253–269
Smulders D, Verbeke G, Mormède P, Geers R (2006) Validation of a behavioral observation tool to assess pig welfare. Physiol Behav 89:438–447
Soler L, Gutiérrez A, Escribano D, Fuentes M, Cerón JJ (2013) Response of salivary haptoglobin and serum amyloid A to social isolation and short road transport stress in pigs. Res Vet Sci 95:298–302
Sordillo LM, Aitken SL (2009) Impact of oxidative stress on the health and immune function of dairy cattle. Vet Immunol Immunopathol 128:104–109
Spiropoulos K, Lymberopoulos D, Garantziotis G, Gogos C (1993) Salivary immunoglobulin A production in chronic bronchitis patients given an orally administered bacterial extract. Respiration 60:313–318
Spruijt BM, van den Bos R, Pijlman FTA (2001) A concept of welfare based on reward evaluating mechanisms in the brain: antic- ipatory behaviour as an indicator for the state of reward systems. Appl Anim Behav Sci 72:145–171
Svobodová I, Chaloupková H, Končel R, Bartoš L, Hradecká L, Jebavý L (2014) Cortisol and secretory immunoglobulin A response to stress in German shepherd dogs. PLoS One 9(3):e90820
Symons F, ElGhazi I, Reilly BG, Barney CC, Hanson L, Panoskaltsis-Mortari A, Armitage IM, Wilcox GL (2015) Can biomarkers differentiate pain and no pain subgroups of nonverbal children with cerebral palsy? A preliminary investigation based on noninvasive saliva sampling. Pain Med 16:249–256
Syrjänen S, Piironen P, Yli-Urpo A (1984) Salivary content of patients with subjective symptoms resembling galvanic pain. Oral Surg Oral Med Oral Pathol 58:387–393
Takatsuji K, Sugimoto Y, Ishizaki S, Ozaki Y, Matsuyama E, Yamaguchi Y (2008) The effects of examination stress on salivary cortisol, immunoglobulin A, and chromogranin A in nursing students. Biomed Res 29:221–224
Takiyyuddin MA, Brown MR, Dinh TQ, Cervenka JH, Braun SD, Parmer RJ, Kennedy B, O'Connor DT (1994) Sympatho–adrenal secretion in humans: factors governing catecholamine and storage vesicle peptide co-release. J Auton Pharmacol 14:187–200
Tecles F, Tvarijonaviciute A, De Torre C, Carrillo JM, Rubio M, García M, Cugat R, Cerón JJ (2016) Total esterase activity in human saliva: validation of an automated assay, characterization and behaviour after physical stress. Scand J Clin Lab Invest 76:324–330
Tecles F, Contreras-Aguilar MD, Martínez-Miró S, Tvarijonaviciute A, Martínez-Subiela S, Escribano D, Cerón JJ (2017) Total esterase measurement in saliva of pigs: validation of an automated assay, characterization and changes in stress and disease conditions. Res Vet Sci 114:170–176
Tecles F, Escribano D, Contreras-Aguilar MD, Rubio CP, Szczubiał M, Cerón JJ, Dąbrowski R, Tvarijonaviciute A (2018) Evaluation of adenosine deaminase in saliva and serum, and salivary α-amylase, in canine pyometra at diagnosis and after ovariohysterectomy. Vet J 236:102–110
Toda M, Den R, Nagasawa S, Kitamura K, Morimoto K (2007) Relationship between lifestyle score and salivary stress markers cortisol and chromogranin A. Arch Environ Occup Health 60:266–269
Tóthová L, Kamodyová N, Červenka T, Celec P (2015) Salivary markers of oxidative stress in oral diseases. Front Cell Infect Microbiol 5:73
Ulmer-Yaniv A, Djalovski A, Priel A, Zagoory-Sharon O, Feldman R (2018) Maternal depression alters stress and immune biomarkers in mother and child. Depress Anxiety 35:1145–1157

Vahedi M, Mazdeh M, Hajilooi M, Farhadian M, Barakian Y, Sadr P (2018) The Relationship between salivary alpha amylase activity and score of McGill pain questionnaire in patients with tension type headache. Basic Clin Neurosci 9:59–64

van der Staay FJ, de Groot J, Schuurman T, Korte SM (2008) Repeated social defeat in female pigs does not induce neuroendocrine symptoms of depression, but behavioral adaptation. Physiol Behav 93:453–460

van Stegeren AH, Wolf OT, Kindt M (2008) Salivary alpha amylase and cortisol responses to different stress tasks: impact of sex. Int J Psychophysiol 69:33–40

Walker LJ, Aldhous MC, Drummond HE, Smith BR, Nimmo ER, Arnott ID, Satsangi J (2004) Anti-Saccharomyces cerevisiae antibodies (ASCA) in Crohn's disease are associated with disease severity but not NOD2/CARD15 mutations. Clin Exp Immunol 135:490–496

Weisman O, Zagoory-Sharon O, Feldman R (2012) Intranasal oxytocin administration is reflected in human saliva. Psychoneuroendocrinology 37:1582–1586

Wetherell MA, Hyland ME, Harris JE (2004) Secretory immunoglobulin A reactivity to acute and cumulative acute multi-tasking stress: relationships between reactivity and perceived workload. Biol Psychol 66:257–270

White-Traut R, Watanabe K, Pournajafi-Nazarloo H, Schwertz D, Bell A, Carter CS (2009) Detection of salivary oxytocin levels in lactating women. Dev Psychobiol 51:367–373

Windle RJ, Kershaw YM, Shanks N, Wood SA, Lightman SL, Ingram CD (2004) Oxytocin attenuates stress-induced c-fos mRNA expression in specific forebrain regions associated with modulation of hypothalamo-pituitary-adrenal activity. J Neurosci 24(12):2974–2982

Woolston JL, Gainfredi S, Gertner JM, Paugus JA, Mason JW (1983) Salivary cortisol: a non traumatic sampling technique for assaying cortisol dynamics. J Am Acad Child Psychiatry 22:474–476

Wozniak KM, Rojas C, Wu Y, Slusher BS (2012) The role of glutamate signaling in pain processes and its regulation by GCP II inhibition. Curr Med Chem 19:1323–1334

Yeates JW, Main DCJ (2008) Assessment of positive welfare: a review. Vet J 175:293–300

Yoto A, Fukui N, Kaneda C, Torita S, Goto K, Nanjo F, Yokogoshi H (2018) Black tea aroma inhibited increase of salivary chromogranin-A after arithmetic tasks. J Physiol Anthropol 37(1):3

Yun J, Björkman S, Pöytäkangas M, Peltoniemi O (2017) The effects of ovarian biopsy and blood sampling methods on salivary cortisol and behaviour in sows. Res Vet Sci 114:80–85

Zidverc-Trajkovic J, Stanimirovic D, Obrenovic R, Tajti J, Vécsei L, Gardi J, Németh J, Mijajlovic M, Sternic N, Jankovic L (2009) Calcitonin gene-related peptide levels in saliva of patients with burning mouth syndrome. J Oral Pathol Med 38:29–33

Chapter 15
The Future of Saliva as an Analytical Sample

Asta Tvarijonaviciute, Silvia Martínez-Subiela, Pia López-Jornet, and Elsa Lamy

Saliva has gained attention in research mainly because of the noninvasive way in which the samples are obtained, allowing repeated sampling even in very susceptible populations such as children (Hartman et al. 2016). Moreover, increasing scientific evidence confirms and highlights the potential of saliva as an analytical sample in application to both local and systemic diseases (see previous chapters). Several aspects of saliva have been studied in the search of biomarkers for a broad range of disease conditions. Likewise, different techniques have been used, from simple evaluation of salivary flow and colorimetry to more complex omics approaches. Furthermore, saliva as a fluid biopsy has been assessed, with very promising results (Aro et al. 2017; Khan et al. 2017). However, saliva is still little used in clinical settings, and the main reasons for this could be:

1. Lack of knowledge
2. Lack of uniform guidelines
3. Lack of highly sensitive affordable methods

A. Tvarijonaviciute (✉) · S. Martínez-Subiela
Interdisciplinary Laboratory of Clinical Analysis (Interlab-UMU), Regional Campus of International Excellence Campus Mare Nostrum, University of Murcia, Murcia, Spain
e-mail: asta@um.es; silviams@um.es

P. López-Jornet
Department of Oral Medicine, Faculty of Medicine and Dentistry, Regional Campus of International Excellence Campus Mare Nostrum, University of Murcia, Murcia, Spain
e-mail: majornet@um.es

E. Lamy
MED - Mediterranean Institute for Agriculture, Environment and Development, IIFA - Instituto de Investigação e Formação Avançada, University of Évora, Évora, Portugal
e-mail: ecsl@uevora.pt

A. Tvarijonaviciute et al. (eds.), *Saliva in Health and Disease*,
https://doi.org/10.1007/978-3-030-37681-9_15

15.1 Lack of Knowledge

One of the reasons why saliva is not routinely used in clinical practice is because it is still relatively "unknown" compared to other body fluids such as blood.

Furthermore, the concrete conditions under which testing is performed pose problems, since saliva composition experiences significant variations as a result of different internal and external factors such as:

- Sex, age and ethnicity
- Circadian and circumannual rhythms
- Food consumption and sensory stimuli
- Psychological aspects, e.g., stress
- Physical activity
- Blood leakage

Sex and age are well known causes of variations in saliva composition. As an example, elements like zinc, copper and magnesium have been reported to present differences in relation to gender and in different age groups (Bales et al. 1990). The same can be said of other parameters, including salivary pH, buffering capacity and protein content (Prodan et al. 2015). Besides age and sex, ethnicity can also represent a source of variation in saliva composition, and more in-depth research is required in this field. For example, differences in cortisol levels have been observed among black, Caucasian and Hispanic adults (Hajat et al. 2010). In this same line, differences in saliva microbiome have been reported among people from different climate zones (Li et al. 2014).

Some salivary constituents such as salivary amylase activity, IgA, leptin or cortisol have been shown to present circadian rhythms (Shinkai et al. 1993; Randeva et al. 2003; Ivars et al. 2015; Wada et al. 2017). However, information is lacking regarding other salivary molecules. Such information is of particular importance in order to obtain homogeneous results and minimize bias, with a view to securing more accurate data and improved interpretation of the results.

Factors related to food intake and dietary habits can also result in inter-individual variability in saliva (see Chap. 2). Depending on the type, timing and amount of food previously eaten, as well as on different dietary patterns (e.g., vegetarian, omnivorous diets, etc.), the levels of a given biomarker may differ, and this needs to be taken into account in establishing a diagnosis. Moreover, psychological and physical stress, including physical activity, have been shown to alter analytes in saliva, including cortisol, amylase and IgA, as well as total flow rate and protein concentration (see Chaps. 13 and 14). Further research is also needed in order to assess potential changes in other less studied molecules.

A particularly important aspect, for which knowledge is essential, but which has been not been so widely examined, is blood leakage. Blood contains analytes at concentrations about 1000-fold higher than in saliva. Therefore, even minor blood leakage can significantly alter the concentrations of target biomarkers. Moreover, if the amount of blood is large enough to change the color of the sample, it can interfere with different methods, yielding erroneous analyte values (Kamodyová et al. 2015). Therefore, adequate oral health and the collection of saliva without abrasive methods and before teeth brushing are advised in order to avoid blood contamination of saliva samples.

Apart from the above, a number of studies searching for salivary biomarkers of disease or physiological conditions have adopted so-called "omic" approaches, which allow the simultaneous identification of a large number of analytes. This offers the advantage of affording a large amount of information about potential variations corresponding to a large number of molecules. However, the techniques employed only allow analysis in a limited number of individuals, and validation of each proposed biomarker is necessary before it can be used in practice. This is one of the reasons why such markers are not used on a routine basis, despite the number of studies describing potential biomarkers for different conditions.

Overall, the lack of knowledge related to salivary biomarkers is even more accentuated in veterinary medicine. Although this body fluid has generated interest in recent years in application to many species, much work remains to be done before this noninvasive fluid can be taken advantage of for analytical purposes.

15.2 Lack of Uniform Guidelines

Guidelines for saliva sampling and storage should be developed and reported for accurate sample management, thereby minimizing bias within and between studies. For this purpose, the following critical points related to saliva obtainment and storage should be taken into account:

Sample Collection Although many sampling methods have been described, no standard technique has been unanimously accepted to date. In general, the most widely used methods in humans are drainage and the use of absorbent devices, while the preferred method in animals is mechanical stimulation by chewing. But even in these cases there is a lack of standardization of the procedure used. Furthermore, not only the selected sample collection method but also several factors such as fasting, the ingestion of substances such as alcohol or medications, or the presence of periodontal disease must be taken into account for adequate interpretation of the results (Bhattarai et al. 2018; Malamud 2011). It is important to underscore that different collection methods result in different saliva composition, making it important to gain more in-depth knowledge of the changes in saliva composition associated with each technique.

In this regard, a list of standard procedures should be proposed to ensure proper collection of the samples, including information on the factors that need to be avoided and on the procedures and time required to obtain an adequate sample.

Sample Storage It is important to keep in mind that the saliva components can be significantly affected if the sample is not properly stored, since certain unstable analytes can change rapidly at room temperature (such as oxidative stress markers or some proteins). Therefore, it is essential to investigate the effects of the storage conditions upon the tested analytes, and if this is not possible, to follow basic recommendations for the storage of samples. As a general rule, it is advisable for all samples to be kept refrigerated or on ice after collection, and frozen as soon as possible (preferably at −80 °C). In addition, multiple freezing and thawing cycles should be avoided (Barranco et al. 2019).

The availability of guidelines with standard procedures can minimize variability of the results and make the data obtained in different studies more comparable. Such guidelines could make salivary biomarker determinations more easily applicable in clinical practice.

15.3 Lack of Highly Sensitive Affordable Methods

Saliva contains a huge number of potential biomarkers, though at very low concentrations; as a result, expensive and sophisticated technologies are required that only major laboratories can afford. Saliva analysis-related problems arise, including stability, transport and location-related issues that complicate the use of saliva in clinical settings. In this regard, there has been growing interest in recent years among researchers in developing easy-to-perform and sensitive point-of-care testing systems (Hartman et al. 2016; Khan et al. 2017; Pappa et al. 2018). These systems usually integrate microfluidics, electrochemical sensing or bio-nanochips, among other elements, that allow:

- Reduction of sample and reagent volume, and thus lesser resource consumption
- Obviation of sample storage and transportation-related issues
- Point-of-care disease screening
- Increased patient compliance

Overall, such systems would allow both personalized medicine and the screening of large populations in the context of epidemiological studies.

A number of such systems have been designed and validated for different local (e.g., periodontitis) and systemic disorders (stress, diabetes mellitus, cardiovascular or kidney disease, cancer) and conditions (smoking, drug consumption) (Khan et al. 2017). However, in addition to a need for further large-scale studies in order to contribute to existing knowledge and help correctly understand and thus interpret the data obtained in each specific situation, there is a need for reducing the associated equipment and hardware costs. In this regard, open-source hardware and software, 3D printing, low-cost electronic boards, and the use of mobile phone image processing tools are increasingly gaining attention among researchers. As an example, open-source toolkits for ultrasound-guided intervention systems (Lasso et al. 2014), scientific microscopes (Gualda et al. 2013) or electrochemistry-based analytical techniques (Rowe et al. 2011) have already been developed and shared. In this way, anyone interested can benefit and contribute to advancement, not only by accessing these resources in an easy and inexpensive way, but also by improving them through collaboration and cooperation (Erny and Tvarijonaviciute 2019). Overall, despite the need for more studies before open-source, low-cost hardware and point-of-care devices can be integrated into clinical practice, their future potential applications in human and veterinary medicine are clear.

15.4 General Conclusions

Saliva is a fluid of enormous potential. Interest in the study and applications of saliva is increasing, though several points need to be addressed before this body fluid can be routinely used for diagnostic purposes. On one hand, saliva composition changes in different situations, making it a good indicator of such changes, though on the other hand, its dynamic nature makes it difficult to establish reference values for many molecules. Research needs to continue in order to increase our knowledge about the exact modifications produced by each source of variation. Improved knowledge would allow the definition of guidelines for sample collection and analysis, and the development of practical and sensitive methodologies – defining saliva as a fluid with successful future applications to diagnosis, as well as a good source of information about different biological processes.

References

Aro K, Wei F, Wong DT, Tu M (2017) Saliva liquid biopsy for point-of-care applications. Front Public Health 5:77. https://doi.org/10.3389/fpubh.2017.00077

Bales CW, Freeland-Graves JH, Askey S et al (1990) Zinc, magnesium, copper, and protein concentrations in human saliva: age- and sex-related differences. Am J Clin Nutr 51(3):462–469. https://doi.org/10.1093/ajcn/51.3.462

Barranco T, Rubio CP, Tvarijonaviciute A et al (2019) Changes of salivary biomarkers under different storage conditions: effects of temperature and length of storage. Biochem Med 29:010706. https://doi.org/10.11613/BM.2019.010706

Bhattarai KR, Kim HR, Chae HJ (2018) Compliance with saliva collection protocol in healthy volunteers: strategies for managing risk and errors. Int J Medl Sci 15(8):823–831

Erny GL, Tvarijonaviciute A (2019) PortASAP – supporting and encouraging an open source labs. Inženjerstvo okoliša 6:6

Gualda EJ, Vale T, Almada P et al (2013) OpenSpinMicroscopy: an open-source integrated microscopy platform. Nat Methods 10:599–600. https://doi.org/10.1038/nmeth.2508

Hajat A, Diez-Roux A, Franklin TG et al (2010) Socioeconomic and race/ethnic differences in daily salivary cortisol profiles: the multi-ethnic study of atherosclerosis. Psychoneuroendocrinology 35:932–943. https://doi.org/10.1016/J.PSYNEUEN.2009.12.009

Hartman M-L, Goodson JM, Barake R et al (2016) Salivary biomarkers in pediatric metabolic disease research. Pediatr Endocrinol Rev 13:602–611

Ivars K, Nelson N, Theodorsson A et al (2015) Development of salivary cortisol circadian rhythm and reference intervals in full-term infants. PLoS One 10(6):e0129502. https://doi.org/10.1371/journal.pone.0129502

Kamodyová N, Baňasová L, Janšáková K et al (2015) Blood contamination in saliva: impact on the measurement of salivary oxidative stress markers. Dis Markers 2015:479251. https://doi.org/10.1155/2015/479251

Khan RS, Khurshid Z, Yahya Ibrahim Asiri F (2017) Advancing point-of-care (PoC) testing using human saliva as liquid biopsy. Diagnostics (Basel) 7:pii: E39. https://doi.org/10.3390/diagnostics7030039

Lasso A, Heffter T, Rankin A et al (2014) PLUS: open-source toolkit for ultrasound-guided intervention systems. IEEE Trans Biomed Eng 61:2527–2537. https://doi.org/10.1109/TBME.2014.2322864

Li J, Quinque D, Horz HP et al (2014) Comparative analysis of the human saliva microbiome from different climate zones: Alaska, Germany, and Africa. BMC Microbiol 14:316. https://doi.org/10.1186/s12866-014-0316-1

Malamud D (2011) Saliva as a diagnostic fluid. Dent Clin N Am 55(1):159–178

Pappa A-M, Parlak O, Scheiblin G et al (2018) Organic electronics for point-of-care metabolite monitoring. Trends Biotechnol 36:45–59. https://doi.org/10.1016/J.TIBTECH.2017.10.022

Prodan A, Brand HS, Ligtenberg AJM et al (2015) Interindividual variation, correlations, and sex-related differences in the salivary biochemistry of young healthy adults. Eur J Oral Sci 123:149–157. https://doi.org/10.1111/eos.12182

Randeva H, Karteris E, Lewandowski KC et al (2003) Circadian rhythmicity of salivary leptin in healthy subjects. Mol Genet Metab 78:229–235. https://doi.org/10.1016/S1096-7192(03)00004-0

Rowe AA, Bonham AJ, White RJ et al (2011) CheapStat: an open-source, "do-it-yourself" potentiostat for analytical and educational applications. PLoS One 6:e23783. https://doi.org/10.1371/journal.pone.0023783

Shinkai S, Watanabe S, Kurokawa Y, Torii J (1993) Salivary cortisol for monitoring circadian rhythm variation in adrenal activity during shiftwork. Int Arch Occup Environ Health 64:499–502. https://doi.org/10.1007/BF00381098

Wada M, Orihara K, Kamagata M et al (2017) Circadian clock-dependent increase in salivary IgA secretion modulated by sympathetic receptor activation in mice. Sci Rep 7(1):8802. https://doi.org/10.1038/s41598-017-09438-0

The manufacturer's authorised representative in the EU is Springer Nature Customer Service Centre GmbH, Europaplatz 3, 69115 Heidelberg, Germany. If you have any concerns regarding our products, please contact ProductSafety@springernature.com

Printed and bound by CPI Group (UK) Ltd, Croydon, CR0 4YY
05/10/2026
02228120-0014